BIRDS OF ARIZONA & NEW MEXICO

For Nick, whose insatiable curiosity and enthusiasm for all things winged, scaled, slime-coated, and furred has changed the course of our lives in unimaginable ways.

And for Hog Island and your spellbinding moonlit shores.

Copyright © 2024 by Melissa Fratello and Steven Prager. All rights reserved.

Photo and illustration credits appear on page 523.

Hachette Book Group supports the right to free expression and the value of copyright. The purpose of copyright is to encourage writers and artists to produce the creative works that enrich our culture. The scanning, uploading, and distribution of this book without permission is a theft of the author's intellectual property. If you would like permission to use material from the book (other than for review purposes), please contact permissions@hbgusa.com. Thank you for your support of the author's rights.

Timber Press
Workman Publishing, Hachette Book Group, Inc.
1290 Avenue of the Americas
New York, New York 10104
timberpress.com

Timber Press is an imprint of Workman Publishing, a division of Hachette Book Group, Inc. The Timber Press name and logo are registered trademarks of Hachette Book Group, Inc.

Printed in China on responsibly sourced paper
Text design based on series design by Adrianna Sutton
The publisher is not responsible for websites (or their content) that are not owned by the publisher.
The Hachette Speakers Bureau provides a wide range of authors for speaking events. To find out more, go to hachettespeakersbureau.com or email hachettespeakers@hbgusa.com.

ISBN 978-1-64326-198-0
A catalog record for this book is available from the Library of Congress.

OPPOSITE: Mexican Jay

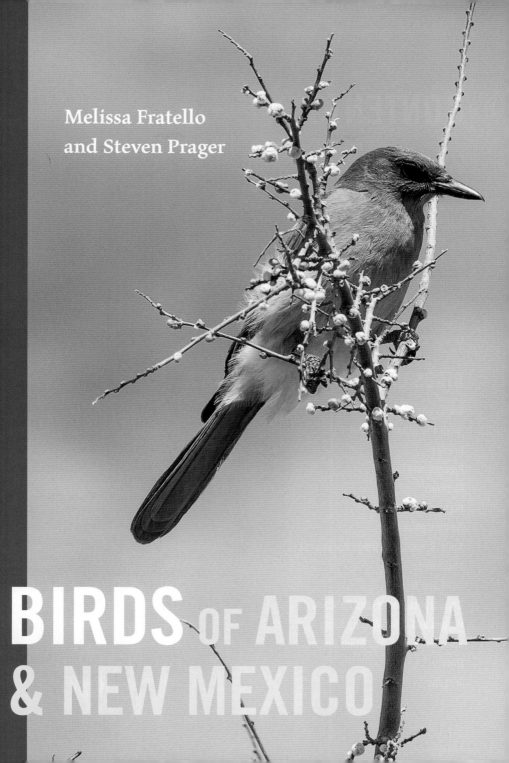

Melissa Fratello and Steven Prager

BIRDS OF ARIZONA & NEW MEXICO

CONTENTS

Birdwatching in the Borderlands ... 7
An Acknowledgement of Occupancy ... 11
What Do You Need to Bird? ... 12
Bird Identification ... 15
Finding Birds ... 27
Birding Hotspots ... 31
Bringing Birds to You ... 45
Inclusivity in Birding ... 58
Conservation and Community Science ... 62
How to Use This Book to Identify Birds ... 65

BIRDS OF ARIZONA & NEW MEXICO

Ducks and Other Waterfowl ... 72
Quail and Other Landfowl ... 105
Grebes ... 117
Pigeons and Doves ... 124
Cuckoos ... 133
Nightjars and Swifts ... 137
Hummingbirds ... 145
Rails, Cranes, and Other Marshbirds ... 161
Shorebirds ... 169
Gulls and Terns ... 201
Cormorants, Loons, and Pelicans ... 213
Herons and Other Wading Birds ... 218
Vultures, Hawks, and Other Raptors ... 228
Owls ... 251
Trogons ... 266

Kingfishers	269
Woodpeckers	272
Falcons	287
Parrots	293
Tyrant Flycatchers and Becard	295
Shrikes and Vireos	322
Corvids	332
Larks and Swallows	344
Chickadees, Titmice, and Others	354
Wrens	366
Gnatcatchers, Dipper, and Kinglets	374
Thrushes	381
Thrashers and Other Mimids	390
Starlings, Waxwings, Silky-flycatchers, and Olive Warbler	399
Old World Sparrows, Pipits, Finches, and Longspurs	404
New World Sparrows	423
Blackbirds, Orioles, and Others	457
New World Warblers	473
Tanagers, Cardinals, and Others	493
Rarities	509
Recommended Resources	515
Acknowledgments	521
Photography and Illustration Credits	523
Index	529
About the Authors	544

BIRDWATCHING IN THE BORDERLANDS

Ditch your assumptions—and bring water.

UNLESS YOU LIVE in the desert, forget what you think you know about it. Throw away that vision of a monotonous sandscape strewn with tumbleweeds. In reality, few places offer such a vast array of habitats as Arizona and New Mexico, where aspen-tipped mountains meet saguaro sentinels of the low desert, canyons and caves meet vast grassland, and riparian corridors rich with cottonwoods meander like green ribbons through the arid, rocky landscape. In the southernmost reaches of the states, the tropics of Central and South America creep into the Madrean Sky Islands, while farther north, you meet the Rocky Mountains and the Kaibab Plateau. Among these mind-bendingly gorgeous landscapes, nearly 600 bird species occur in any given year. Of those, roughly 300 are resident breeders—a testament to the region's astounding biodiversity.

As residents of Arizona (and admirers of our neighbors to the east) and as active members of the birding and conservation communities in our personal, professional, and volunteering lives, we are regularly reminded of the resilience required of the flora and fauna that call these seasonally inhospitable states home, and birds are no exception. Birding in the Southwest can be extremely accessible, from the benches at the lush Paton Center for Hummingbirds in Patagonia and the feeders at the famed Santa Rita Lodge in Madera Canyon, to what some might call, only half-jokingly, a death wish expedition to the rugged peaks and flash flood–prone canyons of our wilderness areas.

The signs of a changing climate are acute here. Invasive grasses lay claim to much of the landscape, exacerbated by unfettered grazing and irresponsible landscaping. Drought begets wildfire, now a regular occurrence in areas not historically prone or adapted to recover from it. To witness the loss of a swath of century-old saguaros is deeply unsettling. Water resources are alarmingly finite, and monsoon patterns, once as predictable as my Grandma Helen falling asleep within twenty minutes of a movie's opening credits, have become inconsistent and nearly impossible to predict. Yet development rages on, and the birds are hard-pressed to adapt on the fly, pun entirely intended.

In short, this landscape—and the weather that accompanies it—will humble you. Spend enough time here, and you will experience bone-chilling cold and skin-sizzling heat, alpine peaks and sandstone canyons, all in a matter of hours. You might also spot a Golden Eagle and a Green Kingfisher. The Southwest is a birdwatcher's paradise, no matter your skill level or birding style. We are constantly reminded of how lucky we are to live here, how fragile our existence on this land has become, and how the fractures of our relationship with the land must be healed for future generations of birds and birders.

OPPOSITE Red-tailed Hawk

We want this book to be a tool for birders of every skill and interest level, and we hope that our approach to this guide provokes thought, sparks joy, and helps you figure out what the heck that bird was. So grab this book, pack your sunscreen (along with more water than you think you can drink—throw an electrolyte pack in there while you're at it), remember whose homelands you are visiting, and Leave No Trace as you venture out into our very birdy backyard.

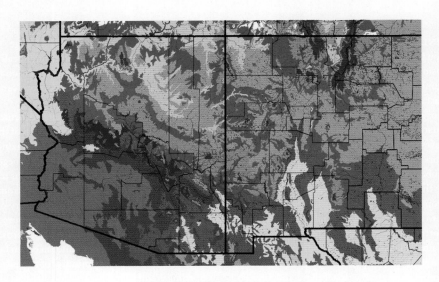

HABITAT MAP KEY

- Alpine Tundra
- Petran Subalpine Conifer Forest
- Petran Montane Conifer Forest
- Great Basin Conifer Woodland
- Madrean Evergreen Woodland
- Chaparral
- Subalpine Grassland
- Plains and Great Basin Grassland
- Semidesert Grassland
- Great Basin Desertscrub
- Mohave Desertscrub
- Chihuahuan Desertscrub
- Sonoran Desertscrub: Lower Colorado River Sub
- Sonoran Desertscrub: Upland Subdivision
- River/water

ABOVE This map, commonly referenced by sciencey types in the region, shows the incredible habitat diversity present in Arizona and New Mexico.

OPPOSITE Adult Harris's Hawk lifting off

AN ACKNOWLEDGMENT OF OCCUPANCY

AS YOU CELEBRATE the birds of this region, please take a moment to reflect on, revere, and express gratitude for the deep history and stewardship of the dozens of Indigenous communities that have lived in the region for time immemorial and that continue to live and work throughout this now-occupied land. Members of at least forty-five tribes in Arizona and New Mexico moved throughout this unforgiving but bountiful landscape, working in harmony with its cycles of winter and summer rain, often migrating through its canyons and peaks with the seasons.

Land acknowledgements in and of themselves are gestures, nothing more. We believe, however, that these gestures are worth expressing, with hopes that, collectively, we may see an acknowledgement as a starting place for action—for supporting cultural easements and landback movements; for resisting the suppression of truth-telling as it relates to natural, cultural, and political history; and for assessing our own relationship with the land as part of, not separate from, the ecosystem.

While you are birding in Arizona and New Mexico, we hope you'll take note of our guidance for birding on tribal lands.

OPPOSITE Upland Sonoran Desert near Florence, Arizona

WHAT DO YOU NEED TO BIRD?

YOU REALLY DON'T need anything special to enjoy birds. Go look (or listen) out your nearest window, no matter where you are, and you'll likely be greeted by birds. Whether you're in a rural or an urban setting, a natural habitat or a human-influenced ecosystem, your neighborhood or a national park, there's no shortage of opportunity to spot a bird. That said, there's also no shortage of gear that can enhance your experience and comfort while birding.

CLOTHING

Although there is no official birding dress code, it's worth remembering that the Southwest holds a variety of rugged, harsh habitats, and your clothing is your first defense against the elements. Long sleeves and pants can be your best friends in thorny scrub, layered clothing will enable you to deal with drastic shifts in temperature, sturdy boots can protect you from unseen hazards in the brush, and a broad-brimmed hat can shield you from the often oppressive sun. Most importantly, know the areas you will be birding, watch the weather, and dress accordingly.

It's also worth considering the impact your clothing choices can have on the birds you hope to observe. For example, bright or noisy clothing can alert birds to your presence, robbing you of the chance to see them before they see you. Lastly, consider your own comfort. If you're not comfortable, you'll be wasting energy that you could be using to find birds!

OPTICS

A pair of binoculars is the go-to tool for birding, and there are a few things you should know. First, binoculars range widely in price. The things that come with a price tag are durability, water- and vapor-proofing image quality, and an increased ability to gather light. So before you purchase binoculars, ask yourself some questions about how you want to use them. Will they take a beating in the field or will they be safely stored at home? Will you be birding in dimly lit forests or at a local park? Depending on their use, expensive binoculars can make anywhere from a huge to a tiny difference in your birding experience.

Next, it's important to understand the numbers that accompany all pairs of binoculars. The first number is the magnification factor. The second number is the diameter of the front lens in millimeters. So, for example, a pair of 7×35 binoculars will magnify your view by 700 percent compared to the naked eye and will have a 35mm front lens, and 8×42 binoculars will magnify your view by 800 percent and will have a 42mm front lens. Both of these variations are popular among birders.

Both magnification and lens size come with trade-offs. High magnification can give you a shot at identifying distant birds, but it can make it difficult to track those that are close by. Also, high magnification amplifies the effect of shaky hands, and because high magnification binoculars are often heavy and bulky, at some point this becomes a problem for even the most steady handed birders.

A large front lens increases the binoculars' ability to gather light, but the larger the lens, the narrower the field of view (the size of the area you can see through the binoculars) becomes. Large front lenses can help when you're looking for subtle field marks in dim lighting, but the narrow field of view they offer can make birds in motion difficult to spot. When looking for high-magnification optics for identifying distant, slow-moving birds such as waterfowl and raptors, birders usually stop at 10×50 binoculars before moving onto tripod-mounted spotting scopes, which can overcome the limitations of high magnification and provide increased accessibility for people for whom binoculars are not a good fit.

IDENTIFICATION TOOLS

Last on the list of helpful birding tools to consider are bird identification resources. While observing birds, you are bound to run into identification puzzles; questions about range, habitat, and seasonality; and countless other wonders. To satisfy your curiosity, field guides (like this one), electronic apps, and bird checklists for your area are great tools to bring along. Also, bring a writing utensil and a pad—your own field notes can be invaluable down the road.

Most importantly, remember that the birds don't care how you're dressed or through what equipment you're viewing them. They don't even care if you know what species they are. Just be safe, get out there, and see some birds!

ABOVE Binoculars are nice for details and spying on shy sparrows, but charismatic species like this Vermilion Flycatcher are easy to enjoy sans optics.

BIRD IDENTIFICATION

SPOILER ALERT: Sometimes it's really difficult. Sometimes, you just can't. And that is absolutely fine, but these tips will help.

The value of a bird doesn't depend on your ability to identify it. Similarly, your identity as a birder doesn't hinge on your ability or desire to identify every bird you encounter. Nevertheless, if you're watching birds, at some point you're likely to be curious about exactly what you're observing, and that brings us to identification, which requires a process of observation, analysis, and deduction.

Iconic birds such as the Greater Roadrunner and Cactus Wren (the state birds of New Mexico and Arizona, respectively) are easy to identify, being emblematic of the region and well known even by the nonbirding public. Others, such as the Vermilion Flycatcher and Yellow-headed Blackbird, are easy to identify because of their exceptional uniqueness and descriptive monikers. If only they could all be so distinct! Species of sparrows, shorebirds (or "peeps"), gulls, flycatchers, and a handful of lookalikes spread across other groups can be frustratingly difficult to differentiate. There's a reason sparrows have been not-so-affectionately referred to as "little brown jobs." When you encounter these challenges, you'll want to gather every available clue.

BIRDING BY EYE

It probably goes without saying, but a bird's physical appearance can offer many clues about its identity. But you have to know what to look for, and sometimes that can be complicated. It helps to become familiar with the basic parts of birds as well as field marks, or physical traits observable in the field that are helpful in distinguishing individual species. It's not uncommon for a beginning birder to record a long list of a mystery bird's features, only to return home and realize they failed to record the one feature that would've led to successful identification.

It also helps to start with broad observations before honing in on smaller details. General features such as overall shape and structure; bill, body, wing, and tail shape; posture; size; flight pattern; and behavior can help you organize birds into taxonomic, or evolutionarily related, groups, such as hummingbirds, woodpeckers, sandpipers, and so on. A slightly closer look can reveal groups within groups—*Empidonax* versus *Myiarchus* flycatchers, for example. The birds featured in this book are organized by these groups, or in taxonomic order, so starting with these categories will help you find the appropriate section in which to search. Once you've found the correct group, start paying attention to more species-specific field marks—pattern, color, and more nuanced observations of size, shape, and structure. Each group will have its own set of field marks that are of outsized importance in identification, such as head and tail shape in accipiters or leg color and bill shape in shorebirds.

But what do you do when you can't get a good look?

OPPOSITE The underbelly of a tricksy *Myiarchus* flycatcher in flight doesn't give you much to go on; refer to the species accounts and then come back to identify it, if you dare!

BIRDING BY EAR

You may not be able to get a great look at a bird for many reasons. It could be a particularly skulky bird, hiding behind dense branches and never offering you a full view. The bird could be extremely far away. You might be distracted by something else, such as another bird or a conversation with fellow birders. Or maybe you did get a good look but you're trying to choose between two visually indistinguishable birds (those pesky flycatchers!). Whatever the case may be, your ears can offer you a whole new set of clues. Sounds can be helpful in identifying unseen or poorly seen birds, they are critical to differentiating visually indistinguishable species, and they enhance birding accessibility for people with impaired sight.

With many species having their own unique song (often complex vocalizations used in courtship) and with all having their own array of calls (usually simpler vocalizations used for contact, alarm, territoriality, or other announcements beyond a desire to breed), birding by ear can seem daunting. However, training your ears is really no different from training your eyes. First, familiarize yourself with the basic types of calls and songs you may encounter, such as the sharp chip and melodious song of a warbler, the rattling laughter of a woodpecker, or the buzzy, rolling song of a bunting. Again, knowing the basic types of calls and songs you may observe will help you file your observation into a taxonomic group of birds so you can narrow down your options when listening to them. From there, as with visual observation, it comes down to more minute details: Was that vireo's call three parts or two? Was that warbler's call ascending or descending in pitch? It's also important to remember that both calls and songs can vary regionally and among individuals. One thing that is helpful to pay attention to beyond the specific pattern of a bird's call or song is its general voice: Summer Tanagers, for example, have a smooth voice, while Western Tanagers have a gravelly one. Learning the general voice of an individual species can help you identify it even when it's making an unusual vocalization that you can't find in a recording.

THE LOOKALIKE DILEMMA

Was that a Cooper's Hawk or a Sharp-shinned Hawk? A Rufous-crowned Sparrow or a Rufous-winged Sparrow? The abundance of lookalike species can seem endlessly maddening as you're trying to identify a bird. Worry not, however, for this is where you really get to hone your birding skills.

BEHAVIOR

So you think you're close to nailing that identification, but you just can't get a good enough look at the bird's shoulder patch, malar stripe, or rump to come to a decisive conclusion. Luckily, birds and their surroundings offer us a bit more information to work with. Often, their behavior offers a "tell." Cue the Spotted Sandpiper. You're watching at a distance, and it seems impossible to decipher which darn peep you're observing. Then you notice its tail, bobbing up and down as it quickly skitters along the shore, all by its lonesome. Now you're equipped with the tools you need to make that identification. When we're beginners, it's easy to get caught up in appearances, but it's often these subtle behavioral clues that tell us so much more. Is the bird flying solo or in a flock? Foraging on the forest floor or plucking berries from a shrub? Once you've watched the distinct behaviors of a species, you'll find that behaviors are not only a joy to observe, but they're also a great way to commit a species, and your experience watching them, to memory. When you're out in the field, take your time observing, even if you've already got that identification all figured out.

ABOVE Rufous-winged Sparrows don't always show off their rufous wings, which are really just tiny shoulder patches. A singing sparrow can make identification much easier.

LEFT It's easy to mistake this Brown-crested Flycatcher for its relative, the Ash-throated Flycatcher—they often dwell in the same habitat and are nearly identical.

SEASON, RANGE, AND HABITAT

The season, range, and habitat in which you make your observations are your next clues, and they are often the most helpful tools in your identification toolbox when differentiating between lookalike species, particularly when your subject doesn't vocalize. Our human seasons, in birding terms, coincide with the seasons of a bird's migratory cycle. In short, there is a breeding season typically in late spring to late summer, migration seasons in spring and fall, and overwintering/nonbreeding seasons. Resident species remain in their range year round, though this term can be an overgeneralization. The Phainopepla, for instance, breeds in the low desert in early spring, only to migrate to higher elevations for a second breeding cycle in summer.

Bird migration is spectacular to behold, and there are still many mysteries to be unraveled. There's nothing quite like sitting under the moonlight as thousands of warblers or thrushes fly silently overhead (with the exception of their inconspicuous migration calls) or, conversely, being awakened at 2:00 a.m. to a cacophony of Sandhill Cranes. But the once reliable migration patterns of many species are beginning to shift in response to climate change—both in timing and in course. Sightings of rarities are becoming more commonplace as storms become more intense and frequent, knocking birds off course, and as ranges of species shift northward.

Range refers to the geographic boundaries of a species, as opposed to habitat, which represents the features within a bird's range where it forages or feeds, roosts, or nests. For example, the range of Yellow-breasted Chat spans much of the North American continent and they can be found throughout Arizona and New Mexico, but a sighting is highly unlikely in arid lands, as they prefer brushy habitats along rivers and streams.

OPPOSITE Adult male Phainopepla

BELOW Sandhill Cranes

When you're in the field trying to distinguish between a little brown job and another little brown job, remember to ask yourself a couple of questions: Is it within season? Is this the species' preferred habitat? Each bird species account in this guide includes range and habitat descriptions and, except in accounts of the rarest species, a range map. These tools will help you determine whether the bird you're observing is within range and season.

TIPS AND PITFALLS

If you've never misidentified a bird, you haven't spent much time birding. Though these mistakes can be (and are) made by even the most experienced birders, a few common blunders can trick beginning birders into a misidentification.

The first is not recognizing how subjective size can be. All sorts of variables, such as distance, perspective, motion, and even state of mind, can affect a birder's perception. Reference points such as objects or birds of known size can be helpful but are not always available. For these reasons, be wary of using size as a definitive field mark. A species' color can also vary, influenced by lighting, individual variation, feather wear, and time of year. Hence, you may notice that the colors you see on birds you observe may not match exactly the shades shown in this book's photos. To avoid the trappings of size and color, you should identify birds with multiple field marks and observations about seasonality, behavior, and habitat. The more information you can gather to confirm one species and rule out others, the better.

Another way to avoid misidentifications is to consider the most common birds first. Rare birds are just that, after all. It's smart, then, to start with the likely suspects and work your way out, considering rarities only once you've ruled out the birds common to the area (and in season). Studying ahead and becoming familiar with common species can help you avoid spending time digging through a field guide while there are birds to be watching. Local bird songs are a great addition to any playlist, and thumbing through this guide is a good way to pass the time between birding trips.

Alas, studying will get you only so far. The best way to improve your identification skills is through experience. Get out and bird as much as you can. Take advantage of others' experience by connecting with local birders, or take your friends birding and learn together. Most importantly, don't get discouraged! Even the most impressive birders, whether they admit it or not, had to start somewhere, and the struggle to identify a bird is half the fun.

PLUMAGE VARIATION

No matter how hard you study, no matter how well you know every species' identifying features, birds can throw you a curveball.

Birds of the same species (conspecifics) can differ greatly in physical appearance for a multitude of reasons, the first being simple individual variation. Just as humans in the same family can vary physically, so can birds within a population. Within a species, slight variations in size, color, and pattern should be expected, and this makes it especially important to focus on field marks, which tend to remain consistent across individuals (with some being more reliable than others). Field marks won't always help you with juvenile birds, though—just ask anyone who has mistaken a juvenile Verdin, which lacks the bright yellow head of the adult, for a wildly out-of-range Bushtit. It pays to consider other identifiers, such as overall shape and posture, habitat, and behavior; to think about what other species are in the

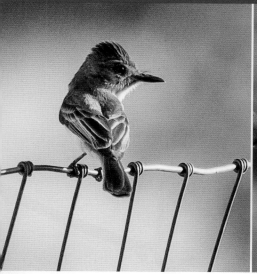 

area to which a young bird could belong; to watch for adults; and to become familiar with juvenile plumages, especially for birds like Bald Eagles and Summer Tanagers that retain their youthful looks for multiple seasons.

Another reason conspecific birds can differ in physical appearance is their sex, a feature referred to as sexual dimorphism. Such dimorphism can be minimal or lacking (as in doves and thrashers), subtle but apparent (as in woodpeckers and sparrows), or dramatic (as in hummingbirds and buntings). Historically, birding and avian science have been biased toward male birds. Perhaps this is because sexism is inherent in even the most niche corners of our society, or perhaps it is because in the majority of (but not all) avian species, the males are the flashier and more vocal of the two sexes, making them easier to observe and study. In the most egregious cases, birders and scientists have paid no mind to females except to note them as the males' counterparts. Not only does this have implications for bird identification, but because female birds often exhibit behaviors and have habitat requirements

LEFT A Dusky-capped Flycatcher has a big noggin in relation to its body, which helps as a subtle field mark, but using only size to identify it can be tricky in the field because it resembles several doppelganger *Myiarchus* species.

RIGHT Juvenile Verdin

that differ from those of males, it's to the detriment of science and conservation as well.

Birds can also vary within a species geographically. In species such as the White-crowned Sparrow, individuals vary depending upon which breeding population they belong to. In these species, the variation can be subtle. In others species, such as the Dark-eyed Junco, the differences can be dramatic. Relating to geography (but also timing), individuals can undergo drastic changes in appearance as they shift between spring/summer (breeding) and fall/winter (nonbreeding) plumage, a process of feather loss and regrowth known as molting. Molting serves several functions for birds. It enables brightly colored breeders to blend in more easily during the nonbreeding season, it

The adult male Anna's Hummingbird in the image above shows little color in its gorget when in the shade, but in the right light—bedazzling!

BIRD IDENTIFICATION | 23

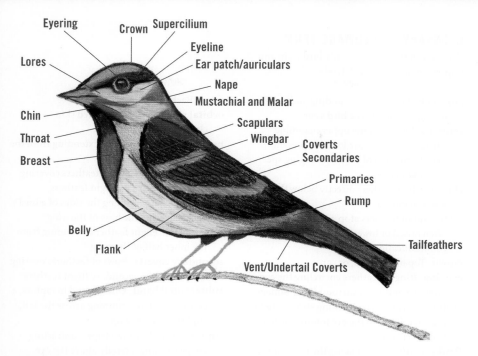

helps migratory birds prepare for and recover from their travels, and, most basically, it replaces old, worn feathers with shiny new ones.

Lastly, lighting can drastically alter a bird's appearance. Some birds, such as hummingbirds that show off their brightly colored throats, or gorgets, by turning their heads to catch the sun just right, use lighting to their advantage. All birds, however, can vary in brightness and color depending upon the available light. Don't use color as your sole field mark when birding in excessively dim or bright conditions, and keep in mind that the sparrow you spotted during a brilliant Southwestern sunset might not actually have been pink.

Though we wish it were possible to capture and curate photographs for this guide that fully encompass all possible variations, it's not—unless you'd like to carry around

ABOVE Although not all of these features are distinct or present on every bird, they often serve as diagnostic field marks that are critical for identification when a bird is reluctant to offer other clues through vocalization and other species-specific behaviors.

multiple volumes, or an entire book about Red-tailed Hawks. For this reason, avoid trying to match your observed bird exactly with photos in this book, pay attention to consistent field marks, and consider all available clues, not just physical appearance, when identifying birds.

PARTS OF A BIRD

The diagram above should help you when fellow birders start throwing around words that sound like they belong on an ingredients list.

GLOSSARY OF PLUMAGE TERMS

You too can throw out a nerdy bird term and impress, or annoy, your friends.

auricular Feathers surrounding and channeling sound into the bird's ear

ceres Fleshy area of the upper mandible where the nostrils are located, present mainly on birds of prey, scavenging species, and parrots

chin Patch of feathers directly below the bill/lower mandible

crest Tuft of feathers atop a bird's crown, often raised or lowered as a means of communication

crown Top of the head, or cap

eye ring Ring of feathers encircling the eye, often in contrast to surrounding plumage

eyeline Feathered, contrasting stripe extending laterally from the eye toward the nape/back of crown

flanks Feathered area along the bird's sides, extending from the legs to under the wings

forehead Area directly above the bill

gorget Bib of iridescent feathers on the throats of many hummingbirds, particularly showy in adult males

lores Small area between the eye and the base of the bill

malar stripe Contrasting feathered stripe that runs downward and backward from the base of the bill along the sides of the throat

mandible Bill, lower and upper

mantle Upper back

mustachial stripe Contrasting feathered stripe that runs backward and downward from the base of the bill between the malar area and auriculars

nape Back of neck

orbital ring Bare, unfeathered skin circling the eye

primaries Flight feathers extending from the outer half of the wing

primary coverts Rows of feathers covering the base of primary flight feathers

scapulars Feathers along the sides of a bird's back, covering the base of the wing

secondaries Flight feathers extending from the inner half of the wing

secondary coverts Rows of feathers covering the base of the secondary flight feathers

subterminal band Often found in raptors, a contrasting band running across the tail, slightly above the tip

supercilium Eyebrow stripe; contrasting stripe running laterally above the eye

supraloral Area directly above the lores

tail coverts Small feathers at the base of the tail

tertials Innermost secondary flight feathers between the rest of the secondaries and the scapulars

throat Area below the chin and above the upper breast

undertail coverts Feathered patch covering base of tail feathers/vent area

wingbars Contrasting stripes on the wing surface, usually present on wing covert tips

wingtips Tips of primary flight feathers

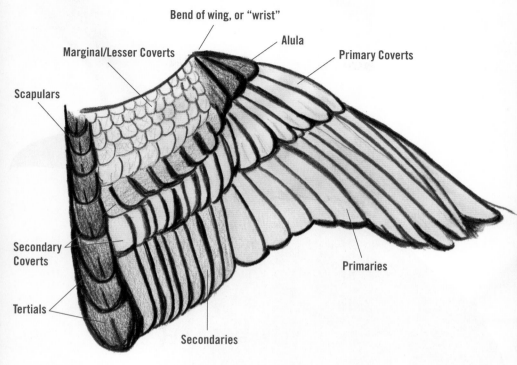

Anatomy of a wing

FINDING BIRDS

Be warned: You can't turn off your birding senses once they've been activated!

Once your eyes and ears are open to birds, you'll never stop seeing them. Maybe you'll be gardening or walking the dog when you notice a species you've never seen before. Perhaps you'll be enjoying another outdoor activity when an unfamiliar song or call catches your ear. Or maybe you'll be sitting in a work meeting when that much needed distraction swoops past a window. Birds can bring new joy to old activities, connect you more deeply to places you already know and love, and add excitement to otherwise banal tasks. Once you're hooked on birds, they tend to find you whether you're looking or not.

That said, many birders enjoy the challenge of seeking out birds, building long and meticulous "life lists," finding rarities and local specialties. It's easy enough to consult the birding community's data through online resources like eBird or to glean information from them directly by joining the numerous social media groups and organizations facilitated by birders, but it can be far more rewarding to do the work yourself. If you're up to the challenge, the task of finding birds comes down to four things: being in the right place, being there at the right time, honing your knowledge and observation skills, and finding birds before they find you.

RIGHT PLACE, RIGHT TIME

"The right place" can be a place with a large overall number of birds, a place with a large number of species, or a place likely to support the specific species you're looking for. When you're looking for large numbers of birds, aim for areas of healthy habitats that offer the four things that all animals need: food, shelter, space, and water. The Southwest's scarcity of water can have an outsized impact on bird productivity, so wetlands, lakes, ponds, rivers, streams, and washes, as well as places where we add water to the landscape such as agricultural areas and parks, should be high on your list. When looking for these areas of high productivity, pay attention to other flora and fauna. Are the plant species you'd expect to see in a given habitat present? Are you seeing a good selection of other wildlife such as insects, reptiles, and mammals?

When looking for a high species-diversity, continue to watch for the signs of overall high productivity, but also look for areas where multiple habitat types come together. These areas of habitat convergence, known as ecotones, can be particularly diverse and can offer you the opportunity to observe multiple communities of birds in a single visit. Areas with this sort of diversity include mountains, where changes in elevation

OPPOSITE Yellow-billed Cuckoos are a rare sight for most birders in the Southwest, but in areas rich with cottonwoods, willows, and oaks, these "rain crows" can be heard calling at dawn and dusk during the breeding season.

Remember, if you know you're heading out to look for birds, it's best to blend in and adopt a low profile look for the day. In our region, wearing plain colors, long sleeves, and closed-toed shoes is a better idea than wearing bright colors and sandals.

result in changes in habitat; drainages, where the distance from water informs habitat type; and human-influenced areas, where differing land uses create artificial habitat diversity.

When seeking a particular species of bird, continue to watch for overall productivity and diversity, but also pay attention to your target species' natural history. In what habitat types does it live? Across what geography does it range? What does it eat? With what other species does it associate? We've made your work easy by including much of this information in this guide, and taking the time to read up on your target species will help you narrow your search, saving you much time and effort.

To be successful, you'll have to visit the right place at the right time, and, again, this can mean different things. It can refer to the time of day. Songbirds are typically most active in the early morning and just before dusk, raptors soar later in the day once the sun begins creating the updrafts of warm air they depend on for lift, and nightjars (such as nighthawks) and owls tend to be most active between sunset and sunrise.

Being in a place at the right time can also refer to seasonality. Many of the birds described in this book are present in the Southwest year round, while others may visit the region only during the spring and summer breeding season, winter, or spring and/or fall migration. Spring and summer can be a great time to visit riparian (streamside) areas for breeding neotropical migrants that have migrated north from South and Central America. Winter can be a great time to view waterfowl and shorebirds that have flown in from the north. Consider visiting the lower Colorado River in winter to see incredible concentrations of ducks and grebes, stopping by Arizona's Willcox Playa, or visiting New Mexico's Bosque del Apache National Wildlife Refuge to view thousands upon thousands of Sandhill Cranes and Snow Geese—truly one of North America's greatest wildlife spectacles.

During migration season, birders are on the lookout for birds traveling between their wintering and breeding grounds. At this time of year, landscape features such as mountains and rivers can be great places to bird, because they provide pathways of productive habitat through the Southwest's often harsh and arid lowlands. During migration, it's also worth paying attention to isolated urban habitat patches. For example, a patch of pine trees in an urban park can be exactly what a high-elevation migrant is looking for when passing through the desert.

Birding at the right time can also refer to the weather. Conditions such as heavy precipitation, strong winds, and extreme temperatures can make birds difficult to observe—but don't let poor weather deter you entirely. The short breaks between storms during the Southwest's rainy seasons can offer opportunities to see birds frantically feeding before the next storm hits, and hurricanes traveling northward from Mexico can push rare birds our way.

Even if you're in the right place at the right time, you'll need keen observation skills and knowledge to find birds and identify what you are seeing or hearing. Before heading out to look for birds, take some time to learn what species are prevalent where you're going and during the season you're visiting, and then study their appearance, habitat preferences, behavior, songs, and calls.

Lastly, finding birds often comes down to observing them before they observe you. Try to make your approach as quiet as possible by keeping your voice down, being careful where you step, and wearing neutral-colored and quiet clothing. Try to be as unobtrusive as possible by not standing out in the open or getting too close, consider the size of your group, and above all else, be patient. Sitting still and quietly in the right place at the right time is often the best approach to spotting birds.

BIRDING HOTSPOTS

NO MATTER THE TIME of year, birds provide a great excuse to get out and see something—and somewhere—new, and there is no shortage of destinations throughout Arizona and New Mexico.

OPPOSITE Saguaro National Park

BIRDING HOTSPOTS | 33

ABOVE Sandhill Cranes lower their landing gear at Whitewater Draw Wildlife Area.

OPPOSITE, TOP The Catalina Highway provides astounding habitat diversity, from saguaro forest to spruce-fir woodlands, in less than 30 paved miles.

OPPOSITE, BOTTOM Sweetwater Wetlands in Tucson

ARIZONA

Southeast Arizona

Sweetwater Wetlands This is an urban birder's paradise, especially for a desert dweller. With the creative use of reclaimed wastewater, the City of Tucson has developed a wetlands complex in the heart of the city that serves as an oasis for birds and people alike. Boasting more than 300 observed species, this is an easy stop off Interstate 10 that will bring you desert classics such as the Greater Roadrunner and the Abert's Towhee alongside Black-bellied Whistling-Ducks and Cinnamon Teals. Bonus: It is entirely accessible, with wide, paved pathways and restrooms.

Catalina Highway/Mount Lemmon
A drive into the Catalina Mountains leads up to Mount Lemmon, which towers over the city of Tucson on its northeastern boundary. Mount Lemmon provides a whirlwind tour of seven distinct biomes, from desert saguaro stands to spruce, pine, and aspen at its peak. The diversity of bird species is as astounding as the scenery along the way, and in the desert spring and summer, the higher elevations bring relief from the heat. Olive and Red-faced warblers, Western Flycatchers, Golden Eagles, and Steller's Jays await at the higher elevations.

Whitewater Draw Wildlife Area This seasonal wetland is known for its throngs of overwintering Sandhill Cranes. Flocks of 30,000 or more cranes mingle with migrating Snow Geese and Yellow-headed, Red-winged, and Brewer's blackbirds for sunrise and sunset spectacles, making it worth the two-hour drive southeast from Tucson.

Madera Canyon Less than an hour's drive south of Tucson, this is a relatively accessible gateway into the Santa Rita Mountains. Madera Canyon offers several paved paths, well-maintained hiking trails, and a paved road leading to a busy feeder station notorious for Magnificent Hummingbirds, Arizona Woodpeckers, Gould's Wild Turkeys, and the entertaining antics of Mexican Jays. Birders flock to the canyon in search of the Elegant Trogon—but even if you don't find one, you'll not be disappointed in the many consolation prizes you're bound to observe.

Patagonia This is not the scenic area in South America, but a delightful little town 60 miles southeast of Tucson in the foothills of its namesake Sky Island, the Patagonia Mountains. Patagonia acts as a quirky, delightful home base for the famed Paton Center for Hummingbirds, where you can witness the splendor of Violet-crowned Hummingbirds and Northern Beardless-Tyrannulets, and then head over to Patagonia Lake State Park

OPPOSITE, TOP Violet-crowned Hummingbirds are a mainstay at Tucson Audubon's Paton Center for Hummingbirds in Patagonia.

OPPOSITE, BOTTOM The South Fork Trail at Cave Creek Canyon offers an imposing view of the Chiricahua Mountains.

to seek the diminutive Green Kingfisher for an easy, laid-back day of Southeast Arizona specialties.

Chiricahua Wilderness/Portal Within the Chiricahua Mountains at the extreme southeast corner of Arizona, this dramatic landscape teeming with life is a well-known destination for birders, plant nerds, bug enthusiasts, and herpers (reptile and amphibian seekers). Hit up Cave Creek Canyon for a scenic and birdy hike. In Portal, visit the Southwestern Research Station's hummingbird feeders during monsoon season for a chance to see a Blue-throated Mountain-gem.

A Gould's Wild Turkey forages under the feeder station at Santa Rita Lodge in Madera Canyon.

BIRDING HOTSPOTS | 35

Southwest Arizona

Lower Colorado River wildlife refuges
Scattered along the lower Colorado River on the southwest border of Arizona are a string of national wildlife refuges including Bill Williams River, Cibola, Havasu, and Imperial. Together, they provide a haven for species that rely on riparian habitat, freshwater marsh, and open water.

Central Arizona

Riparian Preserve at Water Ranch In the suburb of Gilbert just east of Phoenix, this preserve offers area residents and visitors an accessible birding hotspot that draws more than 300 species to its wetlands. The preserve uses treated wastewater to create a complex of wetland habitats, which often draws rarities and vagrants, such as the Roseate Spoonbill, seeking water among a sea of suburbia and asphalt.

Boyce Thompson Arboretum An hour west of Phoenix, the arboretum is a great spot for spring and fall migrants, offering a verdant hideaway in an otherwise rocky, arid landscape. The oldest and largest botanical garden in Arizona, known as BTA to frequent visitors, is ADA accessible, with guided walks, native plant sales, and unique habitats to explore.

Tres Rios Wetlands Thirty minutes west of the Phoenix metro area in the shadow of the Sierra Estrella Mountains, this large complex of wetlands is adjacent to agricultural fields, offering birding opportunities along the way, and is frequented by several species of conservation concern, including the Ridgway's Rail, Willow Flycatcher, and Yellow-billed Cuckoo.

Northern Arizona

Watson and Willow Lakes Ecosystem
Within Prescott city limits, the lakes of this Important Bird Area (IBA) are notable for large concentrations of waterfowl and shorebirds during fall and winter, with an abundance of Wood Ducks, Yellow Warblers, and Green Herons during the breeding season.

Grand Canyon National Park The spectacle of the big ditch aside, the canyon offers a range of habitats—from pine forest to desertscrub and riparian—with more than 370 species present. The North Rim offers the greatest habitat diversity with a broader elevation range, and it rewards visitors with fewer crowds, while the South Rim offers an easy road trip from Phoenix or Flagstaff.

Anderson Mesa/Mormon Lake About 20 miles southeast of Flagstaff, Anderson Mesa is an IBA with global significance, primarily for its population of Pinyon Jays. Nearby coniferous and deciduous forests, open grasslands, and Lake Mary and Mormon Lake offer birders a diverse suite of species, including White Pelicans, Purple Martins, Bald Eagles, and Virginia's and Red-faced warblers.

OPPOSITE The Riparian Preserve at Water Ranch is a birder's playground and a bird's lifeline—an oasis within the sprawling suburbs of Phoenix.

NEW MEXICO

Western/Southwest New Mexico

Percha Dam State Park An easy stop along the Rio Grande, about 20 miles northwest of Hatch, Percha Dam State Park is relatively accessible and offers camping accommodations. The combination of riparian habitat and cottonwood bosque makes for outstanding birdwatching during migration, with American Goshawks and Golden Eagles mingling with migrating warblers and waterfowl.

Bosque del Apache National Wildlife Refuge A short drive south from Socorro, this refuge is a must-stop on your winter birding checklist. The Bosque serves as wintering grounds for tens of thousands of Sandhill Cranes, Snow Geese, and waterfowl. As winter fades, springtime warblers filter through the bosque and shorebirds abound in the wetlands, welcoming the change of seasons.

Gila National Forest Where the high Chihuahuan Desert meets the Rocky Mountains, Silver City serves as an ideal home base for birding adventures throughout this largely undeveloped national forest. Find Red-faced Warblers, Painted Redstarts, Montezuma Quail, Greater Pewees, and American Goshawks among absolutely stunning landscapes and cliff dwellings, and perhaps take a break to soak in a hot spring!

OPPOSITE, TOP Bosque del Apache NWR draws tens of thousands of waterfowl and Sandhill Cranes.

OPPOSITE, BOTTOM Adjacent to charming Silver City, the Gila National Forest will please birders and nonbirders alike with its awe-inspiring Gila Cliff Dwellings

BIRDING HOTSPOTS | 39

Mesilla Valley Bosque State Park Though the habitat here has taken a beating from wildfire and drought in recent years, the park still warrants a stop when visiting the Las Cruces area. It remains a popular site for wintering raptors and Long-eared Owls and will not fail to serve up many desert classics such as the Pyrrhuloxia.

Central New Mexico/Albuquerque/Santa Fe

Rio Grande Nature Center State Park This true gem within the Albuquerque city limits shines brightest during migration seasons, providing winged visitors an urban oasis of riparian corridor and bosque. Of the more than 300 species that frequent the park, only 40 or so are residents, offering birders seasonal variety in an accessible wildland setting.

Las Vegas National Wildlife Refuge Clearly the superior Las Vegas, this New Mexico refuge at the intersection of the Great Plains, Rocky Mountains, and Chihuahuan Desert offers a variety of habitat including riparian corridor, grassland, meadow, marsh, seasonal mudflats, and woodland. Sandhill Cranes arrive en masse in autumn, where they join migratory shorebirds and waterfowl, Bald Eagles, Long-billed Curlews, and Burrowing Owls.

Urban refuges such as Rio Grande Nature Center State Park offer some of the best birding opportunities around, especially during peak migration, when they are nearly always stopovers for hungry birds needing to rest.

Maxwell NWR offers up specialties galore, such as the elusive Black Tern, a species in steep decline across its range.

Northern New Mexico

Rio Fernando Wetlands Straddling the eastern- and westernmost range boundaries for many bird species, this relatively tiny, 30-acre wetland habitat surrounded by upland fields and forests offers a chance to catch Black-capped Chickadees mingling with Western Tanagers. This worthwhile pit stop for Taos County visitors yields great reward for little effort.

Maxwell National Wildlife Refuge A bit more than two hours northeast of Albuquerque, this refuge's picturesque rolling prairie aptly caters to grassland species such as Chestnut-collared Longspur, while its lakes draw Black Terns, Wilson's Phalaropes, and Western and Clark's grebes. Burrowing Owls are known to inhabit old rodent dens, so keep your eyes peeled for those two-legged potatoes with golden eyes!

Eastern/Southeast New Mexico

Milnesand Prairie Preserve Owned and managed by The Nature Conservancy, this vast, 28,000-acre property on New Mexico's eastern boundary is home to a remarkable density of Lesser Prairie-Chickens and their leks, where males display elaborate courtship rituals. Note that there is limited access to this preserve, and it is recommended that potential visitors contact The Nature Conservancy's New Mexico office for more details before visiting.

Melrose Woods migrant trap Just west of the village of Melrose in a sea of grasslands stands a small remnant patch of cottonwood and poplar that act as a magnet for birds seeking refuge and respite during migration. A year-round home to Barn and Great Horned owls, this unassuming stand can be truly aflutter in spring and fall with warblers, flycatchers, vireos, orioles, and sparrows.

Bitter Lake National Wildlife Refuge

Some of the richest biodiversity and birding in the state lies just a few miles from famed Roswell. This 24,000-acre refuge along the Pecos River draws thousands of wintering Snow and Ross's geese, Sandhill Cranes, a plethora of duck species, and hundreds of Wilson's Phalaropes to its shores, while desert specialties such as the Harris's Hawk and Pyrrhuloxia lurk in the bosque.

Bitter Lake NWR draws thousands of wintering birds.

A migrant trap that draws birds from both east and west, Carlsbad Caverns National Park is a marvel in its own right, where you can spend your nights looking for bats!

Rattlesnake Springs at Carlsbad Caverns National Park A whopping 357 species have been recorded in this relatively small park oasis. Beyond its draw as a migratory stopover point, Rattlesnake Springs is a breeding site for three of New Mexico's threatened species: the Bell's Vireo, the Gray Vireo, and the Varied Bunting, as well as the brightly colorful Painted Bunting. Pay special attention during migration, as spring serves as a migrant trap for wayward warblers and other rarities.

BRINGING BIRDS TO YOU

BIRDING IS A FINE EXCUSE to take grand adventures into the wilderness. It's also a great excuse to go for a walk in a park or community garden, sit on the patio, or pay a visit to a local water treatment plant (some seriously great birding, if not the sexiest locale). One of the biggest perks of the pastime is that you can observe and hear birds virtually anywhere, from garbage dumps to garden groves, and with a little bit of effort, you can bring them right to you.

URBAN BIRDING

City parks are bird magnets, particularly during migration, when they serve as oases in seas of asphalt and development. Often equipped with a combination of water features, forest canopy, and grassy areas, parks can attract surprising species diversity, and their residents are sometimes a bit more tolerant of our gawking than are their counterparts farther afield.

Cemeteries, office parks, and apartment complexes—particularly those with high concentrations of native flora—should be high on every urban birder's list. Most birders realize fairly quickly that it pays to have binoculars within reach at all times, even in places not immediately thought of as likely birding hotspots.

OPPOSITE A Yellow-rumped Warbler inspects the contents of a trash bin at Gene C. Reid Park in central Tucson.

Water treatment plants and water reclamation facilities can offer some spectacular birding opportunities in a desert landscape. Two of Arizona's top eBird hotspots, Sweetwater Wetlands and the Riparian Preserve at Water Ranch in Gilbert, fall into this category, with upwards of 300 species observed.

NO YARD NECESSARY

You don't need a backyard to "backyard bird." A window or a patio works just fine, and in Arizona and New Mexico, you can attract a variety of hummingbird species year round with a simple feeder and some sugar water (don't buy that red stuff, sugar and water is all you need, and it's better for the birds). Just be sure to keep feeders clean, and treat windows with decals or other deterrents to avoid avian collisions. If properly situated, birdhouses are often appreciated by cavity-nesting species—no yard necessary, but do your research before placing them.

PLANTS FOR BIRDS

If you have a yard, consider creating habitat rather than relying on feeders to draw in birds. The two most impactful things you can do to attract birds are to provide water and native plants. The thrill of watching a hummingbird or an oriole treat itself to the nectar from the flowers you grew from seed, or hosting nesting birds year after year, is worth the effort. You'll be increasing habitat connectivity for birds while benefiting pollinator species

OPPOSITE, TOP A male Costa's Hummingbird visits the flowers of a chuparosa.

OPPOSITE, BOTTOM This little house has been occupied for several consecutive breeding seasons by a family of Black-capped Chickadees. This proud parent is bringing home a mouthful of caterpillars, one of hundreds of trips made in a single day.

to boot. You can often collect seeds for free if you know what to look for.

Brush piles provide shelter and safety for sparrows, finches, and lizards (not birds, but fun bycatch—and why not share the love?). They cost nothing to create and might just get you to trim that tree you've neglected for too long. Water features provide a critical resource in the arid Southwest and draw everything from warblers to owls. Just keep the water clean, shallow, and, preferably, moving.

ETHICS AND ETIQUETTE

No one wants to be that jerk who gets up in the middle of the game, blocking someone's view of the winning goal. We don't cut in line, we don't use our cell phones at the movies, and we certainly do not disturb owls! As an activity enjoyed by millions, birding has relatively universal community guidelines, for better or worse. Because you won't find birder etiquette conspicuously posted at every birding site, here are a few guidelines that may help you avoid any unnecessary shame on future outings and help you offer gentle reminders to others during your birding adventures.

Proximity

As a general rule, when birding, your behavior should not impact the birds' behavior, disturb their nests or roosts, or cause harm to their habitat. If you are flushing birds, you're too close. If you're trampling plant life, you're destroying habitat and food sources. If you're forcing a bird away from its nest or stopping it from courting, feeding, calling, or sleeping, that's right, you're too close!

Do stay on the trail to view birds, plants, and other wildlife. Do use optics if you have them, giving birds their personal space. Otherwise, use those ears! Vocalization is far and away the best mode of identification and an impressive skill to acquire.

Callbacks

Callbacks, in this context, are recordings of bird calls that are broadcast via audio device in the field—an enormously useful tool for those conducting bird surveys or otherwise providing species data intended to discover or protect habitat for birds. Occasionally, group leaders will use recorded calls or mimic bird vocalizations to draw in a target species (an owl prowl, for instance) for the enjoyment of the group. Scientists and guides use these methods judiciously and infrequently, and they cease playback as soon as the target bird is detected.

Using callbacks as a mode of birdwatching is, quite frankly, lazy and potentially destructive. Birds will often respond out of curiosity or defense and will expend needless energy trying to locate the bird living in your cell phone rather than going about their day. If you'd like to confirm your observation by playing back a song or call, keep your phone or device at a low volume so as not to alarm surrounding birds.

Photography

This is a touchy subject for many of us. Sharing your love of birds through photos is a gift, and photographing birds is an endlessly challenging endeavor. However, the moment your determination to photograph a bird results in its changed behavior, you've crossed

LEFT This terrific shot of an active Harris's Hawk nest in a giant saguaro is appropriately blurry, meaning the photographer kept their distance.

OPPOSITE Members of the Tucson chapter of the Feminist Bird Club hike through a forest at a monthly birding outing.

the line from admirer to antagonist, and you are too close.

As a general rule, avoid using flash photography, which can temporarily blind the subject, preventing it from hunting and/or feeding or avoiding obstacles or even predators. And for a multitude of reasons, do not use drones to photograph birds or bird colonies. A drone can not only cause severe stress for birds and other wildlife, but it is a nuisance to fellow humans seeking solace in nature.

Sharing observations

Be mindful of sharing locations, particularly those of threatened or nocturnal species whose feeding and roosting patterns can be dramatically altered by the presence of humans. Owls are particularly sensitive, and there are far too many accounts of them meeting their demise after throngs of photographers and unruly birders have disrupted their sleep and hunting cycles. Better to view an owl's quirky magnificence from a distance, and keep its location to yourself.

Be inclusive

Birding has a bad, but admittedly deserved, rap for being elitist, exclusive, snooty, patriarchal—all of those icky words. Let's change that by being better birders. In an ideal birders' world, all people would revel in observing birds and their behaviors and would be inspired to protect their habitat. If we desire that, then we,

as a community, should strive to make birding an inclusive and accessible activity. There are many ways to discourage folks from getting into birding, and perhaps the most effective deterrent is giving them a negative experience that makes them feel like outsiders. It's just as easy to create an environment that sparks curiosity and discovery, without the elitism.

- Refrain from judging a person's birding ability or interest by their lack of equipment, gender, skin color, age, disability, or any combination thereof.
- Respect and use preferred personal pronouns.
- Optics are a luxury for many, and some don't like to use them or lug them around on a hike (we often opt out of optics and choose to put our ears to use). Avoid offering unsolicited critiques on the quality or absence of someone's optics.
- Respect differences. In the culinary world, there's a delightful saying: "Don't yuck my yum." Some of us enjoy the antics of starlings, while others can't get past their introduced status. Some of us aren't rare species chasers, and some of us love the thrill of adding another bird to a life list. Whatever your birding flavor, don't assume the same from other birders, and learn to appreciate varying perspectives.
- Be kind, listen more than you speak, and respect boundaries. Like birds, some birders prefer solitude, and some are gregarious. Be mindful of others' body language when you're on the trail, and give space to those who desire it.
- Personally acknowledge and endeavor to understand the Indigenous histories of the land you are birding on, and be a grateful steward.
- Openly consider the observations of others, even if you question their validity. Instead of questioning them, help them verify their observations.

- When birding in a group, be aware of others and make space for those with disabilities without making a production of it.
- If you're a group leader organizing a birding event, ensure that you're leading an inclusive, accessible outing that is welcoming to folks of all backgrounds and abilities.

Trail etiquette

Birding will often send us on wild goose chases (pun intended) through hiking areas shared with recreators of all varieties. Not all of them will be interested in the Gray-crowned Rosy-Finch or Elegant Trogon you're trying to track down. These simple guidelines will ensure your consideration for others on the trail, and for the wildlife that surrounds it.

Right of way and shared-use trails Downhill hikers yield to uphill hikers. When hiking in a group, hike single file to enable others to pass. Horse riding is common on trails in the Southwest. Yield to oncoming riders by stepping aside or safely off trail, if necessary. If passing a rider, gently announce yourself to avoid frightening the horse.

Dogs Keep them on leash in urban or shared spaces and hold them away from passersby. Be vigilant in avoiding venomous wildlife your dog may encounter.

Photography Be sure you aren't holding up traffic on the trail while you're getting that shot.

Preparedness Know where you're going, and tell someone where you'll be.

Leave it better

Public lands, for better or for worse, have experienced a drastic uptick in visitors in recent years, and this is sadly apparent as an encounter with toilet paper, plastic waste, food waste, and fishing line is as likely as an encounter with wildlife. The Southwest offers free access to hundreds of thousands of acres of wilderness to explore and recreate within. In a perfect world, we would in turn revere and steward these unique habitats. In reality, public lands in the Southwest face the persistent perils of overgrazing, extreme drought, intense fires, invasive grasses, extractive industries such as logging and mining, and irresponsible recreation. We'll be doing ourselves, our land, and our fellow birders a favor by following a few simple principles.

Flipping rocks or logs Many birders are also general naturalists who are tempted to overturn logs and rocks to discover the hidden worlds beneath. If, like us, you can't resist the urge, use caution and care: there could be a rattlesnake, a scorpion, or a giant centipede hiding below, and they may not appreciate the disturbance. Always gently place these little habitats back in the position you found them, taking great care not to disturb or damage the creatures that live there.

Building "ornamental" cairns Conspicuously stacked rocks used for wayfinding in the most remote locations are tools for backcountry travelers, not entertainment for casual visitors. Rocks are habitat and should be left in place, especially those in streams and riverbeds.

TOP Cows can't bury their poop, but we can—do your part to lessen your impact on the wilderness.

BOTTOM By moving rocks, you are moving the shelter on which wildlife relies.

Making noise Step lightly, keep your voices down, and listen for those bird songs!

Staying on the trail Even if puddles are present, stay on the trail, and don't cut corners on switchbacks. Staying on the trail preserves the integrity of the trail itself and of the flora and fauna surrounding it.

Disposing of waste You know the motto: pack it in, pack it out. That goes for food, packaging, pet waste, and, for the love of Pete, if you can't bury it, pack out your toilet paper. Relieving yourself in the wilderness is just fine, but please do so at least 200 feet from any water source. Do not leave your waste under a rock with toilet paper strewn about; this is both detrimental to the environment and extremely unpleasant for fellow hikers to discover. Dig a hole at least 6 inches deep and cover your solid waste, and pack away wipes in a sealed bag.

Making fires If you're camping out to catch the dawn chorus or a particularly early bird such as the Elegant Trogon, be aware of and adhere to fire restrictions. The dry conditions of Southwestern deserts are a perfect recipe for wildfires. You do not want to be responsible for starting one! Always, without exception, extinguish fires completely before leaving them.

Respecting wildlife Don't feed it, don't approach it, don't harass it.

Leaving everything in place Don't take souvenirs, and scrape those boots clean before and after your visit. Invasive grass and other plant seeds love to hitch rides on your shoes and clothing.

Birding on public lands

Public lands are federally managed lands to which all Americans share access. (Note that state lands are not public lands and access requirements for these areas can vary.) Public lands come with many designations, including national forests, national monuments, parks, wildlife areas, and more, and they are managed by federal agencies including the US Forest Service, Bureau of Land Management, US Fish and Wildlife Service, and National Park Service. In Arizona and New Mexico, roughly 35 percent (52.7 million acres) of the landscape is public land.

Public lands are managed differently depending upon their type. For example, national parks are managed primarily for recreation and wildlife resources, while national forests are managed for multiple uses, including recreation and wildlife resources, but also for grazing, extractive uses like mining and logging, hunting, off-road vehicle use, and other uses. It's important to remember that these lands belong equally to all of us and to respect all public land uses and users.

That said, it's critical that these uses are managed with a balanced approach that ensures that one use doesn't exclude another. As users of public land, it's our responsibility to be active in its management, and you can do this by engaging with public comment processes through managing agencies and by participating with and supporting friends groups, local nonprofits that work to connect communities to their nearby public lands. Through this participation, birders can help protect our public lands and the habitats, wildlife, and birding opportunities they support.

Birding on Indigenous lands

If you are birding in Arizona or New Mexico, you are birding on what was, before

colonization by Europeans, Indigenous land. The wildlife, the habitats, and the land itself all have deep ties to native communities and cultures, and it is important that birders acknowledge the history of and give due respect to the lands on which we bird. Some of the most stunning bird habitats in the Southwest are on land belonging to tribal nations.

In Arizona and New Mexico today, roughly 19 percent of the landscape (some 27.9 million acres) is designated Indigenous land that belongs to sovereign nations with independent governments, their own laws and regulations, and unique cultural connections to wildlife and habitat. Though some Indigenous communities openly welcome noncommunity members, others require special permits; many allow visitors only in certain areas, and others do not allow outside visitation.

LEFT Exploring wilderness areas is always an exciting adventure.

RIGHT The Appleton-Whittell Research Ranch of the National Audubon Society, in Elgin, Arizona, allows visitors by appointment only; other landowners may not be so welcoming.

Indigenous lands are not public lands. Before planning to bird on any Indigenous lands belonging to communities of which you are not a part, prepare as you would when visiting any other sovereign country. Educate yourself on laws, regulations, and culture; familiarize yourself with important boundaries; and acquire necessary permits. As a visitor on tribal lands, your behavior can affect both the community you're visiting and accessibility for future birders.

Birding on private land

"No Trespassing" means just that: your presence on private land without permission from the landowner is trespassing and can be treated as such, both legally and by landowners who are often ready and willing to defend their property. Ignorance of land ownership is not an excuse when you're caught trespassing on private land without permission. It is the Wild West, after all.

Though some landowners do not offer access to their land, some do and levels of access vary. Some landowners allow the public to pass through their property only on main roads on their way to other destinations, while others allow folks to stick around and bird. In either case, it's important to follow some basic rules.

- Obtain any needed permissions and be sure to bring proof with you. If you're visiting a property that isn't often visited by the public, try to contact the landowner directly before entering the land.
- Always stay on designated roads and trails, stay out of restricted areas, and respect the property's wildlife, habitat, and infrastructure.
- Leave all gates as you find them. Landowners with livestock often use gates to manage their animals' grazing and water access, and you won't be doing anyone any favors by closing a gate intentionally left open.

Remember, your behavior while on private land can directly influence future access. In Arizona and New Mexico, landowners take private property rights seriously and so should birders. Both safety and access to some of the states' best birding locations depend on it.

BIRDING SAFELY

Ignoring safety can ruin a day of birding, and in a region rich with hazards, those who venture outdoors unprepared can easily cut their life (and life list) short. Whether you're birding in an urban area or deep in the wilderness, the uniqueness of the Southwest brings with it unique dangers. Experiencing the region and its birds safely requires that you understand the risks, be prepared, and plan ahead.

The most ever-present danger in the Southwest is the sun. Even during the cooler months, the combination of cloudless skies and low humidity can lead to rapid overheating and dehydration. Try and plan your birding for the early morning or near dusk, and no matter where or when you're birding, be sure to bring plenty of water—at least one liter for each hour you'll spend outside, and more if you have room in your pack or vehicle (you never know when your trip may become longer than anticipated). You should also be prepared to protect yourself from the sun's harmful rays, as the Southwest leads the nation in skin cancer rates. Sunscreen, a hat, neck protection, and long sleeves can all help keep the sun from damaging your skin. Lastly, remember that even the most rugged, prepared birder can be bested by the desert sun. If you are spending time outdoors in the Southwest, you should know the signs of heat exhaustion, heat stroke, and dehydration and how to respond to each.

Although it may seem that the Southwestern sun never rests, the reality is that the region's weather is both diverse and fickle. Particularly in the desert, daytime and nighttime temperatures can vary drastically. Temperatures can also vary dramatically with gains and losses in elevation. Check the weather ahead of your visit and come prepared with multiple layers of clothing. Just as the thermometer bounces up and down, precipitation can come with little warning. Storms typically occur in winter and summer. Winter storms bring gentle rain to the lowlands and sometimes heavy snow to high elevations. Summer storms, which typically occur between early July and late September during our monsoon season, are characterized by sudden downpours and violent thunderstorms. Throughout the monsoon season, flash floods are common, when typically dry drainages transform into roaring streams. These floods don't present as steadily rising water, but as walls of water that crash downstream with enough force to uproot trees and bring boulders, debris, and even vehicles along for the ride. Never attempt to cross a flooded wash, never set up camp or park in a wash, and pay attention to washes you cross on your way to your birding destination. Even if you're not anticipating rain, a storm during your outing could render your return route impassable. Storms tend to roll in just before dusk, and lightning strikes are extremely common. If you find yourself at high elevations during the monsoon season, especially if you're birding above the tree line, head downhill by mid-afternoon.

Other natural hazards in the Southwest come in the form of flora and fauna. Our vegetation is among the spiniest in North America, and it's not just the cacti. Our trees, shrubs, and even annual flora are especially thorny, providing another reason to wear long sleeves and pants. Wildlife can also pose risks. Ticks and chiggers can be extremely common (you'll know they're active when you see folks with their pants tucked into their boots), and you may encounter larger invertebrates including scorpions, centipedes, and spiders. These animals should be appreciated and respected. You should never put your hands into holes or crevices you can't see into, and you should always check your shoes for critters before slipping them on.

You also may be fortunate enough to catch a glimpse of one of the Southwest's venomous reptiles—rattlesnakes, coral snakes, or Gila monsters. Enjoy them from a distance and do not attempt to handle them. To ensure that you always see them before they see you, avoid hiking through brush that prevents you from seeing your feet and legs (or the reptiles) and near the edges of trails, dense vegetation, and rock walls, where a coiled snake could be hiding. If you do get surprised by a snake or other reptile, sturdy boots that cover your ankles can be a lifesaver if it decides to strike. Other threatening wildlife may be present as well, including javelina (typically only a threat if you have a dog in tow), mountain lions, and black bears.

In addition to threats from the natural environment, some humans you encounter while birding may pose risks. Always be aware of other people in the area where you're birding, and remember that your safety is influenced by your identity. Just because a place is safe for one birder does not necessarily mean it's safe for another. Always be sure to know who owns the land where you're birding, and get permission before you enter the area; confrontations with private property owners can escalate quickly. If you're birding along the Mexican border, be aware that you may encounter law enforcement (Border Patrol) at a much greater frequency than you would elsewhere. Drug and human trafficking can be common in certain areas along the border, but encounters with them are infrequent, and you are more likely to be approached by traveling migrants seeking help. Remember that offering help in the form of water and food is not against the law and may save a life. Transporting migrants in your vehicle, however, is a crime.

If you do come across activity, it is best to keep to yourself, avoid initiating contact, and move along as quickly as possible. Do

ABOVE Heat exhaustion and heat stroke are serious issues; stay hydrated and wear proper clothing.

OPPOSITE, TOP Gila monsters are a rare delight to encounter, but they pack a punch; do not touch.

OPPOSITE, BOTTOM Rattlesnakes are to be respected, but the hot sun can actually prove more dangerous if you fail to prepare appropriately for a day spent in the desert.

not engage. In areas where border-crossing attempts are common, be aware that desperation can sometimes motivate theft. As you would anywhere else, take precautions and don't tempt fate by leaving your valuables under your seat. Other precautions include camping only in established campgrounds, carrying identification, and traveling with a companion.

Last on the list of potential risks are bad roads and bad luck. Backroads often aren't mapped on GPS platforms; if you're navigating by cell phone GPS, keep in mind that you are likely to lose the cell signal in remote areas. Up-to-date paper maps are far more reliable. Four-wheel-drive vehicles may be necessary, and a full-size spare tire (or two) and a can of extra gasoline may be lifesavers if you're traveling in the boonies and have a flat or run out of gas. Check ahead to see what services are or are not available in the rural areas through which you are passing, bring emergency supplies, avoid traveling alone, and let someone know where you are going and when you expect to return.

Almost as impressive as the Southwest's avian diversity is the diversity of risks faced by those seeking to explore the region. With the right knowledge and a bit of preparedness, you can help ensure that you return home from each trip with many more birding adventures ahead of you.

INCLUSIVITY IN BIRDING

THE BIRDING COMMUNITY, by and large, has consisted of wealthy, older, white men and women. Birders, however, span a broad range of demographics and identities. When you think of a birder, you should include in your mental image a person waiting for a train in Brooklyn, New York, observing Rock Pigeons with fervor; or a family taking a weekend walk in their local park, enjoying the birds that add color and song to their morning, not worrying themselves with identifying each and every bird that flies their way.

Birding is for everyone, everywhere and anywhere, and the way people observe and appreciate birds is deeply individual. For generations, however, this intimate experience of discovering bird behavior, song, and appearance has seemingly been reserved for those without physical disability and with income to support optics and gear habits, the privilege of leisure time, and, often, skin color and/or gender that isn't perceived as out of place in the outdoors. Thankfully, those perceptions and demographics are rapidly shifting, as are the accessibility features of birding hotspots throughout the Southwest.

ACCESSIBILITY

Barriers to accessibility include lack of inclusivity, lack of awareness that birding can be an accessible activity, and physical barriers to birding hotspots. Some of these barriers are beginning to break down as the birding community at large redefines itself with growing networks and initiatives. The Feminist Bird Club, Black Birders Week, Queer Birders of North America, and Birdability, as well as movements within established conservation organizations, are shifting to a birding culture that embraces a new definition of what being a birder means.

FINDING ACCESSIBLE BIRDING LOCATIONS

If you or one of your fellow birders has a disability, is a wheelchair user, or is unable to use optics as a birding tool, finding information about accessible birding locations and the amenities they offer can be a challenge. You certainly do not want to go through the trouble of traveling to a hotspot, just to find it "unbirdable" for some individuals. Birdability, an organization dedicated to creating inclusive and accessible birding locations around the globe, has partnered with the National Audubon Society to support a crowdsourced online map of birdable locations, including many within Arizona and New Mexico.

Stellar examples of accessibility efforts exist in some of the best birding spots in the Southwest. Sweetwater Wetlands, a water treatment facility and urban habitat complex in Tucson, is fully accessible and boasts more than 300 observed species. Just northwest of Santa Fe, the Los Alamos Nature Center (often referred to as the Pajarito Environmental

OPPOSITE The bird blind overlooking the wetlands and bosque at Rio Grande Nature Center State Park provides various levels of viewing openings that are accessible to kids and those in wheelchairs.

Education Center, or PEEC) rates highly for accessibility and offers visitors front row seats to their feeders, frequented by Red Crossbills and the frantic antics of Pygmy Nuthatches. If you are unsure about a site's accessibility, call ahead to confirm before you visit.

A WELCOME SHIFT IN BIRDING CULTURE

Racism and (to a lesser, yet still frustrating, extent) sexism have been embedded in this pastime since its inception, despite both the hobby and field of study realizing great benefit from the contributions of people of color and women. In modern birding history, white men have had a monopoly on the birding and ornithology market, as field guide authors, trip leaders, and chairs and CEOs of prominent birding and ornithological organizations have created a lopsided image of authority on the topic. We shouldn't diminish the contributions of these men to the field, but to assume they've forged the path alone is wildly inaccurate.

WOMEN IN BIRDING AND CONSERVATION

A rich documented history of women in birding goes back many generations—from those who have adopted birding as a hobby, to those who have actively influenced landmark legislation protecting migratory birds. An

Florence Augusta Merriam Bailey penned a series of field guides on North American bird life with birders in mind and helped popularized the birding movement.

important early contributor was ornithologist Florence Augusta Merriam Bailey, who penned one of the first field guides to birds of Eastern America, *Birds Through an Opera Glass*, which was published in 1889. Bailey's book showed a departure from the "shotgun ornithology" practiced by the likes of John James Audubon (who killed birds with a shotgun as part of his study of them) by rendering poetic expressions of species' behavior in addition to vivid, artful descriptions of their identifying features.

Throughout the eighteenth, nineteenth, and early twentieth centuries, when it was common for women's hats to feature feathers, bird conservation in the United States grew in popularity. Created by milliners from Paris to New York, some hats in the late nineteenth century were actually adorned by an entire stuffed bird! As the trend peaked, the unlikely intersection of feminism, fashion, and conservation brought about one of the most critical milestones in bird protection to date. Millions of birds, including egrets, birds-of-paradise, herons, and even hummingbirds, had met their demise from the insatiable demand for avian adornments.

It was early female conservation activists, led by Harriet Hemenway and Minna Hall of Massachusetts Audubon, who rallied to end the wholesale slaughter of wild birds for the sake of fashion. Their advocacy resulted in the passing of the Lacey Act in 1900, which outlawed the transport and sale of illegally taken wildlife across state lines (though it was ultimately ineffective with the feather trade), and later the formation of the National Audubon Society and the passing of the 1913 Weeks-Mclean Act. This act was the precursor to the bedrock Migratory Bird Treaty Act of 1918, which to this day makes it a crime to "pursue, hunt, take, capture, kill [or] offer for sale . . . any migratory bird [or] any part, nest, or egg of any such bird."

Today, women account for well more than half of the birding community, and they are blazing trails in avian research and habitat protection; they also author and co-author field guides (like this one) and compile breeding bird atlases for sites around the world. Going beyond creating a space for women in birding, the thirty-five active chapters of the international Feminist Bird Club, founded by Molly Adams in 2016, take an intersectional approach to birding through the mission of promoting inclusivity in birding while fundraising and providing a safe opportunity for members of the LGBTQIA+ community, BIPOC, and women to connect with the natural world.

BRINGING IN NEW, NEEDED PERSPECTIVES

The field of conservation and its institutions have not only contributed to the insidious history of institutionalized racism in the United States; in fact, they have upheld it. These have been some of the slowest organizations to make real change when it comes to internal power structures and cultures. What does this have to do with birding? A lot. Not only do these organizations have difficulty retaining racially diverse staff, but they have perpetuated the image that birding and ornithological research belongs primarily to white people.

While institutions take their time loosening their grip, BIPOC (Black, Indigenous, People of Color) birding communities and scientists are taking flight on their own, and the results are exactly what the birding community needed but could not do in a vacuum. The BIPOC birding community is thriving, and social media has become a powerful tool in showing both BIPOC and LGBTQIA (Lesbian, Gay, Bisexual, Transgender, Queer, Intersex, and Asexual) communities that they are represented in and add value to the birding and conservation space. Individuals—including the amazing, auburn-haired Pattie Gonia and the utterly delightful Alexis Nikole Nelson—use social media platforms not only to amplify the voices of other queer and BIPOC scientists, photographers, and naturalists, but they speak to issues facing wildlife habitats, women, and the LGBTQIA and BIPOC communities. Specific to birding, the Black AF in STEM (which stands for science, technology, engineering, and mathematics) collective, along with the Smithsonian Institution, founded the awesome and continually growing Black Birders Week, which unites and celebrates a robust national community of Black scientists and birders.

There is much to celebrate as sorely needed diversity in representation increases in the birding world, but to assume that birding is always a safe activity for people of color and the LGBTQIA community is naive. It is the responsibility of all of us to keep turning up that amplification dial, not only in representation in birding, but in accepting how people are birding and their personal expression of it. For the birds (and people), this can only be a good thing.

Representation among birders, scientists, and naturalists is expanding, to the benefit of all.

CONSERVATION AND COMMUNITY SCIENCE

THE SOUTHWEST SUPPORTS an incredible diversity of habitats, birds, and other wildlife, but an almost equally diverse list of conservation challenges puts it all at risk.

ARIDIFICATION AND CHANGING PATTERNS

As climate changes, average temperatures across the American West are rising, the region is becoming increasingly arid, and the precipitation we do get is falling outside historical patterns, leaving habitats stressed and communities racing to adapt. As reliable water sources dry up, people are turning to untapped groundwater and surface water to make up the difference, which puts water-dependent habitats and the birds that rely on them in an increasingly precarious situation.

Climate change is also increasing the risk of catastrophic wildfire. In our forests, poor management has led to an overdensity of trees, which carries these fires much farther and with much greater intensity than in the past, leaving massive scars that are slow to recover—and since these areas can no longer absorb rainfall, this puts communities at risk of flooding. Nonnative plants are carrying flames into desert habitats. These areas are poorly adapted to the impacts of fire, and we likely will not see their recovery in our lifetime.

Already stressed, natural areas have become less able to withstand disruptive land uses such as urban encroachment; inappropriate cattle grazing; irresponsible agricultural practices; mining, oil, and gas development; and careless outdoor recreation, to name a few. Fortunately, as someone who cares about the Southwest's birds (you bought this book, after all), you can help. Opportunities to give back come in all shapes and sizes, and you don't have to be an expert naturalist or a high-dollar donor to have a significant impact. Regardless of your skill set or resources, there is a place for you in conservation.

YOU CAN TAKE ACTION

If you're inspired to take action, you can start by connecting with people who are already working to protect the places you care most about, such as conservation nonprofits, friends groups, birding and other nature-centric clubs, and governmental agencies such as the US Forest Service or state wildlife departments. In addition, use your voice to support policies that prioritize conservation and the policymakers who create them. Whether you take action at the local, regional, or national level, your voice can make a difference for birds.

Lastly, as the proud owner of this field guide, you are one step closer to taking action for birds in yet another capacity—as a community scientist! Community science projects send people of all stripes, regardless of occupation or other identifiers, into the field to collect data that, on their own, professional

collect. With provided training, mentorship from experienced observers, and projects designed with public accessibility in mind, anyone can be a community scientist. In fact, you don't even have to engage with an established project to become a community scientist. Online and phone-based applications such as eBird and iNaturalist enable you to enter your observations in real time, turning even the most casual of outings into a scientific expedition.

As birders, it is our responsibility to give back to the birds and habitats that provide us so much joy. The Southwest's habitats, birds and other wildlife, and human communities depend on it.

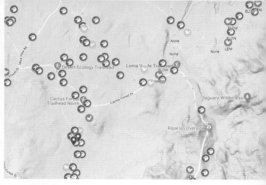

TOP Fire is a natural occurrence, but human activity has increased fuels and intensity.

BOTTOM Volunteer community scientists collected data to track Elf Owl populations in Southeast Arizona.

HOW TO USE THIS BOOK TO IDENTIFY BIRDS

Birds of Arizona and New Mexico provides everything you need to explore birds and their habitats across both states. Included are all the birds you are most likely to see in the region—common wintering, breeding, and migrant birds, as well as rarities that occur with some regularity. With about 600 species recorded across the two states, we had to make some tough calls about which birds to include. When deciding which rarities to include, we prioritized species that occur in small numbers most years (such as the Rose-breasted Grosbeak) and species emblematic of the Southwest that are unlikely to be found in other guides to North American birds (such as the Elegant Trogon).

The organization of species in this guide closely follows the most recent American Ornithological Society (AOS) Checklist, which organizes birds taxonomically or by evolutionary relatedness. The book also follows AOS conventions for species' common and scientific names, and for each species account, both names are included, along with suggested alternatives for honorific common names.

OPPOSITE The Scott's Oriole is likely on its way to a more suitable name: the Yucca Oriole.

PHOTOS

Each species account includes one or more photos that highlight key identifying features, or field marks, of the species being described. Beyond basic editing for lighting and color, we have not significantly altered any of the included photographs. Although we did our best to curate photos that fully represent each species, appearances can vary between individual birds—geographically, during different times of year, by age, by sex, and depending on environmental lighting. When you're using these photos as identification tools, you may have difficulty identifying a bird if you simply try to find a perfect photographic match; instead, use a combination of field marks shown in photos along with range, habitat, behavior, vocalizations, and other identification clues to reach a conclusion.

Each account offers a succinct physical description of the bird. Length is measured from beak tip to tail feather, and wingspan measures the distance between the tips of the primary flight feathers. Each description covers the bird's appearance in terms of stature, body shape, distinguishing features, and colors. If sexual dimorphism is present in the species, descriptions of male, female, and juvenile features are included. If only subtle differences in appearance occur between sexes, the description mentions that sexes are similar and known subtleties are noted. In some species, such as the Red-tailed Hawk and Dark-eyed Junco, descriptions may refer to various subspecies, or morphs, that occur within Arizona and/or New Mexico, such as the "dark morph" Red-tailed Hawk or "pink-sided" Dark-eyed Junco.

HABITAT

In the Desert Southwest, habitat types vary drastically, often within a relatively small geographic area, as a result of topography and elevation. Within 15 miles, one can experience a classic low- and high-desert saguaro forest or harsh chaparral; oak, juniper, and pine forests; and riparian areas lined with cottonwood and sycamore. With each shift in habitat comes a new suite of birds. For habitat generalists such as grackles and Mallards, the preferred habitat is highlighted, and for habitat specialists such as the Botteri's Sparrow or Elegant Trogon, a description of habitat is accompanied by key site descriptions. Keep in mind that birds can often be found outside of their traditional habitat and range, particularly during migration.

BEHAVIOR

Observing bird behavior can be delightful, awe-inspiring, and sometimes terror-inducing, but it is always fascinating. In some cases, behavior is diagnostic of a species. The telltale rump-bobbing of a Spotted Sandpiper or tail-pumping of a Gray Flycatcher is enormously helpful for identification, particularly with those notoriously difficult-to-identify shorebirds and *Empidonax* flycatchers. These behavioral descriptions aid in identifying as well as in developing a deeper understanding of the species itself, including foraging and flight patterns—such as the rustling hop-kick of a foraging Abert's Towhee or the slow-pumping wingbeats of a Cooper's Hawk versus the erratic flap of the Sharp-shinned Hawk.

VOCALIZATIONS

Oh, the challenges of birding "by ear" and of translating bird language within the confines of the Latin-script alphabet! Birdsong, and the way we hear and describe it, is both subjective and variable. Just like people, birds adopt regional dialects and sometimes sing out of tune, and many a mimic can fool even the best of us. That said, birding by ear is one of the most useful skills one can develop as a birder, because, as all birders know, birds are fickle, inconspicuous creatures, and not often do they wish to make their presence known, unless—case in point—they are trying to mate, in which case you'll hear their songs ad nauseam. Breeding season is the best time to get out and immerse yourself in birdsong and to start picking up on the sound quality of a thrush versus a warbler, blackbird, jay, or flycatcher.

Having listened to these species both in the field and through recordings, we have done our best to represent the vocalizations of each one phonetically to aid your identification. Although we have not adhered to any hard and fast rules in our interpretations, we use vowels to communicate pitch and capital letters to communicate emphasis. In some instances, we introduce mnemonic devices that have aided us in the field.

ABUNDANCE

Extra information about each species' relative abundance in Arizona and New Mexico will help you determine which birds are more likely to be seen than others. For example, although the ranges of the Golden Eagle and House Finch span the entirety of Arizona and New Mexico, you are much more likely to see a flock of House Finches than you are to see even a single Golden Eagle. In terms of relative abundance, House Finches are

A Scaled Quail poses conspicuously while calling for a mate.

widespread and common and Golden Eagles are widespread and rare.

Abundance information will help you zero in on species that are uncommon to rare at the regional level but common in a particular habitat or at a particular site. For example, Sandhill Cranes can be incredibly abundant in their wintering grounds at New Mexico's Bosque del Apache National Wildlife Refuge or Arizona's Whitewater Draw Wildlife Area, but these are among only a few sites where cranes winter in the Southwest. This species' abundance is described as "extremely numerous at key winter roosts and a common winter resident in surrounding areas." When this information is combined with information in the "Key Sites" section of the account, you will be able to determine exactly when and where to go to observe these large birds.

Abundance information is also useful when discussing rare vagrants to the region. For example, though Rose-breasted Grosbeaks are extremely rare visitors from the Eastern United States, they do occasionally show up in areas across Arizona and New Mexico. For this species and others, abundance may be described as "rare/uncommon but regularly observed across both states." This information also helps clarify instances in which a species' conservation status is seemingly in conflict with its abundance in our region. For example, the Yellow-billed Cuckoo is listed by the International Union for the Conservation of Nature (IUCN) as a species of Least Concern, but it is also considered a federally threatened species in the Western United States.

Wilson's Warbler, aka Black-capped Warbler

KEY SITES

For species that regularly occur but with limited range, or for those whose ranges only slightly overlap with Arizona/New Mexico boundaries, descriptions of specific sites where they are commonly observed are included.

LOOKALIKES

In the cases of lookalike species, both a reference to the species and highlighted distinguishing factors are included for identification.

ALTERNATIVE NAMES

Black-throated Gray Warbler, Hooded Oriole, Blue Grosbeak, and Scott's Oriole. One of these things is not like the other, and one of these things just doesn't belong. For a multitude of practical reasons, birds should not be saddled with the surnames of people, and throughout this guide we offer suggested alternative names to eponymous and honorific bird names.

From an ethical perspective, many eponymous and honorific names act as a memorial to colonizers with historically documented, horrific pasts. The Scott's Oriole, for example, is a particularly blighted species because it refers to its namesake, General Winfield Scott, who both handed out orders to commit and directly committed acts of genocide against Indigenous peoples—specifically, the systematic and brutal removal of Cherokee peoples from their land, an atrocity widely known as the Trail of Tears. To boot, Scott had no documented affiliation with or affinity for birds or nature expeditions. Conversely, the poet and naturalist Alexander Wilson, for whom the Wilson's Warbler is named, was of seemingly upstanding character. Inarguably, neither offers anything of value in defining a bird's features. In June 2021, a proposal was submitted by Robert Driver of East Carolina University and Jessica McLaughlin of University of Oklahoma to the American Ornithological Society (AOS) to change the name of the Scott's Oriole to the befitting Yucca Oriole or Golden-tailed Oriole. In November 2023, the AOS announced that it will be creating a special committee to guide the process of removing and replacing all honorific or otherwise offensive bird names. No matter, as birders have no need to wait on a naming committee, and many of us have readily invited Yucca Oriole into our birding lexicon.

CONSERVATION STATUS

Each account includes the bird's present conservation status as classified by the IUCN Red List of Threatened Species. Widely accepted as the industry standard in the international conservation and life sciences community, the Red List classifies a species in accordance to its vitality, factoring in habitat availability and ecology, climate threats, current populations, and use and/or trade, and then assigns it a level of concern. Categories of concern include Extinct, Extinct in the Wild, Critically Endangered, Endangered, Vulnerable, Near Threatened, and Least Concern, with all categories but the first two acting as a warning system and indicator of the level of coordinated conservation effort necessary to protect a species from extinction. Because the Red List assesses trends at a global level, the level of conservation concern included in this book may conflict with classifications in lists organized at the state level, such as Arizona and New Mexico's Species of Greatest Conservation Need, or lists organized at the regional and national levels, such as the US Fish and Wildlife Service's Species of Greatest Conservation Need.

RANGE MAP

A bird's range can be defined as the geographical area in which it regularly occurs, whether year round, during migration, during the breeding season (summer), or while overwintering. Note that there is often variability within any given species' range. As beginning birders, we often overlook range maps as the essential tool that they are, both in terms of understanding a species and its patterns on a deeper level and to aid in identification. When you've boiled down your bird identification to two lookalike species, the range map will likely offer you a final, decisive clue.

Although the range information in this guide is as accurate as possible, our changing climate means you are more likely than ever to see birds outside of the areas shown in the maps. Rapid changes in ranges and migration patterns are being observed as birds are forced to adapt swiftly. Although we aimed to represent the most recently observed ranges for each species, we recognize that they are likely to shift—for some species more drastically than others, as grassland and desert riparian habitats face more imminent threats than oak scrub and chaparral, and habitat specialists may be forced to travel farther than generalists

Range map key

in their search for suitable breeding or wintering grounds.

As you explore the species in this guide, you will find a handful of accounts for which maps were intentionally left out. For rarities with broadly scattered records, such as the Rose-breasted Grosbeak, maps were not included because mapping these sporadic observations as points across the region would be (at best) of minimal use and (at worst) misleading—especially given the scale of the maps in this book. Maps were also omitted for rarities, like the Black-capped Gnatcatcher, that occur reliably in specific locations, as these pinpoint occurrences are better expressed verbally than they are visually. For introduced species with limited ranges, like the Rosy-faced Lovebird, maps were omitted for the same reason. For these accounts without range maps, consult the abundance section for range information.

Without further ado, folks, let's meet the birds of Arizona and New Mexico.

OPPOSITE Lazuli Bunting

DUCKS AND OTHER WATERFOWL

When the Southwest's warm seasons retreat along with the region's migratory, neotropical specialties, the ducks, geese, and swans (family *Anatidae*) fly in to keep birders busy. While some species linger here year round, these bulky and water-bound birds are most abundant in winter, when impressive congregations gather on waterbodies from large reservoirs to tiny urban ponds. Offering easy-to-identify classics like the Mallard and American Widgeon, as well as challenges like spotting a Greater Scaup in a raft of Lessers, this family has something for beginning and expert birders alike.

ABOVE Female Common Merganser

Adult

Juvenile

Adult

Black-bellied Whistling-Duck

Dendrocygna autumnalis

Shallow, open freshwater wetlands and marshes; cultivated land with abundant vegetation
LENGTH 19–21 in. WINGSPAN 30–36 in.

Statuesque and slender, with both sexes sporting a vibrant, bubblegum-pink bill and legs to match. Deep chestnut-colored body and cap contrasts with black belly and white wing patch; pale, dusky gray face; and distinct white eye ring. Elongated silhouette with slightly downturned head and pink feet extending beyond the tail make them distinct in flight. Juvenile lacks black belly and deep chestnut color, with gray-brown plumage and dark gray bill. BEHAVIOR Unique among ducks, this gregarious species forms monogamous pair bonds. While primarily granivorous, these opportunistic, nocturnal feeders will also dine on aquatic and terrestrial vegetation, arthropods, and aquatic invertebrates. This cavity nester once known as the "tree duck" will make use of nest boxes and other human-made structures and, if necessary, will nest on the ground. Often nests in colonies. VOCALIZATIONS Soft, airy whistling in flight and while swimming and standing. Alarm and distraction *hoo-eek* calls reminiscent of the Wood Duck. ABUNDANCE Uncommon but regularly observed breeders with limited range. Found primarily in Central and Southeast AZ wetlands. CONSERVATION STATUS Least Concern.

Adult (white morph)

Adults

Adult (blue morph)

Snow Goose

Anser caerulescens

Agricultural fields, open water, wetlands
LENGTH 27–32 in. WINGSPAN 53–56 in.

Chunky, medium-sized goose. Two morphs: each has pink legs, black primaries, and pink bill with curved base and black "grin patch." More common white morph is all white with black wingtips and sometimes buffy staining on the head. Juvenile has grayish wash. Less common blue morph is a variable mix of white and brownish gray overall with white underwing coverts and head. Juvenile lacks white head. All juveniles have dark bill and legs. BEHAVIOR Forages along edges of large waterbodies and in wetlands and agricultural fields. At night, flocks roost in large groups on water, producing quite the spectacle at dawn and dusk, often alongside Sandhill Cranes. VOCALIZATIONS Single or series of loud, harsh honks; louder than other geese. In flight, flocks explode with honks, sharp quacks, and cries. Honks vary greatly in pitch. ABUNDANCE Locally abundant at key wintering sites; rare elsewhere. KEY SITES AZ: Cibola National Wildlife Refuge and Whitewater Draw Wildlife Area; NM: Bosque del Apache National Wildlife Refuge. LOOKALIKE Ross's Goose. Snow Goose is larger, has a black grin patch and a longer bill, and sometimes shows buffy staining on a white head. CONSERVATION STATUS Least Concern.

Ross's Goose

Anser rossii

Agricultural fields, open water, wetlands, meadows
LENGTH 23–25.5 in. WINGSPAN 43.5–45.5 in.

"Honey, I shrunk the Snow Goose" would largely sum up the appearance of this small, snow-white goose with jet-black primaries/wingtips, pink legs, and a stubby pink bill. Immature geese are white with gray wash and darker bill. A dark, or blue, morph is extremely rare. BEHAVIOR Forages on agricultural fields, shallow wetlands, meadows, and grasslands, often in mixed flocks of Snow Geese and other waterfowl. VOCALIZATIONS Higher pitch than the Snow Goose; repetitive, cackly *keek-keek* calls as well as lower, grunting *hoink* sounds. ABUNDANCE Overwinters in relatively small but increasing numbers throughout much of NM. A regular rarity in AZ. LOOKALIKE Snow Goose. Ross's Goose is significantly smaller, has a shorter bill that lacks contrasting black pattern, and usually lacks buffy staining. The Snow Goose roots for tubers and the Ross's prefers surface vegetation. ALTERNATIVE NAMES Lesser Snow Goose, Stubby-billed White Goose. CONSERVATION STATUS Least Concern.

Greater White-fronted Goose

Anser albifrons

Agricultural fields, open water, wetlands
LENGTH 25–32 in. WINGSPAN 53–60 in.

Medium-sized, stocky goose with pink-orange bill, thick orange legs, white feathers around the base of the bill, white vent and rump, and white tip on black-banded tail. Brownish gray overall with irregular black bars on belly. Juvenile less distinctly patterned, lacks white in face, and has paler legs. BEHAVIOR A shy goose; migrating and wintering flocks are quick to flush upon being spotted. VOCALIZATIONS High pitched, laughlike yelp often consisting of two or three rapidly delivered syllables; higher and less hoarse than other geese. Also a variety of high-pitched single notes. ABUNDANCE An uncommon migrant in Eastern and from Central to Southern NM and in extreme Southwest AZ along the lower Colorado River. LOOKALIKES Domestic Graylag Goose, blue morph Snow Goose. Greater White-fronted distinguished from domestic Graylag by orange versus pinkish legs and lack of neck stripes. Distinguished from blue morph Snow Goose by lack of black grin patch and less white on head. CONSERVATION STATUS Least Concern.

Cackling Geese (small, at left) with Canada Goose (larger, at right)

Cackling Goose

Branta hutchinsii

Agricultural fields, open water, wetlands
LENGTH 24.5–26 in. WINGSPAN 42–44 in.

A duck-sized Canada Goose lookalike (previously thought to be a subspecies) with a stubby, steeply angled bill—much like the Ross's Goose is the miniature version of the Snow Goose. Compact with elongated wings, short legs, and short neck. Plumage identical to Canada Goose—gray-brown overall, with black head, white chinstrap, and white rump. Subspecies *taverneri* present in the Southwest. BEHAVIOR Grazes in large, open areas of grasses or agricultural fields, usually in large, mixed flocks, often with Canada Geese or other waterfowl. Forming strong family bonds, it is common for young to remain with parents for far more than twelve months. VOCALIZATIONS High-pitched cackling flight call, markedly distinct from the classic honk of the Canada Goose. ABUNDANCE Uncommon, regularly winters throughout Eastern NM. LOOKALIKE Canada Goose. Cackling Goose distinguished by shorter neck, smaller size, shorter bill, and cackling versus honking vocalizations. CONSERVATION STATUS Least Concern.

Adult with young

Adults

Adult

Canada Goose

Branta canadensis

Agricultural fields, open water, wetlands, urban areas such as golf courses, parks, sports fields
LENGTH 25–40 in. WINGSPAN 45–60 in.

An overall grayish brown goose, often lightest on the breast and belly, with a long, black neck and head; dark legs; black bill; and white chinstrap. Subspecies ranging in size from medium to large. Rump and vent are white. Black tail has a broad white band. BEHAVIOR In winter, large flocks fly in V-shaped formations and forage in open fields for grains and grass. VOCALIZATIONS A low, resonant honk. Flocks produce low honks at a relaxed pace with no harsh quacks or cries. ABUNDANCE Common and abundant in winter. Year-round populations exist in Northwest AZ, Northern NM, and urban areas across both states. Populations have increased in the Southwest on golf courses, lawns, and human-made waterbodies, where their presence is often considered a nuisance. LOOKALIKE Cackling Goose. Canada Goose is larger with a longer bill and neck. CONSERVATION STATUS Least Concern.

Juvenile

Adult

Tundra Swan

Cygnus columbianus

Open lakes and rivers; in winter, feeds in agricultural fields
LENGTH 47–58 in. WINGSPAN 64.5–67 in.

A slender, dainty swan compared to the bulky Trumpeter Swan or Mute Swan. Adult is pure white with a jet-black bill, sometimes displaying a small yellow skin patch near eyes. Immature swans are downy gray or mottled white and gray with black-bordered, pink bill. BEHAVIOR A swan with good posture, the Tundra keeps its neck straight and outstretched both in flight and while swimming, with the exception of during foraging, when it dabbles like a duck. In flight, wings produce a characteristic whistling sound and during landing resemble an open, floating parasol. Typically found in large flocks, but in AZ or NM, most sightings are of a single bird separated from its flock. Monogamous, usually mated by two or three years of age. VOCALIZATIONS Bugling hoots in flight and while feeding, often single- or double-syllable *ooh-ooh* or *ooo-uh*. Slapping feet on the water's surface, reminiscent of the tail-slapping of a beaver, serves as an alarm call or calls attention to another nearby flock. ABUNDANCE Uncommon winter visitor along the lower Colorado River; rare in migration throughout AZ and NM in agricultural fields and urban parks with water features. KEY SITES Uncommon but regularly observed in winter along the lower Colorado River. CONSERVATION STATUS Least Concern.

Adult male

Adult female

Wood Duck

Aix sponsa

Open water with plentiful adjacent or overhanging trees
LENGTH 18–21 in. **WINGSPAN** 26–30 in.

Both sexes have small bills, long necks, and long, wide tails. Male has brilliant green and purple crest, two white stripes on face and neck that meet below chin, buffy flanks bordered by black-and-white bars, black back, rufous breast speckled with white, orange legs, red eyes, and orange, yellow, and black bill. After breeding, male molts and transitions to eclipse plumage—dull, female-like plumage that "eclipses" his usual bright feathers, just before a second molting. At that time, male loses color but retains faded white facial stripes, dull green cap, speckling in the breast, and orange bill. Female is grayish brown overall, with speckled flanks, bold white eye ring tapering toward the rear, muted orange bill, and yellow legs. Juvenile similar to female with muted speckling, limited white eye ring, and dark eye stripe. **BEHAVIOR** Shy, often lurking among dense emergent vegetation or water-adjacent trees. Cavity nester. **VOCALIZATIONS** Both sexes deliver a high-pitched, two-note *ooWEEP*, rising in pitch on the second note. **ABUNDANCE** Uncommon local winter resident throughout NM and from Central to Southeast AZ. Year-round populations exist in NM along the Rio Grande near Albuquerque and along the Animas River near Farmington, and in AZ in the Verde Valley near Prescott. **CONSERVATION STATUS** Least Concern.

Blue-winged Teal

Spatula discors

Calm wetlands, flooded agricultural fields
LENGTH 14–15.5 in. **WINGSPAN** 22–24 in.

Petite and understated, with nonbreeding male looking very similar to female: mottled, dusky brown with scalloped pattern and deep brown eyeline and matching cap, often with white feathers at the base of the bill, extending to the throat. In flight, namesake icy blue upper coverts are revealed along with emerald-green secondary feathers, the two colors middled by a white patch. Bills are black and light brown; legs are yellow-orange. Breeding male sports black speckles on dusky brown body, slate-blue head, and bright white crescent facial markings. **BEHAVIOR** A long-distance migrant, often first to arrive in winter and last to depart in spring. Diminutive duck is a dabbler, not a diver. Usually in pairs or small flocks. When not feeding, rests along the water's edge. **VOCALIZATIONS** Female utters a low-pitched quack akin to the *ruff ruff* of a barking dog, and male uses a loud, whistling *peep* and nasally, airy *chuk chuk*. **ABUNDANCE** A common breeding resident across much of NM. In winter, an uncommon resident across extreme Southern NM and from Central to Southeast AZ. **CONSERVATION STATUS** Least Concern.

Cinnamon Teal

Spatula cyanoptera

Open water with plentiful emergent vegetation
LENGTH 15–16 in. WINGSPAN 21–22 in.

Small duck with an oversized bill. Both sexes have large, flattened, black bills and light blue wing patches. Male has red eyes and is deep brownish red (cinnamon) overall. Female is light brown to tan overall, with scalloped flanks and lighter feathers around base of bill. Juvenile and eclipse male resemble female. BEHAVIOR Dabbles for food in shallow water, often with rump in the air in an attempt to reach deeper. Commonly observed in small groups among other dabbling ducks. VOCALIZATIONS Male delivers short, gruntlike rattles, distinct from whistles of Blue-winged Teal. Female delivers short quacks similar to other teal. ABUNDANCE Common to uncommon breeder throughout much of the region. Uncommon winter resident in Southern AZ. Year-round resident along the lower Colorado River and Rio Grande. LOOKALIKES Other teal, Northern Shoveler. Cinnamon distinguished from other teal by larger, flatter bill, and in female and nonbreeding male, a more plain face lacking eyeline. Distinguished from Northern Shoveler by darker bill. CONSERVATION STATUS Least Concern (Decreasing).

Northern Shoveler

Spatula clypeata

Marshes, wetlands, flooded agricultural fields
LENGTH 17.5–20 in. WINGSPAN 27.5–32.5 in.

Exaggerated shovel-shaped bill appears slightly out of proportion with medium-sized body. Female and immature plumage is understated but beautifully patterned, with pronounced brown-and-white scalloped pattern on light brown body, sky-blue wing patch, and bright orange bill and legs. Breeding male displays chestnut-colored body contrasting with bright white chest, deep green head, slate-blue upper coverts, and deep green secondaries separated by a band of white. Bill is black, legs are orange. Nonbreeding male lacks contrasting white chest and is awash with deep brown. BEHAVIOR Distinctive, side-to-side sweeping dabbler. Sometimes feeds in large groups with heads submerged while collectively swimming in a circular motion to stir up plants, invertebrates, and crustaceans. Monogamous, often mating for life. When flushed off the nest, females known to spray feces on eggs to deter predation. VOCALIZATIONS Female gives nasally, classic quack. Male gives soft, understated cluck. ABUNDANCE Abundant and widespread in winter across the region. Lingers year round along the Rio Grande north of Albuquerque in NM. CONSERVATION STATUS Least Concern.

Adult female (left) and male

Gadwall

Mareca strepera

Open water with emergent vegetation, flooded fields, lawns
LENGTH 18–22 in. WINGSPAN 30–33 in.

Medium-sized duck, similar in size but slighter than a Mallard, with understated colors and fine patterning. Both sexes have square heads with steep foreheads, slender bills, yellow legs, and white primaries that show as small triangles on flanks when swimming. Male is brownish gray with black bill and tail coverts. Beautifully detailed pattern in breast and flanks visible at close range. Female is mottled brown, tan, and white with plain gray face, orange-and-black bill, and white belly, resembling female Mallard. Juvenile and nonbreeding male resemble female. BEHAVIOR Dabbles in small flocks. Often observed in mixed flocks with Mallards and wigeons. VOCALIZATIONS Males produce a series of low burps and high whistles. Female quack much like Mallard's, but higher and hoarser. ABUNDANCE Common winter resident. Year-round populations exist in Northwest AZ and from Central to Northern NM along the Rio Grande. Occurs as a migrant in Northeast AZ and Northwest NM. LOOKALIKE Mallard. Gadwall distinguished by lack of blue wing patch, slighter bill in both sexes, black bill in male, and less extensively orange bill in female. CONSERVATION STATUS Least Concern.

Adult males

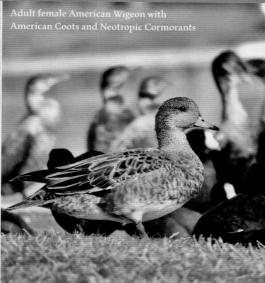
Adult female American Wigeon with American Coots and Neotropic Cormorants

American Wigeon

Mareca americana

Wetlands, marshes, calm lakes, reservoirs, parks, golf courses
LENGTH 16–23 in. WINGSPAN 31–33 in.

Petite to medium-sized dabbling duck. Breeding male displays vibrant, wide green stripe extending from eye down to nape, on gray-brown head flecked in steel gray contrasted by a blazing white cap. Brown body with pronounced white wing patches and black rump flanked by white side patches. Rare "Storm Wigeon" mutation results in fully cream-colored face and neck. Female and nonbreeding male are awash in mottled brown-gray plumage. Bill on both sexes is pale blue-gray with black tip. BEHAVIOR With short, gooselike bills, they feed primarily on aquatic vegetation, gathering in large, remarkably vocal flocks on ponds, lakes, and marshes. VOCALIZATIONS Very vocal duck, with males producing a three-part, nasally whistle akin to a plastic recorder. Females lack whistle but produce a gravelly grunt during courtship. ABUNDANCE Abundant on calm wetlands, lakes, and ponds and in golf courses throughout range from early fall to late spring. Year-round populations exist in extreme Northern NM. LOOKALIKE Eurasian Wigeon (rare). American distinguished by gray-brown versus dark brown head, white underwing coverts, and, in males, green head stripe and whiter crown patch. CONSERVATION STATUS Least Concern.

Adult female (right) with males

Adult male (top) and female

Mallard

Anas platyrhynchos

Open water, ponds, lakes, wetlands, rivers
LENGTH 20–25 in. WINGSPAN 32–37 in.

Medium-sized to large duck. Male has yellow bill, iridescent green head, white collar, chestnut-colored breast, gray flanks, black tail coverts with uppers winding into a tight curl, and white edging on tail. Female is mottled brown and tan overall with brownish gray face, dark cap and eye stripe, and orange bill with dark center. Both sexes have orange legs and blue wing patches bordered in white. Juvenile and eclipse male resemble female. BEHAVIOR Mallards frequently hybridize with other dabbling ducks and share traits with many domestic varieties, causing great confusion in the field. Males' aggressive courtship tactics can be unsettling. VOCALIZATIONS Male produces short grunts, quacks, and whistles. Often mistaken for a male, female delivers a delightful series of laughlike, cackling quacks. ABUNDANCE Common year round in most of the region and frequently the most numerous duck. Common winter resident in southern portions of both states. LOOKALIKE Mexican Duck. Female and eclipse male Mallard similar but distinguished by paler plumage, white in tail and undertail coverts, thicker white borders to wing patch, and (in males) curled uppertail coverts. CONSERVATION STATUS Least Concern.

Adult male

Adult female

Mexican Duck

Anas diazi

Freshwater marshes, lakes, ponds
LENGTH 22–26 in. WINGSPAN 33–38 in.

Difficult to distinguish from female and eclipse male Mallard, and until 2021, was classified as a subspecies. Sexes are similar, both rich brown overall and darker than Mallard, with a dark cap and eyeline, blue wing patches thinly bordered by white, a brown tail and undertail coverts lacking white highlights, and a distinct tan throat, appearing as if it is wearing a dark crew-neck sweater. Male has a bright to olive-yellow bill; female's bill is dull orange mottled with black. Will mix with Mallard flocks and the two frequently hybridize, resulting in animals with traits of both species. BEHAVIOR A dabbler and grazer, frequently submerging head and up-ending, or dabbling, at the surface. Will use its bill to disturb soil, unearthing roots and seeds. Pairs will nest surprisingly far from water, sometimes more than 100 feet away, usually in an inconspicuous, vegetated area in meadow or high grass. VOCALIZATIONS Similar to Mallard, with female giving a boisterous, classic quack, though lower pitched, truncated, and a bit more frog-like than the mocking laughter of a Mallard. Male gives a raspy, gravelly *rek rek rek*. ABUNDANCE Residents and local migrants from Central to Southeast AZ, in Southwest NM, and northward along the Rio Grande, though population may be decreasing in response to a changing climate. LOOKALIKES Female and eclipse male Mallard. Mexican distinguished by darker plumage, lack of white in tail and undertail coverts, thinner border to wing patch, and lack of curled uppertail coverts. Hybrid may show mixed traits including partially green head. CONSERVATION STATUS New designation as species, status not yet determined.

Adult male (right) and female

Northern Pintail

Anas acuta

Ponds, lakes, wetlands, flooded fields
LENGTH 20–30 in. WINGSPAN 33–35 in.

Among our "best-dressed" ducks. Male has blue-gray bill with black central stripe, brown head and nape, narrow white stripe curving from back of head into white breast and belly, gray flanks with buffy white patch toward rear, black rump, green wing patch, and long, pointed tail feathers. Female is mottled brown and tan overall with long and slender neck, bronze wing patch, and light blue bill with black central stripe. Juvenile is similar to female but with dark bill. Male in eclipse plumage retains blue bill. BEHAVIOR Dabbles in open water and forages along wet shorelines for insects. Occasionally gathers in flooded fields foraging for insects and grain. VOCALIZATIONS Male produces a reedy *ze ze ze zoEEo* and abrupt quacks. Female makes a series of quacks reminiscent of Mallard, but hoarser.

ABUNDANCE Common winter resident. Year-round resident in extreme Northern NM and within a limited patch in East Central AZ. CONSERVATION STATUS Least Concern (Decreasing).

Breeding male

Female/nonbreeding male

Green-winged Teal

Anas crecca

Shallow wetlands, marshes, flooded agricultural fields
LENGTH 12–15 in. WINGSPAN 20–23 in.

Both sexes of this impeccably dressed, pint-sized dabbling duck—the smallest in North America—sport namesake emerald-green wing patches. Breeding male has a black bill, a deep cinnamon-colored head and eye, and a wide green stripe extending from eye to nape. Vertical white stripe along shoulder contrasts with intricately patterned and flecked gray-brown body, and buffy rump patches contrast with black tail coverts. Tan breast is speckled in black. Female is gray-brown with a dark gray bill and a dark brown eyeline and cap. BEHAVIOR Feeds primarily in shallow waters or flooded fields. Unique from many migratory waterfowl species, they are known to lack fidelity to their wintering sites, sometimes migrating to another region altogether. VOCALIZATIONS Male makes a soft whistle during courtship, reminiscent of a chirping cricket. Female issues a *HEE-haw* of a quack on wintering grounds, during courtship, and as an alarm call. ABUNDANCE Common and abundant in winter throughout AZ and NM, with small resident populations in Southeast AZ and Central NM. LOOKALIKE American Wigeon and other teal. Green-winged distinguished from wigeon by darker bill and lack of white cap; from other teal by smaller size and slighter bill. CONSERVATION STATUS Least Concern.

Adult male

Adult female

Canvasback

Aythya valisineria

Ponds, lakes, wetlands, often with heavy shoreline and emergent vegetation
LENGTH 18–22 in. WINGSPAN 30–34 in.

Large diving duck most easily recognized by its unique bill. Both sexes have long, wedge-shaped black bill that matches the forehead closely in slope; black legs; gray to white underwings; long, sturdy neck; and low, flat back. Male shows deep rufous head and neck, ruby-red eyes, white flanks and back, and black breast and rump. Female has light grayish brown flanks and tan head, neck, and breast, darkest in the nape and crown. BEHAVIOR Dives for vegetation and invertebrates, specializing in the rhizomes and tubers of aquatic plants. Seldom seen on dry land. Runs along the water's surface before taking flight. Takeoff and feeding style makes larger bodies of water the most suitable habitat. VOCALIZATIONS Usually silent, but female gives low, throaty grunts and male produces gobbling bursts of hoots and coos. ABUNDANCE Uncommon winter resident. LOOKALIKE Redhead. Canvasback distinguished by wedge-shaped bill and head, red eyes, white back, and gray-black versus blue-gray bill. CONSERVATION STATUS Least Concern.

Adult male

Breeding male (left) with female

Redhead

Aythya americana

Lakes, rivers, reservoirs, coastal bays, estuaries
LENGTH 17–21 in. WINGSPAN 28.5–31 in.

Medium-sized diving duck. Female has black bill and eyes and a gray-brown body. Wings are slightly paler, a solid shade, with pale gray wing stripe visible in flight. Breeding male sports marigold-hued eyes with bright cinnamon-colored head, lacy-patterned gray body, and black chest and rear. Bill is gray with black tip. Nonbreeding male appears similar to breeding but with somewhat washed-out/muted tones. BEHAVIOR Often flocks in remarkably large numbers, known as rafts when floating together on water. Females regularly engage in brood parasitism, laying their eggs in the nests of several other duck species. They feed on aquatic vegetation by shallow surface diving. VOCALIZATIONS Female gives a raspy, gurgly, repetitious quack, while male's call is a wheezy weew. ABUNDANCE Uncommon but regular winter resident throughout AZ and NM, with small year-round populations along the Rio Grande and Colorado River corridors. Breeds in extreme Northern NM and Northwest AZ. LOOKALIKE Canvasback. Redhead distinguished by angle of forehead—Canvasback is like a rolling hillside, while Redhead is like a steep cliff. Redhead lacks red eye, has a gray versus white back, and has a blue-gray versus gray-black bill. CONSERVATION STATUS Least Concern.

Adult male

Adult female (left) with male

Ring-necked Duck

Aythya collaris

Ponds, lakes, wetlands; may inhabit smaller waterbodies than other diving ducks
LENGTH 16–18 in. WINGSPAN 25–26 in.

Among our more misleadingly named ducks: the ring around this duck's bill is far more prominent than the one around its neck. Both sexes have bluish gray bill with white ring and black tip, grayish primaries and underwings, dark legs, short tail, and slight crown that is highest toward the back (more exaggerated in male). Male has yellow eyes; white ring around bill; black head, back, breast, and rump; and gray flanks with white patch toward front. Reddish ring around base of neck, originally described from a deceased specimen, is rarely seen in the field. Female is mottled brownish gray with gray face, brown cap, white eye ring, and white feathers at base of bill. BEHAVIOR Dives for plants and invertebrates, often making an upward jump before disappearing underwater. VOCALIZATIONS Male makes short whistles and toots. Female makes low, short growl. ABUNDANCE Common winter resident across the region. LOOKALIKES Greater, Lesser scaups. Ring-necked Duck distinguished by white ring on bill. CONSERVATION STATUS Least Concern.

Adult male

Female/juvenile

Male Greater Scaup (left) and Lesser Scaup

Greater Scaup

Aythya marila

Lakes, reservoirs, ponds
LENGTH 16–22 in. WINGSPAN 28–31 in.

Medium-sized diving duck with a smoothly rounded head. Female sports shades of warm cocoa-brown, with a solid milk-chocolatey head, golden eye, and creamy white at the base of her sturdy, slate-colored and black-tipped bill. Breeding male displays iridescent green-black head and golden eyes. Black chest and rear contrast with pure white flanks and patterned gray back. Bill is blue-gray. Nonbreeding male lacks bright white flanks. BEHAVIOR Much like the Redhead, these sociable diving ducks form large, raftlike flocks on large bodies of water in winter and during migration. Often seen in mixed flocks. VOCALIZATIONS Typically silent, though males give repetitive, wheezy whistle during courtship, and females croak an *urr urr urr* as an alarm call. ABUNDANCE Uncommon but regularly observed along the Colorado River corridor on AZ's western boundary. LOOKALIKE Lesser Scaup. Greater distinguished by a rounder head shape with more pronounced green (never purple) iridescence, visible only when in the right light. CONSERVATION STATUS Least Concern.

Adult male

Female/juvenile

Lesser Scaup

Aythya affinis

Open water including lakes, ponds, wetlands
LENGTH 15–18 in. **WINGSPAN** 27–30 in.

Tricky diving duck with a few key field marks. In both sexes, head is flattened leading into neck, with peak toward the rear, producing a boxy look. Male has black head with purple/green iridescence, yellow eyes, blue bill with black nail, black breast and rump, grayish white flanks, and grayish back with coarse black barring. Female is brownish overall, darkest in head and lightest in flanks, with white feathers at base of gray bill. Juvenile and nonbreeding male resemble female. In flight, note white flight feathers fading to gray toward wingtip. **BEHAVIOR** Dives for mostly invertebrates in smaller waterbodies than Greater Scaup. **VOCALIZATIONS** Usually silent outside of breeding range. Male produces wobbly *weh-eh-ew*. Female produces harsh grunts and growls. **ABUNDANCE** Common winter resident. More commonly observed in the Southwest than is Greater Scaup. **LOOKALIKES** Ring-necked Duck, Greater Scaup. Lesser Scaup distinguished from Ring-necked by lack of white bill ring; from Greater Scaup by purple head iridescence (Greater shows only green), boxier head, and more white in flight feathers. **CONSERVATION STATUS** Least Concern.

Male | Female

Surf Scoter

Melanitta perspicillata

Oceans, lakes, estuaries, reservoirs
LENGTH 19–24 in. WINGSPAN 29–31 in.

Female is a cool, rich dark brown, with white patches on either side of the bill and behind each deep brown eye. Adult male is jet-black with bright white patches at forehead and extending from nape to base of neck, bright orange legs matching a red-orange bill accented by black-and-white patches, and blue-white eyes. Immature male has faded orange bill and legs. Bill on both sexes is large with an exaggerated slope, resulting in an overall triangular appearance to the head. BEHAVIOR Flocks of thousands of these diving ducks will amass upon coastal waters, but in AZ and NM, you're more likely to see individuals or small groups. Some immature scoters opt out of migration and become temporary year-round residents at lakes and reservoirs. Feeds primarily on mollusks, crustaceans, small fish, and some plant material and aquatic insects.

VOCALIZATIONS Relatively quiet unless courting or defending young, though wings give off a soft, rapid whistle in flight. ABUNDANCE Rare, but regularly observed winter vagrant across both states. Restricted to the region's largest bodies of water. KEY SITES AZ: lakes throughout White Mountains and along the Colorado River corridor; NM: lakes and reservoirs, Bosque del Apache National Wildlife Refuge, and along the Rio Grande River corridor. CONSERVATION STATUS Least Concern.

Male

Female

White-winged Scoter

Melanitta deglandi

Oceans, lakes, estuaries, reservoirs
LENGTH 18–22 in. WINGSPAN 32–34 in.

Large, stocky duck with a more wedge-shaped bill and head compared to other scoters. Both sexes are very dark with broad, sloping bills, heads reminiscent of a Canvasback's, and bold white patches on secondaries (often but not always visible when swimming or at rest). Short tail often held upward. Male is deep black with orange-red legs and bill; small, pointed white patch behind the eye; and a fleshy lump at base of bill. Female is dark brown with white patch in front of and behind eye. Juvenile and nonbreeding male resemble female. BEHAVIOR More common coastally in winter; usually seen singly or in very small groups inland. A deep water diver specializing in prying mollusks from underwater surfaces. VOCALIZATIONS Usually silent in winter range; male produces high whistles during courtship and both sexes give low grunts. Wings whistle in flight. ABUNDANCE Vagrant; rare winter visitor in largest waterbodies only. CONSERVATION STATUS Least Concern (Decreasing).

Male

Female

Black Scoter

Melanitta americana

Oceans, lakes, estuaries, reservoirs
LENGTH 17–19 in. WINGSPAN 27–28 in.

Medium-sized sea duck. Female is gray-brown with a pale gray face and neatly worn, dark brown cap that extends from above the eyeline and down the nape. Wide black bill is relatively small and straight. Male is inky black with dark brown eyes and black bill, accented by an unmistakable deep sunflower-yellow knob, like someone draped a bit of modeling clay over it. Legs are black on both sexes. BEHAVIOR Surface-diving sea duck gathers in linear flocks along coastlines and feeds primarily on aquatic invertebrates. More vocal than their Surf Scoter relatives, males can be heard giving frequent but soft, haunting whistles. VOCALIZATIONS Male has descending, woodwind-like whistle, and female in courtship may respond in kind. As in other scoter species, wings produce a rapid whistling sound in flight. ABUNDANCE Rare, but regularly observed winter vagrant across AZ and NM. Restricted to the region's largest bodies of water. KEY SITES AZ: lakes throughout White Mountains and along the Colorado River corridor; NM: throughout lakes and reservoirs, Bosque del Apache National Wildlife Refuge, and along the Rio Grande River corridor. CONSERVATION STATUS Near Threatened.

Breeding male / Female / Female/juvenile

Bufflehead

Bucephala albeola

Saltwater bays, lakes, ponds, rivers, reservoirs
LENGTH 12–15.5 in. WINGSPAN 21–22 in.

Petite and dapper little bobber. Adult male has a snow-white body, silky black back, and large, ovular white patch on striking, iridescent head with gradients of green, blue, and purple visible in sunlight; otherwise may appear black. Wide, white wing stripe and orange legs in flight. Bill is blunt, pale blue-gray on male, darker on female and first-year male. Female and young male are various shades of charcoal-gray–brown, with a white chest, white cheeks, and white wing patches. BEHAVIOR Impressive diver can remain submerged for more than ten seconds, consuming mollusks and other invertebrates while underwater. Males in courtship offer a delightful head-bobbing display. In flight, rapid wingbeats combine with a diagnostic rocking side-to-side flight pattern. VOCALIZATIONS Not particularly vocal, the female offers a throaty, gravelly *ruk ruk* as she identifies potential nesting sites or when summoning her young. Male's courtship display sometimes accompanied by rapid chattering. ABUNDANCE Common throughout AZ and NM in winter, making use of everything from large lakes and reservoirs to smaller urban ponds. CONSERVATION STATUS Least Concern.

Adult male

Female

Common Goldeneye

Bucephala clangula

Large lakes, wetlands, rivers
LENGTH 16–20 in. WINGSPAN 26–30 in.

Bulky headed, bright-eyed diving duck. Both sexes have large, peaked head with sloping forehead and bill, yellow eyes, and white wing patches in flight. Male has iridescent green head that can appear black in dim light, round white patch behind the bill, black back and rump, and thin black stripes in wings at rest. Female is mostly gray, with brown head and yellow-tipped, usually mostly black bill. First winter birds are mottled gray overall, with black head and rump and white collar. BEHAVIOR Forages in groups in relatively shallow water and along shorelines. Dives for crustaceans, mollusks, other invertebrates, and fish. VOCALIZATIONS Usually silent. Wings whistle in flight. Female produces hoarse grunts and laughlike quacks. Male makes buzzy *reh REH*. ABUNDANCE Uncommon winter resident across NM and all but South Central AZ. Most abundant along the Colorado and Gila Rivers in AZ and from Santa Fe north into the Carson National Forest in NM. LOOKALIKE Barrow's Goldeneye. Common distinguished by longer bill and more sloping forehead in both sexes. Male Common distinguished by round facial spot, whiter flanks, lack of a black spur near shoulder, and black stripes in folded wing; female Common distinguished by darker bill and less richly colored head. CONSERVATION STATUS Least Concern.

Adult male (right) and female

Group

Barrow's Goldeneye

Bucephala islandica

Lakes, ponds, mountain sloughs
LENGTH 17–29 in. WINGSPAN 27–29 in.

Medium-sized, compact diving duck. Both sexes sport namesake golden eyes. Female is mottled gray and brown, fading to pale gray on the chest and throat, contrasting with milk-chocolatey, steeply angled head and short, carrot-orange to yellow (sometimes faded) bill. Adult male has a deep blue-purple head accented by a crescent-shaped white patch on either side of a short black bill. White neck, chest, and belly contrast with jet-black back, decorated with a row of elongated white spots along the wing. BEHAVIOR Dives for extended periods, foraging on aquatic invertebrates. Agile in flight even in heavily forested areas, with powerful, rapid wingbeats. Cavity nesters, sometimes using the nests of other cavity-nesting duck species. VOCALIZATIONS Kazoo-like grunts and *kek-kek* calls. In flight, wings produce rather raucous, rapid whistle. ABUNDANCE Uncommon but regularly observed in winter on the Colorado River along the western border of AZ. LOOKALIKE Common Goldeneye. Barrow's of both sexes distinguished by shorter bill and steeper forehead. Male Barrow's distinguished by white spots on black back, black "spur" at shoulder, crescent-shaped facial spot, and purple head. Female Barrow's distinguished by orange bill and deeper cocoa-brown head. ALTERNATIVE NAMES Crescent-faced Goldeneye, Spotted Goldeneye. CONSERVATION STATUS Least Concern.

Breeding adult male (left) and female

Hooded Merganser

Lophodytes cucullatus

Small to medium-sized ponds, lakes, wetlands
LENGTH 16–19 in. **WINGSPAN** 24–26 in.

Our smallest merganser. Both sexes have thin bills, long tails often cocked upward or held flat on the water, and large crests that are round when raised and oblong with peak at the rear when lowered. Long-tailed and slender-winged in flight. Male has black bill, head, and upper parts with yellow eyes, rufous flanks, thin white streaking on wings, white breast with two black spurs on each side, and large white patch occupying most of the crest. Female is gray overall with rich brown crown and drab yellow-and-black bill. Juvenile and eclipse male resemble female. **BEHAVIOR** Shy and reclusive, quick to vanish into emergent vegetation. Dives for fish, insects, and other aquatic invertebrates. **VOCALIZATIONS** Usually silent. Male produces descending, froglike croak. Female delivers hoarse, croaking quacks. Wings produce high trill in flight. **ABUNDANCE** Uncommon winter resident along the Colorado River, from Central to Southern AZ, and in limited portions of Southeast and Southwest NM. An uncommon migrant across eastern NM. **CONSERVATION STATUS** Least Concern.

Breeding adult male (back) and female

Common Merganser

Mergus merganser

Freshwater lakes, rivers, reservoirs
LENGTH 22–27 in. WINGSPAN 32–24 in.

Everything about this medium-large, sleek diving duck is long and lean, from its slender silhouette to its long, slim, and serrated bill. Cleanly patterned male has a bright white body with black along the back, forest-green head, and deep orange-red, serrated bill tipped with a black hook. Gray-bodied female sports this same bill, with a white throat, rufous head, and a shaggy mullet hairdo. BEHAVIOR With slender bodies made for diving, Common Mergansers feed on small freshwater fish. Females raise young while males gather in flocks, which remain tight-knit through winter months. Often seen in mixed flocks with other diving ducks. Opportunistic cavity nesters, they will nest in everything from nest boxes to tree cavities and rock crevices. VOCALIZATIONS Typically silent, with the exception of various gurgling, grunty croaks given as alarm calls or during courtship. Female will also use a higher pitched, rapid croak to round up her young. ABUNDANCE A winter resident across the region, most commonly in the northern portions of both states. Small resident populations exist in high-elevation forests from Central AZ to extreme West Central NM and in the Southern Rockies north of Santa Fe. LOOKALIKE Red-breasted Merganser. Female Common distinguished by white chin patch on cinnamon-colored throat and white chest, versus overall dusky gray-brown on Red-breasted. CONSERVATION STATUS Least Concern.

Adult male

Female

Red-breasted Merganser

Mergus serrator

Large lakes, reservoirs, rivers
LENGTH 21–25 in. WINGSPAN 26–29 in.

Slender and long-bodied, even for a merganser. Both sexes have long, thin, red-orange bill; thin neck; and shaggy crest. Male has iridescent green head, red eyes, white collar, rufous breast with black mottling, black back, and gray flanks with white upper edge. Leading edge of flank is black with small white spots. In flight, white wing patch shows two thin black stripes. Female is gray overall with a light golden brown head, blurry white throat patch, and light breast. Juvenile and eclipse male resemble female. BEHAVIOR Hunts fish by diving and swimming just below the water's surface. VOCALIZATIONS Usually silent. Male produces a whiny *mew mew*. Female gives short grunts and croaks. ABUNDANCE Uncommon to rare in migration and winter. Most frequently observed on the Colorado River north of Kingman, AZ; near Carlsbad in Southeast NM; farther north from Albuquerque to Tucumcari; and along the lower Rio Grande River. KEY SITES Reservoirs along the lower Colorado, Rio Grande, and Pecos rivers. LOOKALIKE Common Merganser. Red-breasted of both sexes distinguished by thinner bill. Male Red-breasted distinguished by gray flanks and white collar; female Red-breasted distinguished by lighter head with less distinct throat patch. CONSERVATION STATUS Least Concern.

Breeding male

Female/immature male

Nonbreeding male

Ruddy Duck

Oxyura jamaicensis

Marshes, lakes, ponds, rivers, estuaries
LENGTH 14–17 in. **WINGSPAN** 22–25 in.

What they lack in size, they make up for in looks and charisma. Dapper breeding male sports a stocky chestnut-colored body and ashy gray belly; black, rigid, and upright fanned tail with white undertail coverts; and a baseball cap–shaped head. Broad, pale turquoise bill contrasts with a black head and bright white cheeks. Female and nonbreeding male have black bills and mottled brown-gray bodies, and they tend to keep their tails submerged rather than cocked upright. Immature male and female lack bright white cheek patch, instead showing a pale gray patch that contasts with a darker cap extending just below the eye. **BEHAVIOR** As interesting as it is handsome, the Ruddy feeds primarily at night, surface diving for its preferred meal: midge larvae. They nest in vegetated marsh areas, laying large, heavily textured eggs that hatch remarkably well-developed ducklings, and they are extremely aggressive toward just about anyone that dares enter their territory, including mammals. **VOCALIZATIONS** Silent, with the exception of a high-pitched *bleet* or hissing alarm calls produced only by females; courting males slap their bills against the water to produce a clicking and clapping display. **ABUNDANCE** Common breeder across the northern portions of both states, winter resident from Central to Southern AZ and in Southern and Eastern NM, and year-round resident from Central NM to Eastern AZ and along the Colorado River. **CONSERVATION STATUS** Least Concern.

QUAIL AND OTHER LANDFOWL

From the Gambel's Quail of the low desert to the Dusky Grouse of high coniferous forests, quail (family *Odontophoridae*) and pheasants, grouse, and turkeys (family *Phasianidae*) are often emblematic of the habitats they call home. Some, like the Wild Turkey, make themselves easy to spot with their brazen antics, while others, such as the Montezuma Quail, can go undetected until they launch upward from almost underfoot. When birding in the Southwest, be sure to take the occasional break from scanning the trees to check the ground for these chickenlike fowl.

ABOVE Gambel's Quail family

Adult male (masked subspecies)

Adult male (right) and female

Northern Bobwhite

Colinus virginianus

Brushy edges of open habitat, including grasslands, agricultural lands, pastures **LENGTH** 9–11 in. **WINGSPAN** 13–15 in.

Small, strikingly patterned quail. Both sexes have dark caps; pale eyebrows; thick, dark eyelines; and pale throats that create bold head patterns. Male has black-and-white head pattern with a slight rufous crest, black throat, and rufous breast. Female has a brown and buffy head pattern. Both sexes show body intricately patterned with rufous, white, brown, and black. Gray tail. Masked Bobwhite (*C. virginianus ridgwayi*) is a subspecies; male has all black head and rufous body and is less patterned than Northern Bobwhite. **BEHAVIOR** Coveys (groups of quail or other landfowl) forage for foliage and seeds on the ground and in low-hanging vegetation. **VOCALIZATIONS** Both sexes give loud, ascending *bob-whITE*. Calls include squeaky whistles and varied clucks. **ABUNDANCE** Uncommon and local in Eastern NM; populations declining. Masked Bobwhite is listed as endangered and is rare and local in reintroduction area within Southeast AZ grasslands. Escaped domestic birds can be seen regionwide. **KEY SITE** Masked subspecies restricted to Buenos Aires National Wildlife Refuge, AZ. **CONSERVATION STATUS** Near Threatened (Decreasing). The masked subspecies is listed as Endangered by the US Fish and Wildlife Service.

Scaled Quail

Callipepla squamata

Desert grasslands, pinyon-juniper scrub
LENGTH 10–12 in. **WINGSPAN** 13.5–16 in.

Affectionately known as the "cottontop," this small, plump, chickenesque quail has a pronounced scaled pattern that fades from blue-gray at the shoulders to a creamy brown belly, a gray-black beak, and a dusky brown head topped with a cottony white crest. Scaling fades along the back to a solid dusky brown. Juvenile lacks pronounced cottony crest. **BEHAVIOR** Often seen scurrying from shrub to shrub through desert grasslands, this gregarious, granivorous ground forager prefers cover for its covey, with the exception of courting males, who caterwaul conspicuously from perches to attract lady quail. Will sometimes huddle, outward facing, in a circular formation to watch for prey and regulate body temperature. **VOCALIZATIONS** Courting male will conspicuously perch on fence posts, snags, or stumps, offering a repetitive, shrill squawk, as if someone just stepped on its foot. Most common call is a rhythmic, repetitive, truncated *ruh-roo ruh-roo* (which we liken to the chicken song and dance performed on TV's *Arrested Development*). Will also offer high-pitch chips when flushed. **ABUNDANCE** A common year-round resident across most of NM and in limited range in the grasslands of Southeast AZ. **CONSERVATION STATUS** Least Concern (Declining).

Adult male

Adult female

Juvenile

Gambel's Quail

Callipepla gambelii

Brushy areas within desert, chaparral, oak woodland, suburban landscapes; often near water
LENGTH 10 in. **WINGSPAN** 14 in.

The chicken of the desert. Both sexes are light gray overall with white-streaked, rufous flanks; finely streaked breasts; curved black topknots; and unmarked, buffy bellies. Male has a larger topknot, rufous cap, streaked nape, black face bordered by white stripes, and black patch on lower belly. Female has a faintly streaked nape and plain face. **BEHAVIOR** Forages on the ground in coveys for mostly seeds and foliage, moving quickly along the ground between areas of dense cover. **VOCALIZATIONS** A loud, plaintive *chi-CAA-go* or *chi-CAA-ga-go*; a descending *kuaaw*; and various chickenlike clucks and chirps. Flushing birds have loud wingbeats. **ABUNDANCE** Common year-round resident in Western and Central to Southern AZ and in Central to Southwest NM. **LOOKALIKE** California Quail. Gambel's distinguished by lighter tones and lack of scalloped feathers on nape and belly. **ALTERNATIVE NAME** Desert Quail. **CONSERVATION STATUS** Least Concern.

Adult male

Adult female

Montezuma Quail

Cyrtonyx montezumae

Oaky grasslands, canyons, hillsides
LENGTH 7.5–9.5 in. WINGSPAN 16–17 in.

The envy of many a juggalo, adult male shows large, round eyes and a striking black-and-white face pattern that appears as if painted. White spots on black sides meet a deep, mahogany chest. Back is heavily patterned, mottled brown with black spots. Female similarly patterned but in subtle shades of warm and cool brown and tan, with delicate white accents. Both sexes sport a slicked-back rufous or brown mohawk, a rounded, downturned beak, and a distinctly short tail. BEHAVIOR Secretive, hyperlocal, and lacking the gregarious nature of other quail species, "zoomies" will wait until you are extremely close by and then flush suddenly and loudly, scaring the pants off of you, as you'll likely be unaware of their presence. These homebodies hide in dense vegetation, rarely traveling outside of a 150-foot radius throughout their day. Omnivorous, they use their large claws to dig for tubers and invertebrates. VOCALIZATIONS Song a descending, buzzy, one-note whistle reminiscent of a whistling bottle rocket. Calls vary, from soft clucking and cooing to a squealing distress call. ABUNDANCE Fairly common year round in Southeast AZ and Southwest NM. Less common from East Central AZ to West Central NM and in limited patches further east. ALTERNATIVE NAME Oak Quail. CONSERVATION STATUS Least Concern.

Adult male

Adult Female

California Quail

Callipepla californica

Coastal chaparral; birds in AZ frequent brushy areas within high-elevation grasslands, often near water
LENGTH 10 in. WINGSPAN 14 in.

Both sexes are dark gray-brown overall with white-streaked, reddish brown flanks; distinctly scaled napes and bellies; and curved, black topknots. Male has a larger topknot, reddish brown cap, and black face bordered by white stripes. Female has a plain, brownish face. BEHAVIOR Forages on the ground in coveys for mostly seeds and foliage; reluctant to stray far from areas of dense cover. VOCALIZATIONS A loud *chi-CA-go*, higher and shorter than Gambel's Quail. A short *caw*, with various chickenlike clucks and chirps. Flushing birds have loud wingbeats. ABUNDANCE Introduced; uncommon and local. KEY SITES Within the watershed of the Little Colorado River in the White Mountains of AZ near Springerville. LOOKALIKE Gambel's Quail.

California distinguished by darker plumage and scaled feathers on nape and belly. ALTERNATIVE NAME Valley Quail. CONSERVATION STATUS Least Concern.

Wild Turkey

Meleagris gallopavo

Mixed and conifer forests, wooded grasslands, riparian areas, chaparral
FEMALE LENGTH 30–37 in. **WINGSPAN** 50 in.
MALE LENGTH 39–46 in. **WINGSPAN** 64 in.

An enormous and mostly ground-dwelling bird. Both sexes have bald heads and necks, long legs, and long, rounded tails. Male is extremely ornate with widespread iridescence, intricately patterned flight feathers and tail, and blue-and-red facial skin adorned with bright red, fleshy bulges and protrusions. Color fades during nonbreeding season. Female is brown overall with a bluish gray head, bronze iridescence, and subtle yet intricate patterning. Plumage varies slightly between the Southwest's three subspecies: Merriam's, Gould's, and Rio Grande. **BEHAVIOR** Infrequently taking flight, birds in large flocks forage on the ground for nuts, fruit, and invertebrates. Roosts high in trees at night. Males gather in open areas to perform showy courtship displays. **VOCALIZATIONS** Males produce an unmistakable, descending gobble. Both sexes give varied clucks, whistles, and yelps. **ABUNDANCE** Uncommon year-round resident. Most abundant in the region's national forests and in Northeast NM. **CONSERVATION STATUS** Least Concern.

Breeding adult male

Nonbreeding adult

Breeding adult female (front) with young

White-tailed Ptarmigan

Lagopus leucura

Alpine tundra
LENGTH 12–13 in. WINGSPAN 22 in.

Masters of camouflage, both sexes of this petite but rotund grouse show solid snow-white plumage, jet-black beaks and eyes, and heavily feathered feet in winter months, outfitted for the elements. Breeding male has ruddy body with white belly, and breeding female displays deep black-brown barring on a creamy, pale body. Molting adults will appear heavily mottled. BEHAVIOR Slow and steady ground foragers, transitioning into a state of low activity during winter to preserve energy. Often roosts in snowbanks. VOCALIZATIONS Most often silent, but during nesting season they communicate with various clucks and growls. During territorial display, males will give a repetitive, shrill screech in flight. ABUNDANCE Population is isolated within the highest peaks of the Sangre de Cristo Mountains in the Pecos Wilderness, NM. CONSERVATION STATUS Least Concern (Declining).

Breeding male

Nonbreeding male

Female

Dusky Grouse

Dendragapus obscurus

High-elevation forests of fir, aspen, pine, typically with scattered shrubs, clearings
LENGTH 18–23 in. WINGSPAN 25–26 in.

Large, chickenlike bird. Male is dark overall with orange-red eyebrow combs, red throat sac bordered in white when displayed, dark gray breast, mottled black-and-white barring on flanks and undertail, and black tail usually tipped in gray. Female is extremely well camouflaged in mottled browns, whites, and grays. BEHAVIOR Forages on the ground for insects and plant material or in trees for foliage. With eyebrow combs and throat sac displayed, male performs elaborate dance of repetitive strutting, short flights, tail fanning, and vocalizations. VOCALIZATIONS Male makes a short series of hoots so low in pitch that they're easy to miss. Both sexes give short buzzy notes, varied clucks, and barks. ABUNDANCE Uncommon year-round resident from Santa Fe north into the Southern Rockies, and in AZ above the Grand Canyon, near Flagstaff, and in the White Mountains. LOOKALIKE Sooty Grouse. Dusky and Sooty were once considered a single species (Blue Grouse) and are extremely difficult to differentiate. Only Dusky occurs in the Southwest. CONSERVATION STATUS Least Concern.

Male

Female

Ring-necked Pheasant

Phasianus colchicus

Grasslands, agricultural fields
LENGTH 19–27 in. WINGSPAN 23–34 in.

Large, long-necked gamebird. Male is adorned with bottle-green head and neck contrasted with a lipstick-red cheek and namesake white neck ring. Body is a warm, coppery brown, with black spots on sides, white spotting on back, and black barring on long, elegant tail plumes. Female shows more subtle plumage in varying shades of taupe and cinnamon, yet still elaborately patterned with deep black markings on back and wings and delicate black barring on tail. BEHAVIOR Primarily granivorous ground foragers, gleaning seeds and sometimes insects from agricultural fields and prairies. Seldom seen in flight unless flushed by nearby humans or predators, they tend to spend most of the day on foot. VOCALIZATIONS Males are raucous throughout the day and throughout the year, with a cackling, usually two-syllable, truncated, roosterlike *cock-a*, but without the *doodle-doo*. Females give one-note, raspy calls of varying pitch to round up young and warn of predators. ABUNDANCE Uncommon but regularly observed in suitable habitat across NM's northeastern half; rarely observed in AZ, most frequently in the state's southwestern corner. CONSERVATION STATUS Least Concern (Introduced).

Chukar

Alectoris chukar

Dry fields, sagebrush, rocky grasslands; often steep terrain
LENGTH 13–15 in. WINGSPAN 19–20 in.

Boldly patterned pheasant resembles a quail. Gray overall with white throat and lower face bordered by black. Both sexes show black-and-white barred flanks, buffy undertail and belly, red eyes and bill, and pinkish legs. BEHAVIOR Forages on the ground in large groups for plant material and insects. Extremely fast runners are swift and agile on steep, rocky slopes. VOCALIZATIONS Song is a loud, intensifying, high-pitched, and somewhat squeaky *chukah-chukah-chukah-chukahrah-chukahrah*. Calls include short *chuck*s and *chuka-ah*s of similar quality. Flushing groups emit high-pitched squeals. ABUNDANCE Uncommon year-round residents in extreme Northwest AZ. Escaped or intentionally released domestic birds can be seen regionwide. CONSERVATION STATUS Least Concern (Introduced).

Male

Female

Lesser Prairie-Chicken

Tympanuchus pallidicinctus

Oak, sagebrush grasslands
LENGTH 15–16 in. **WINGSPAN** 24 in.

Medium-sized but rotund grouse with short neck and stubby tail, male and female display heavy brown-and-white barring on chest and heavily mottled, ruddy brown-and-white back. Female and juvenile have pale, creamy white eyeline and throat; juvenile is dusty cocoa-brown and white. Male displays bright yellow-orange combs above the eyes, and in display, inflates red, balloonish air sacs on either side of the neck. **BEHAVIOR** Ground foragers dining on both insects and seeds/grains, will forage higher in shrubs for the right treat, such as buds and fruit. Their diverse diet changes with the seasons. Males known for extravagant lekking, when they gather in groups to perform mating dances complete with puffed air sacs, fancy footwork, and booming vocalizations. **VOCALIZATIONS** Little vocalization from either sex when not in display. Displaying male makes a sound with its air sac, called a "boom," like a bubbling gobble. Also uses stomping and feather fluttering to produce sound. Females cackle, generally when in conflict with other females at a lek. **ABUNDANCE** Limited to extreme Eastern NM. Species in drastic decline resulting from overhunting and habitat loss. **KEY SITES** Extremely difficult to observe. Much of its habitat in NM's Milnesand Prairie Preserve is protected and privately owned by The Nature Conservancy. **LOOKALIKE** Greater Prairie Chicken. Lesser distinguished by slightly smaller size and lighter plumage. **CONSERVATION STATUS** Vulnerable.

GREBES

In Latin, *podicis* refers to the rear end and *ped* means "foot." This fact, combined with grebes' oddly oriented legs that are perfect for swimming but of limited use on land, has earned them the less-than-flattering family name of *Podicipedidae* (loosely translated, "buttfeet"). Despite the moniker, these fish-eating specialists are unembarrassed and take full advantage of their form. In the Southwest, birders enjoy the annual gatherings of Clark's and Western grebes on the region's larger reservoirs and are always on the lookout for a Least Grebe, a rare and diminutive visitor from farther south.

ABOVE Nonbreeding adult Pied-billed Grebe

Nonbreeding adult

Breeding adult with young

Least Grebe

Tachybaptus dominicus

Freshwater or brackish marshes, freshwater ponds, lakes
LENGTH 8–11 in. **WINGSPAN** 10–12 in.

Aptly named wee gray grebe with deep golden eyes; a dark gray, sharply tipped, straight bill; and a white, fluffy tuft on its caboose. Breeding plumage is a deep and lustrous gunmetal-gray, while nonbreeding plumage is a duller, paler gray-brown with a pale whitish gray face and throat. **BEHAVIOR** Tiny but effective hunters, Least Grebes surface dive for insects and small vertebrates and glean prey from emergent vegetation. They are known to submerge their bodies in vegetated waterbodies, leaving only their heads above water. They tend to travel solo or in pairs, nesting in shallow water, with young sometimes traveling on parents' backs. **VOCALIZATIONS** Both sexes give a high-pitched, rapid trill, much like a cicada, known as their duet. Female gives a rapid, repetitive chatter in courtship, with male giving a one-note *EEP*. **KEY SITES** AZ: Scattered locations from near the town of Ruby to Patagonia Lake State Park. **ABUNDANCE** A rare but regularly occurring visitor to Southeast AZ. **CONSERVATION STATUS** Least Concern.

Pied-billed Grebe

Podilymbus podiceps

Open water with plentiful emergent vegetation; wetlands
LENGTH 12–15 in. **WINGSPAN** 18–24 in.

Drab waterbird with a colorful personality. Gray and brown overall, with more brown in throat, neck, breast, and underparts. Blocky head; long, stout neck; relatively short, thick bill; puffy white rump; and extremely short tail. Breeding birds show dark crown, face, and throat and a black ring near the tip of white bill. Nonbreeding birds show plain yellowish bill and are more uniformly gray and brown. Juvenile has plain bill and boldly striped face. **BEHAVIOR** Charismatic grebes dive and swim both to catch fish and to harass other waterbirds getting in their way. Can sometimes be seen lurking with just the eyes visible above water. **VOCALIZATIONS** A loud, intensifying *ooh-ooh-ooh-ohwoop-ow-hoop*, descending and fading toward the end. Also hyena-like clattering laughter. **ABUNDANCE** Common year round. **CONSERVATION STATUS** Least Concern.

Breeding adult

Nonbreeding adult

Molting adult

Horned Grebe

Podiceps auritus

Lakes, bays, ponds, usually with emergent vegetation
LENGTH 12–14.5 in. WINGSPAN 17–18 in.

Breeding adult undergoes a drastic molt, transforming from black cap, white cheek, gray-black back, and dingy gray-white belly to a glorious chestnut-red neck, breast, and sides; gunmetal-gray back; clean white belly; and deep gold tufts over the eyes on a rich black head. Sharply tipped, straight bill is black on breeding adult and gray on nonbreeding and immature grebes, always with a white tip. Eyes are bright ruby-red. Molting adult shows varied pattern of mottled gold, chestnut, black, and white. BEHAVIOR Surface divers, diet consists primarily of crustaceans, aquatic insects, and fish, with the occasional tadpole, leech, and, if it must, aquatic vegetation. Almost always seen on water, will form small flocks in winter months; otherwise prefers to live solo or in pairs. Courtship involves both adults extending their bodies vertically above water and quickly shaking their heads from left to right. VOCALIZATIONS Generally silent, with the exception of a raspy, whiny trill during courtship and a soft, high-pitched honking that may be used as an alarm call. ABUNDANCE Uncommon but regularly observed winter visitor. KEY SITES NM: Bosque del Apache National Wildlife Refuge; AZ: Lower Colorado River refuges. LOOKALIKE Eared Grebe. Horned distinguished by heavier bill with white tip and, in breeding plumage, chestnut-colored neck and golden tufts that run back along the head, versus the black neck and lighter colored, more splayed eyelash-like plumes of Eared. In nonbreeding plumage, Horned shows a more distinct border between black cap and white cheek. CONSERVATION STATUS Vulnerable.

Breeding adult

Nonbreeding adult

Eared Grebe

Podiceps nigricollis

Open water in often fishless settings; saline lakes, high-salinity ponds, wetlands fed by briny groundwater and agricultural runoff
LENGTH 12–14 in. WINGSPAN 16–21 in.

Dainty, high-floating waterbird with small, round head; thin bill; orange to bright red eyes; long neck; and extremely short tail. In breeding plumage, is mostly black with copper-brown flanks and thin, golden plumes that radiate from behind the eyes. Nonbreeding birds are gray-and-white overall with black crown fading into smudged ear patch, white throat, and grayish neck. Juvenile similar to nonbreeding but more brownish overall. BEHAVIOR Dives for insects, crustaceans, and fish. Also gleans insects off water's surface. VOCALIZATIONS Mostly silent in nonbreeding range. Breeding male gives high-pitched *ooEEK*. Both sexes give high trills. ABUNDANCE Common winter resident in Western AZ, Eastern NM, and the central to southern portions of both states. Small breeding population exists in East Central AZ. Common migrant and breeder farther north. LOOKALIKE Horned Grebe. In all plumages, Eared distinguished by thinner bill without a pale tip; in breeding, Eared's plumage by black neck and golden feathers displayed as a raised fan, versus the golden tufts that lay flat on the head of Horned. In nonbreeding, Eared distinguished by smudgier border between black cap and white cheek. CONSERVATION STATUS Least Concern.

Western Grebe

Aechmophorus occidentalis

Expansive open water including lakes, ponds, large reservoirs, bays
LENGTH 22–29 in. WINGSPAN 22–26 in.

Large, sleek, long-necked grebe with squarish head and black cap that extends to or slightly past the eyeline and down the nape. White cheek extends down the front of the neck to the breast, belly, and sides. Back is gray. Sharp, finely tipped bill is yellow or olive-yellow, and eyes are ruby-red. Immature appears similar, with faded looking plumage that lessens contrast. BEHAVIOR Spending most of their lives on the open water, these gregarious grebes nest on fresh water in large breeding colonies, diving for small fish and often carrying their young on their backs. Notorious for their courtship dance, both sexes run side by side in a synchronous duet, bodies curved, atop the water's surface, beaks upturned to the sky. VOCALIZATIONS Largely silent in winter but boisterous during the breeding season, using a variety of raspy trills similar to a chirping cricket, lacking rhythm. ABUNDANCE Year-round resident along the Colorado, Rio Grande, and Pecos rivers and on Central AZ's major reservoirs. Regular migrant and winter resident across the region. LOOKALIKE Clark's Grebe. In breeding plumage, Western's extensive black cap leaves eyes surrounded by black (white in Clark's). In all plumages, but most obviously when not breeding, Western distinguished by dingier olive-yellow bill. Western's call is two-parted, while Clark's consists of a single drawn-out syllable. Intermediate and hybrid birds not uncommon. CONSERVATION STATUS Least Concern.

Clark's Grebe

Aechmophorus clarkii

Expansive open water including lakes, wetlands, ponds, large reservoirs, bays
LENGTH 22–29 in. WINGSPAN 23–25 in.

Large, elegant, and neatly patterned waterbird. Black and white with long, pointed, orange-yellow bill; red eyes; black cap, nape, and back; and white chin, cheeks, neck, breast, and underparts. In breeding plumage, black cap recedes and eyes are fully surrounded by white. Juvenile is a plain, dingy white. BEHAVIOR Congregates in large groups on open water in winter, often with Western Grebes. Dives for small fish and invertebrates. VOCALIZATIONS High-pitched, ascending, froglike *kerererEEET*. Also a variety of croaking trills and barks. Mostly silent in winter. ABUNDANCE Most abundant on large waterbodies in winter and migration, with year-round populations on the Colorado, Rio Grande, and Pecos rivers and in Central AZ. LOOKALIKE Western Grebe. In breeding plumage, Clark's reduced black cap leaves eyes surrounded by white (black in Western). In all plumages, Clark's distinguished by brighter orange bill. Clark's call is a single drawn-out syllable, while Western's is two-parted. Intermediate and hybrid birds not uncommon. ALTERNATIVE NAME High-capped Grebe, Carrot-billed Grebe. CONSERVATION STATUS Least Concern (Decreasing).

PIGEONS AND DOVES

Pigeons and doves (family *Columbiadae*) are often overlooked, but that's by design. Subtle plumage, strong group and pair bonds, powerful and agile flight, and feathers that easily break off to fill would-be predators' mouths all help keep these birds off the menus of those who find their chunky physique appealing. Whether you're looking at an urban flock of Rock Pigeons or enjoying the lucky sighting of a Ruddy Ground Dove, throw away your preconceived notions of worthwhileness and give birds in this family a closer look.

ABOVE Rock Pigeon

Rock Pigeon

Columba livia

Cliff dweller in its native range; in North America, roosts and nests on building edges or under bridges within any human-modified landscape; will occasionally use natural cliff faces
LENGTH 12–14 in. WINGSPAN 22–28 in.

A handsome pigeon, but often overlooked or even maligned because of its nonnative and common status. Chunky bird has short, downturned bill with fleshy white base; orange-red eyes; long, pointed wings; long tail; and pink-orange legs. Wild-type adult is light gray overall, with dark head, neck, mantle, and breast; two dark bars on the lower wings; white rump; and dark borders on light-colored underwings in flight. Mantle shows purple-green iridescence. Feral bird colors are varied, including all dark, white, rufous, or mottled black and white. BEHAVIOR Forages on ground in large flocks for seeds, fruit, and dropped food. Showy puffing and cooing is frequent between individuals muscling for rank. When disturbed, flocks often circle repeatedly before resettling. VOCALIZATIONS Low-pitched, resonant *coo*. Wings produce sharp slap on takeoff. LOOKALIKE Band-tailed Pigeon. Rock Pigeon distinguished by urban habits, smaller size, and lack of yellow bill and legs, white tail band, and white neck crescent. CONSERVATION STATUS Least Concern (Introduced).

Band-tailed Pigeon

Patagioenas fasciata

Mountain forests, coniferous and/or oak woodlands, canyons
LENGTH 14–15.5 in. **WINGSPAN** 22–25 in.

Large but slender pigeon bears a slight resemblance to its relative, the Rock Pigeon, with lavender-gray underparts, sooty gray upper parts, black wingtips, and diagnostic white-tipped tail feathers forming a distinct band in flight. Adult shows a distinct iridescent green patch of scaly feathers topped by a bright white crescent on the back of the neck, like a popped collar. Beak is crayon-yellow with a black tip. Juvenile lacks neck patch and crescent. **BEHAVIOR** Defying stereotypes, this gregarious pigeon deftly sweeps through the forest canopy in flocks sometimes numbering in the hundreds, searching for acorns, berries, and other seeds. Will ground forage but is surprisingly agile in the treetops. **VOCALIZATIONS** A deep, hooting, two-note *yoo hoo*. **ABUNDANCE** Breeds throughout much of the region, less commonly in westernmost AZ and easternmost NM. Observed year round in Southern AZ and Southwest NM. **LOOKALIKE** Rock Pigeon. Band-tailed distinguished by its namesake banded tail, white crescent on neck, distinct habitat, and larger size. **CONSERVATION STATUS** Least Concern (Rapidly Decreasing).

Eurasian Collared-Dove

Streptopelia decaocto

Human-modified landscapes with ample bare ground
LENGTH 11–13 in. WINGSPAN 19–22 in.

Large dove, extremely pale gray to grayish brown overall. Short, dark bill; red eyes; black collar bordered in white just above the mantle; relatively short, rounded wings; long tail with black base and white tip; dark primaries; and pinkish legs. BEHAVIOR Forages on the ground for mostly seeds, often among other doves. Habitually flicks tail upon landing. Vocalizes during high, circular flights back to original perch to show off for potential mates. VOCALIZATIONS A loud *hoo-HOOO-hoo*. Display flight call a harsh, nasal, humanlike *EHHH*. ABUNDANCE Increasingly common bird across the Southwest year round, especially around human development. This introduced species' range is expanding rapidly. CONSERVATION STATUS Least Concern (Introduced).

Inca Dove

Columbina inca

Parks, farms, suburban developments; prefers shrubs, palo verde
LENGTH 7–9 in. WINGSPAN 11–12 in.

Petite, subtly patterned dove. When perched or ground foraging, visible plumage is pale, dusty brown with distinct cocoa-brown scaling pattern. In flight, look for bright white outer tail feathers and deep ruddy red underwings. Bill is black. Red eyes are known to brighten when alarmed. BEHAVIOR Soft-footed ground foragers, granivorous and gregarious, they cock their heads forward and backward as they move along the ground. Adults produce crop secretions used to supplement the diets of their young. Relatively cold intolerant, flocks engage in pyramid roosting to keep warm in frigid temperatures. VOCALIZATIONS Extremely vocal at all times of day and year, with a repetitive, cooing *no hope*. ABUNDANCE Common year round in Western AZ and across the central to southern portions of both states. ALTERNATIVE NAMES Mexican Dove, Pessimistic Dove. CONSERVATION STATUS Least Concern.

Common Ground Dove

Columbina passerina

Open, sandy areas; brushy openings within mesquite, riparian, oak, pine, other open woodlands; agricultural fields
LENGTH 6–7 in. WINGSPAN 10–11 in.

Our smallest dove. Brownish gray overall with pinkish undertones, light-colored head, dark bill with pink to red base, rufous primaries, and scaled head, breast, and nape. Black spots on upper wing form messy stripes. Short, square tail is brownish gray with black edges and white corners. Male more pinkish overall with lighter head, red eyes, and deeper red bill. BEHAVIOR Forages on ground for mostly small seeds. Often in pairs or small groups. VOCALIZATIONS A slow, steady *hooope-hooope-hooope*, with notes delivered about once per second. ABUNDANCE Uncommon year round from Southern AZ through to the bootheel of NM. Common year round along the lower Colorado River and in extreme Southwest AZ. LOOKALIKES Ruddy Ground Dove, Inca Dove. Common distinguished from Ruddy by less pinkish coloration, scaled appearance, white tail corners, and reddish to orange bill with dark tip. Common calls at about half the speed of Ruddy. Common distinguished from Inca by smaller size, shorter tail, less heavily scaled appearance, dark wing spots, and pinkish undertones. CONSERVATION STATUS Least Concern.

Female (left) and male

Ruddy Ground Dove

Columbina talpacoti

Riparian areas, parks, farms, suburban developments; prefers open or low shrubby areas
LENGTH 6–7 in. WINGSPAN 11 in.

Attractive, small, short-tailed dove with male showing pale, cool-gray head contrasting with warm-rufous body and sparse black spots and barring on wings. Female is more subtly plumed, with dusty brown rather than rufous body and dusty tan rather than cool-gray head. Both sexes show ruddy underwings in flight. BEHAVIOR Forages on open ground in search of grains, often in mixed flocks with Inca Doves or Rock Pigeons. VOCALIZATIONS Repetitive, cooing *g-guh-WOO*. ABUNDANCE Uncommon to rare, but regularly observed in winter throughout Southern and Central AZ and NM, often in mixed flocks with Inca Doves. Population trends show northward movement from historical range in Northern Mexico. LOOKALIKE Common Ground Dove. Ruddy has more rufous plumage overall, has a darker bill, lacks scaly appearance, and is slightly larger than the Common. CONSERVATION STATUS Least Concern.

Adult

Adult with juvenile

White-winged Dove

Zenaida asiatica

Open, thorny woodlands and deserts; urban and agricultural areas; most abundant in saguaro forest, mesquite woodland, near riparian habitat
LENGTH 11–12 in. WINGSPAN 19–22 in.

Plain at first glance, but finely painted, brownish gray overall, with azure-blue "eyeshadow," reddish-orange eyes, rounded tail with black band and white tip, and black comma shape below cheek. Bold white wing patches visible in flight form white lower edge on folded wing. BEHAVIOR Forages on the ground for mostly seeds and, unlike many other doves, will forage high in vegetation for fruit, specializing in saguaro cactus fruit in the Sonoran Desert. VOCALIZATIONS A loud, raspy *hoo-HOO-hoo-hoo*, similar in cadence to Barred Owl's classic *who cooks for you?* call. Also a steady, repeated *heh-heh-hooa*; other varied coos with similar raspy quality; and a short, nasal *rehh*. Wings whistle when taking flight. ABUNDANCE Common spring and summer visitor to central and southern portions of both states. Year-round populations exist in Southeast AZ, NM's Rio Grande Valley, and irrigated suburban and agricultural areas. CONSERVATION STATUS Least Concern.

Mourning Dove

Zenaida macroura

Open woodlands, farms, towns, suburbs, urban parks, backyards
LENGTH 10–13 in. WINGSPAN 17–18 in.

Understated but beautiful dove with long, slender tail and peachy tan underside subtly contrasting with gray-tan on back and wings. Sparse black spots dot the back, and black edges border a white-tipped tail. Skin around the dark eyes is a lovely sky-blue, on a relatively small, tan head with a short, lead-gray, pointed bill. BEHAVIOR Often resembling a falcon in flight, this dove is agile on the wing. Can be observed in nearly any habitat aside from heavily forested areas and will gather in small to large flocks. Known for nesting in absurdly precarious places, such as on the blade of an outdoor patio fan. VOCALIZATIONS Ubiquitous, repetitive, mournful *cooOOO, coo, hoo, hoo* is produced by air passing through the throat, with a closed bill. Wings produce distinct whistling sound upon takeoff. ABUNDANCE Common and abundant across the region. CONSERVATION STATUS Least Concern.

CUCKOOS

With the European Cuckoo being what most folks envision when they think of a cuckoo, and despite the Greater Roadrunner being New Mexico's state bird, it often comes as a surprise that the Southwest is home to three species in this family (*Cuculidae*). Although cuckoos in the Americas aren't obligate brood parasites like their more famous cousin from across the pond—meaning that they can raise their own young and are not limited to dumping their eggs in the nests of others—they do sport the family's characteristic zygodactyl (X-shaped) feet, long tails, and carnivorous habits. Whether you're after the common but sparsely distributed Greater Roadrunner, the skulking and imperiled Yellow-billed Cuckoo, or the rarely visiting Groove-billed Ani, spotting a Southwestern cuckoo can be a challenging endeavor—but it's well worth the effort.

ABOVE Greater Roadrunner

Groove-billed Ani

Crotophaga sulcirostris

Dense and brushy thickets within open grasslands, savannahs, other open habitats
LENGTH 13–14 in. WINGSPAN 16–18 in.

A shaggy muppet of a cuckoo. Solid black with slight purple-green iridescence, long tail, and extremely heavy bill. Frequently ruffled feathers and drooping wings give an unkempt appearance. Juvenile has smaller bill and is brownish black overall. BEHAVIOR Hops along the ground or in low vegetation in search of insects, lizards, other small animals, and fruit. Warms itself similarly to Greater Roadrunner, with tail pushed downward, wings spread and drooped, and back to the sun. Known to follow both cattle and swarms of army ants, both of which tend to stir up prey. VOCALIZATIONS A clear, high, and repeated *PIK-ooee*. Also a bill rattle similar to Greater Roadrunner. ABUNDANCE Extremely rare visitor to sites scattered across the southern portions of both states. Most frequently observed in midsummer or late fall. CONSERVATION STATUS Least Concern.

Adult

Adult with tiger whiptail lizard

Greater Roadrunner

Geococcyx californianus

Desert scrub; open, shrubby grasslands
LENGTH 20–23 in. WINGSPAN 19–20 in.

Unmistakable bird with a sleek body built for doing what it does best—running. Creamy tan belly contrasts with cocoa-colored back, chest, and wings, accented by creamy white and black streaking. Black crest is spotted with off-white. Bare skin patches behind the eyes are bright blue, sometimes with white and cherry-red. Sturdy, long beak is slightly hooked at the tip. Tail is long and chocolate-brown with blue-green iridescence and a sometimes white tip. BEHAVIOR This charismatic icon of the Southwest is not a cartoon character or a dinosaur, but a member of the cuckoo family, evident in its tail-bobbing displays and soft, cooing vocalizations. Darting swiftly through scrub and brush, seeking out tasty lizards, snakes, insects, birds, and mammals, it holds its body nearly parallel to the ground with tail outstretched, enabling it to move stealthily toward its prey. Nests a few feet off the ground in shrubs, cacti, and scrubby desert trees. One of many adaptations that enable it to thrive in the harshest conditions, it flutters bare gular patches on its throat to dissipate body heat in the scorching hot desert. Leaves a distinctive X-shaped track in its wake when running through loose dirt and sand. VOCALIZATIONS Descending, sorrowful cooing call to attract mates and claim territory. Also produces a clattering sound with the bill. ABUNDANCE Commonly observed in suitable habitat throughout AZ and NM. CONSERVATION STATUS Least Concern.

Yellow-billed Cuckoo

Coccyzus americanus

Low to mid-elevation cottonwood/willow riparian woodlands, mesquite bosques; prefers open floodplains to canyon-bound areas; in Southeast AZ, oak drainages and oak-dominated grasslands
LENGTH 10–12 in. **WINGSPAN** 15–18 in.

Slender, elegant cuckoo emblematic of western rivers. Grayish brown above and white below, with yellow bottom mandible on long, decurved bill; yellow eye rings; long tail with black-and-white–spotted underside; and rufous primaries visible both in flight and when perched. Juvenile lacks eye rings and has muted undertail pattern. **BEHAVIOR** More often heard than seen, this extremely skulky and elusive cuckoo forages in dense canopy for caterpillars, cicadas, and other large insects. Often sits hunched, motionless, and hidden within the densest available canopy. Colloquially referred to as the "rain crow" for its tendency to call before summer monsoon storms. **VOCALIZATIONS** Loud, resonant, and intensifying *kuk-kuk-kuk-KOWLP-KOWLP* that trails off at the end. Also steady, repeated coos at a single pitch. **ABUNDANCE** Uncommon summer resident. Listed as Threatened by the US Fish and Wildlife Service west of the Rio Grande. **CONSERVATION STATUS** Least Concern (Decreasing).

NIGHTJARS AND SWIFTS

Sharing traits with one another and with a mutual relative—the hummingbird—nightjars (family *Caprimulgidae*) and swifts (family *Apodidae*) are built with a single purpose: to catch airborne insects. Their tiny feet render them a bit useless when grounded, but when they take to the skies, their long, pointed wings make them fast and agile flyers, and their gaping, envelope-shaped mouths let few insects escape. When you're birding in the Southwest, use the daylight hours to scan for swifts high in the sky or along prominent cliffs, and come nightfall, thank the nightjars for keeping the mosquitoes at bay.

ABOVE Common Poorwill

Lesser Nighthawk

Chordeiles acutipennis

Open habitats, usually below 4000 feet, including deserts, grasslands, chaparral, agricultural fields
LENGTH 8–9 in. **WINGSPAN** 21–22 in.

Intricately mottled gray, buff, and black overall with white collar below chin and dark primaries with buffy spots. In flight, note long, pointed wings with buffy or white bars near relatively blunt wingtips, buffy spotted underwings, and long, notched tail. When perched, appears legless and heavy bodied, with small, flat head; tiny bill; long tail; and long, upcurved wingtips that just meet or fall short of tail tip. **BEHAVIOR** Most frequently seen at dusk and dawn when they take to the sky with bouncy, effortless wingbeats to forage for flying insects, from mosquitos to beetles. Perches on horizontal branches during the day. **VOCALIZATIONS** A long, trilling *coo* and nasal, monkey-like laugh. Usually silent in flight. **ABUNDANCE** Common summer resident across the southern portions of both states and in Western AZ. **LOOKALIKE** Common Nighthawk. In flight, Lesser shows more blunt-tipped wings and wingbars closer to the wingtips (farther from the wrist); when perched, Lesser's wingtips do not project behind the tail. **CONSERVATION STATUS** Least Concern.

Common Nighthawk

Chordeiles minor

Open areas, forest clearings, urban and suburban areas, grasslands, open wetlands
LENGTH 8–10 in. WINGSPAN 20–23 in.

Medium-sized, dove-like in shape, with long, pointed wings and relatively short, notched tail. Because of its inconspicuous, mottled blend of colors (charcoal-gray, white, buff, and cocoa-brown), it can become virtually invisible when perched in natural surroundings, becoming one with tree bark. Large black eyes and small, wide, black bill just peek out from whiskery plumage on a round head. Very short neck has a white V-shaped pattern at the throat, visible in flight along with white bars on the underwings. At rest, appears legless and heavy bodied with small, flat head; tiny bill; and wingtips that extend beyond the end of the tail. BEHAVIOR Its name being somewhat of a misnomer, this nightjar is crepuscular, taking to the sky at twilight to feed on insects, and is observed most frequently at dusk, though quite active in the morning hours as well. Flight pattern resembles that of a bat or butterfly—erratic and undulating. Historically ground nesters, they have adapted to habitat loss by building nests on flat-roofed structures. VOCALIZATIONS Loud, nasally, buglike *meep* or *peent* in flight. Females cluck while defending nests. In breeding season, males known to dive through the air, positioning their wings to allow air to pass through the primaries, creating a whooshing, booming sound. ABUNDANCE Breeds throughout NM and much of AZ, except southwest and far northwest portions of the state. LOOKALIKE Lesser Nighthawk. Common has white bar on underwing farther from wingtip and, at rest, wingtips that extend beyond the end of the tail. CONSERVATION STATUS Least Concern (Rapidly Decreasing).

Common Poorwill

Phalaenoptilus nuttallii

Open, shrubby habitats with ample open ground and rock, including desert, chaparral, sparse woodlands
LENGTH 7–8 in. WINGSPAN 16–17 in.

Our smallest nightjar is mottled gray and brown overall, with a thin, white collar below chin; buffy underwings; and buffy, dark-banded primaries that are often visible at rest. In flight, note long, rounded wings and relatively short, round-edged tail. At rest, appears legless and heavy bodied with large, flat head; tiny bill; short tail; and straight wingtips that meet the tip of the tail. BEHAVIOR More often heard than seen. At night, sits on bare ground (including dirt roads) and flits upwards in pursuit of passing insects. VOCALIZATIONS A high, fluty *poor-will-wup*, with the final note often difficult to hear, like a droplet of water. Also a short *wep*, sometimes repeated with urgency. ABUNDANCE Common year round in Southern AZ and Southwest NM. Common summer resident farther east and north. LOOKALIKES Nighthawks, Buff-collared Nightjar, Mexican Whip-poor-will. Common Poorwill distinguished from nighthawks by shorter wings and lack of pale bars on underwings in flight, from Buff-collared Nightjar by lack of buffy collar and less white in tail, and from Mexican Whip-poor-will by smaller size and shorter tail with less white. CONSERVATION STATUS Least Concern.

Buff-collared Nightjar

Antrostomus ridgwayi

Rocky hillsides, desert canyons
LENGTH 8.5–9.5 in. WINGSPAN 14–15 in.

Small and slender, mottled from head to toe with dark gray, black, and brown tones above and buffy tones below. Namesake buff collar is conspicuous. Small, round head; large, coppery brown eyes; and plentiful whiskers around small, black bill with downturned tip. When not in flight, has a legless, heavy bodied, and flat-headed appearance. BEHAVIOR Much like a flycatcher, forages primarily by perching on the end of a shrub or tree branch, flying out to ambush insects and returning to its perch. Will do the same from the ground and will sometimes forage on the wing in one- to two-minute intervals. Rarely observed in daylight, most often identified by its distinct vocalizations in the dead of night. Ground nester that lays eggs directly on loose soil, yet very few nests have actually been found. VOCALIZATIONS Repetitive, rapid, ascending staccato clucks, very insectlike but not buzzy like other nightjars. Generally heard extremely late at night. ABUNDANCE Uncommon but regularly breeds in key sites within range. KEY SITES Southeast AZ: Warsaw Canyon/California Gulch in the Atascosa Highlands; NM: Whitmire Canyon Wilderness Study Area in extreme southwest corner. CONSERVATION STATUS Least Concern.

Mexican Whip-poor-will

Antrostomus arizonae

Oak woodland, pine-oak communities at higher elevations than other nightjars
LENGTH 9–10 in. WINGSPAN 18–19 in.

Similar to Eastern Whip-poor-will. Mottled buff, tan, and gray overall with indistinct buffy collar, gray band on upper wing, white to buff tail corners, and dark buff-spotted primaries. In flight, note long, rounded wings and long, rounded tail. At rest, appears legless and heavy bodied with large, flat head; tiny bill; and long tail that extends beyond wingtips. Best identified by vocalizations and habitat. BEHAVIOR Uses a variety of strategies to forage for flying insects—flycatching from a perch, soaring along woodland edges, or flitting off the ground like a Common Poorwill. VOCALIZATIONS A quick, trilled *whip-poor-will*, very similar to Eastern Whip-poor-will. Also a short *whip-whip*. ABUNDANCE Common summer resident in Southeast AZ and Southwest NM. Uncommon summer resident in mountainous areas farther north and east. LOOKALIKES Nighthawks, Common Poorwill, Buff-collared Nightjar. Mexican Whip-poor-will distinguished from nighthawks by shorter wings and lack of pale underwing bars in flight, from Common Poorwill by larger size and longer tail with white corners, and from Buff-collared Nightjar by lack of complete buff collar, lack of dark crown stripe, and longer tail. CONSERVATION STATUS Least Concern (Decreasing).

Vaux's Swift

Chaetura vauxi

Conifer or mixed conifer forest; prefers old growth
LENGTH 4–5 in. WINGSPAN 10–11 in.

A classically cigar-shaped swift, but tiny in comparison to other North American swift species. Gray-brown on back and wings, with paler, buffy gray belly; round head; and a wee bill on wide mandibles perfect for chomping insects on the fly. Boomerang shaped in flight. **BEHAVIOR** Aerial foragers spend much of their day deftly picking off insects in the skies above forests, lakes, and rivers. Gregarious swifts nest and roost communally in large tree cavities, preferring mature conifers, though they will take advantage of a large chimney if need be. Roosting in large groups, they engage in torpor, dropping their body temperatures in the night to endure freezing temperatures. **VOCALIZATIONS** Extremely high-pitched, rapid chittering. **ABUNDANCE** Uncommon in migration across Southern AZ. **LOOKALIKE** Chimney Swift. Vaux's and Chimney distinguished by range—Vaux's to the west and Chimney to the east. **ALTERNATIVE NAMES** Tree-roosting Swift, Charcoal Swift, Least Dusky Swift, Western Chimney Swift. **CONSERVATION STATUS** Least Concern (Decreasing).

White-throated Swift

Aeronautes saxatalis

Nests in crevice-rich rock faces along cliffs and in canyons; forages over any habitat type
LENGTH 6–7 in. WINGSPAN 14–15 in.

Perhaps the most easily identifiable swift. Solid black to dark brown overall with stark white eyebrows, cheeks, and throat, extending into long, white band down belly. Large white "saddlebags" adorn flanks behind wings. Long-tailed with boomerang-shaped wings, long and slightly notched tail, and tiny bill. BEHAVIOR Extremely agile and fast flyer. With a gape that opens far wider than its tiny bill would imply, forages for flying insects high along cliffs and canyon walls. Like other swifts, spends almost the entire day on the wing and lands only to roost. Roosts and nests in large colonies on expansive vertical rock faces with plentiful crevices, but will also take advantage of human-made structures. VOCALIZATIONS A harsh, descending *kee–kee-kee* and high-pitched trills. ABUNDANCE Common breeding resident across much of the central to northern part of the region, with year-round populations in Western AZ and the southern portions of both states. Uncommon in extreme Southwest AZ and Eastern NM. CONSERVATION STATUS Least Concern (Decreasing).

HUMMINGBIRDS

The abundance of hummingbirds (family *Trochilidae*) in the Southwest puts the rest of North America to shame. Although it's a delight to have these nectar-guzzling gems buzzing around in both backyards and far afield, and even though the diversity is enough to inspire a region-wide quest to spot them all, the abundance of species can also create some difficult identification challenges. Especially when dealing with juvenile and female birds sporting subtle field marks, pay special attention to range, habitat, overall structure and size, and vocalizations. Or, to take things a little less seriously, just put up a feeder (or a few) and enjoy the show.

ABOVE Broad-billed Hummingbird

Adult male

Female/immature

Rivoli's Hummingbird

Eugenes fulgens

Madrean pine-oak forests
LENGTH 4.5–5.5 in. WINGSPAN 7–7.5 in.

Our second largest hummingbird after the Blue-throated Mountain-gem. Male is very dark green to black overall with white corners on uppertail and, when well lit, amethyst crown and brilliant blue-green throat. Female recognized by size, with solid green backside and very long, straight bill. Both sexes have small white marks behind eyes and dark undertails. BEHAVIOR Feeds on nectar from flowers and catches small insects, both by flycatching and gleaning from vegetation. May be more reliant on insects than other hummingbirds. Will venture outside habitat to visit feeders. VOCALIZATIONS A very sharp, clear *sip-sip-sip*, sometimes given singly as calls. Also a rattling chatter similar to Anna's Hummingbird but lower in pitch and volume. ABUNDANCE Common breeding resident in the Madrean Sky Islands of extreme Southeast AZ and Southwest NM. Uncommon to rare farther north and east. KEY SITES AZ: Madera Canyon, Paton Center for Hummingbirds, Miller and Ramsey canyons, town of Portal in Cochise County. LOOKALIKE Female Blue-throated Mountain-gem. Rivoli's female lacks bronze rump and has a green uppertail and longer bill. ALTERNATIVE NAME Magnificent Hummingbird. CONSERVATION STATUS Least Concern.

Plain-capped Starthroat

Heliomaster constantii

Wooded streams, canyons, arid to semi-arid forests, forest edges
LENGTH 4–5 in. WINGSPAN 5 in.

Long, slender, and relatively large as hummingbirds go, the starthroat is fancier in name than appearance, with a greenish body, dusky gray wings, white belly with buffy gray flanks, and iridescent red feathers on a white throat that tend to appear blackish unless caught by light. White-tipped, charcoal tail. Long, black bill and white supercilium. BEHAVIOR Canopy feeder, spending much of its day flycatching from high perches; feeds on flower nectar when available. VOCALIZATIONS Monotonic, single-note, whistling chip. ABUNDANCE Uncommon but regularly observed throughout the Madrean Sky Islands of Southeast AZ. KEY SITES Madera Canyon at the Santa Rita Lodge feeder station, Paton Center for Hummingbirds in Patagonia, AZ. CONSERVATION STATUS Least Concern (Decreasing).

Adult male | Female/immature

Blue-throated Mountain-gem

Lampornis clemenciae

Damp, humid canyons within mixed mid- to high-elevation forests of oak, pine, fir
LENGTH 4.5–5 in. WINGSPAN 7.5–8 in.

Our largest hummingbird, bulkier and longer winged than Rivoli's. Both sexes have relatively short bills, green backs, bronze rumps, dark uppertails, gray underparts, and eyes sandwiched between two white stripes. Male has bright blue throat that can appear dark in dim light, and bold white corners on tail, visible from above and below. BEHAVIOR Feeds on nectar from flowers and catches small insects, both by flycatching and gleaning from vegetation. Will visit beyond natural habitats when feeders are present. Nests are covered in green moss, distinct from the usual lichen-covered nests. VOCALIZATIONS A high, squeaky, liquid, rattling *krrikriikrrr*. Also a high, clear *sip* or *seep*, delivered more slowly than Rivoli's. ABUNDANCE Uncommon breeding resident in extremely limited range in Southeast AZ and Southwest NM. Rare elsewhere. KEY SITE Cave Creek Canyon near Portal, AZ. LOOKALIKE Female Rivoli's Hummingbird. Blue-throated Mountain-gem female distinguished by bronze rump, blackish uppertail, and shorter bill. CONSERVATION STATUS Least Concern.

Adult male

Female/immature

Lucifer Hummingbird

Calothorax lucifer

Desert canyons and scrub, hillsides rich with ocotillo and/or agave, occasionally oak-dominated grasslands
LENGTH 3.5–4 in. WINGSPAN 4 in.

Tiny jewel of a bird with heavy, downturned, black bill that is seemingly out of proportion with its petite body. Distinctive tail is narrow and deeply forked. Male is emerald-green with white belly and an exaggerated purple gorget that appears black when not catching the light. Immature male shows sparse purple spotting on throat. Female is green with a buffy rufous wash on belly. BEHAVIOR A lover of agave and ocotillo nectar, feeds on flowering desert plants; sometimes spotted at feeders. VOCALIZATIONS Rapid, buzzy chips and quiet, repetitive *chi-dee* with a metallic quality. ABUNDANCE Uncommon but regularly observed in extreme Southeast AZ and Southwest NM. KEY SITES In AZ, Ash Canyon Bird Sanctuary in the foothills of Huachuca Mountains, Chiricahua Mountains near Portal. CONSERVATION STATUS Least Concern.

Adult male

Female/immature

Black-chinned Hummingbird

Archilochus alexandri

Generalist in wooded areas in low-elevation washes, riparian areas, mid-elevation oak woodlands
LENGTH 3.5 in. WINGSPAN 4.5 in.

A slender, long-bodied hummingbird. Both sexes are green above and grayish white below with greenish flanks; long, dark, slightly downward-curving bills; and tails that extend slightly beyond the wingtips when perched. Male shows dark outer tail feathers, black chin, and iridescent purple throat bordered below by white breast and collar. Female's dark tail feathers are tipped in white. BEHAVIOR Feeds on nectar from flowers and catches small insects by flycatching from a perch. A regular at feeders and in urban landscapes. Male displays by buzzing side to side in front of perched female and by diving from as high as 100 feet. VOCALIZATIONS Common call a high *chew-chew-chew*. Also other high-pitched chips and squeaks given singly or in combination. Wings produce a low buzz. Male produces high trill at low point in display dive. ABUNDANCE Common breeding resident in suitable habitat throughout the region. LOOKALIKES Costa's, Anna's hummingbirds. Female Black-chinned distinguished from both by long, slender shape and longer bill; from Anna's by lack of red central throat patch. CONSERVATION STATUS Least Concern.

Adult male

Female/immature

Anna's Hummingbird

Calypte anna

Landscaped features in urban and suburban areas, open woodlands, low and high desert, chaparral
LENGTH 3.5–4 in. WINGSPAN 4–5 in.

Compact, sturdy little bird with straight, black bill. The male's back is emerald-green, fading to chartreuse on the sides and flanks, with dirty white chest and marvelous deep magenta gorget that flashes in the sunlight, otherwise appearing black in the shade. Female is metallic and paler green on back, sides, and flanks, with dingy white belly, white line (as if someone applied eyeshadow) over the eye, and small, triangular patch of deep rose-colored spots on the throat. Immature male shows spotty, incomplete gorget. BEHAVIOR Feisty and fierce, the Anna's announces its presence with scolding vocalizations and will likely quickly evict competitors at a feeder, even if there are several feeders to choose from. Ubiquitous in landscaped desert areas, look for them hovering over nectar-producing blooms, including chuparosa and desert honeysuckle, and sometimes gleaning small spiders and insects from the air or off human-made structures. VOCALIZATIONS Song is an extended series of metallic, buzzy chips and whistles. Calls consist of a series of short, high-pitched, buzzy chips. ABUNDANCE Very common in Western and from Central to Southern AZ. Common in winter farther east. LOOKALIKES Black-chinned, Broad-tailed, Costa's hummingbirds. Male Anna's distinguished from these by red crown, and female distinguished by reddish spotting on throat. ALTERNATIVE NAMES Magenta-headed Hummingbird, Ruby-headed Hummingbird. CONSERVATION STATUS Least Concern.

Adult male

Female/immature

Costa's Hummingbird

Calypte costae

Sonoran and Mojave deserts, often with ocotillo, chuparosa, other desert blooms **LENGTH** 3–3.5 in. **WINGSPAN** 4.5 in.

Fuzzy ping-pong ball of a hummingbird. Both sexes are round-bodied and short-tailed with short, slightly curved bills; whitish underparts; and grayish green flanks. Male has purple gorget with long tails, extending down either side of the throat nearly to the shoulder and dividing white patches on the throat and neck. Female has white eyebrow stripe and throat bordering a faded gray ear patch. **BEHAVIOR** Feeds on nectar from flowers and catches small insects midair. Visits feeders, but is less common in human landscapes during breeding and seldom strays far from deserts. Male displays include deep, U-shaped dives during courtship. **VOCALIZATIONS** Male produces short whistles when perched and longer whistles that rise and fall in pitch during display dives. Females give short, high chips, either singly or in rapid succession. **ABUNDANCE** Common year-round resident in Southwest AZ. Common breeding resident slightly farther north and east. **LOOKALIKES** Female Black-chinned, Anna's, Calliope, Broad-tailed hummingbirds. Costa's female distinguished from Black-chinned by rounder body, shorter tail, and shorter bill; from Anna's by lack of red central throat patch; and from Calliope and Broad-tailed by lack of buffy flanks. **ALTERNATIVE NAMES** Desert Hummingbird, Purple-mustached Hummingbird. **CONSERVATION STATUS** Least Concern.

Adult male

Female/immature

Calliope Hummingbird

Selasphorus calliope

Mountain meadows, high-elevation open forests and streams, aspen forests
LENGTH 3–3.5 in. **WINGSPAN** 4–4.5 in.

This tiny bird—the smallest in North America—is like a flying tourmaline jewel, with male showing a clean white breast, bronze wash on sides, and a deep rose starburst gorget. Female shows bronze-and-green spotting on the throat. Immature male has an incomplete, stippled gorget. **BEHAVIOR** Unless perched in a willow or an alder to keep watch over its territory (during breeding season), the Calliope forages for nectar very low to the ground in high-elevation meadows or regenerating burn areas. This thimble of a bird steers clear of competing with larger, more aggressive hummingbird species. Males perform impressive U-shaped flight display during courtship. **VOCALIZATIONS** While foraging and in flight, a high-pitched, single-note chip or series of chippy, buzzy *cheeediddydees*. Displaying male gives a quick, high-pitched *honk* during its dive. They sound much like bumblebees in flight. **ABUNDANCE** Uncommon in AZ and NM, but they migrate through both states, primarily in high-elevation areas along the length of the AZ–NM border. **CONSERVATION STATUS** Least Concern.

Adult male

Female/immature

Adult male flaring tail

Rufous Hummingbird

Selasphorus rufus

Brushy, open areas within mid- to high-elevation forests and montane meadows; feeders, parks, yards
LENGTH 3–3.75 in. WINGSPAN 4.5 in.

Small, chunky hummingbird with heavy, straight, black bill and dark wings. Male is fiery copper-orange overall with gold-and-red gorget and white breast. Usually shows no green in back. Female has green back, orange-washed flanks, whitish belly, orange-red central throat spot, rufous-washed rump, and tail with orange base, black margins, and white tips. Outer tail feathers relatively broad. Tail feathers bordering central pair show distinct notch toward tip. BEHAVIOR Very aggressive even in migration and quick to chase other hummingbirds away from feeders. Feeds on nectar and insects captured in flight or gleaned from spiderwebs. VOCALIZATIONS Both sexes give high chips either singly or in a series. Male gives three-part, buzzy *chup-chupity-chew*, very similar to Allen's Hummingbird. Wings produce a loud hum that varies in pitch. ABUNDANCE Common migrant during late summer and early fall, especially at higher elevations. LOOKALIKES Allen's, female Calliope and Broad-tailed hummingbirds. Male Rufous sometimes distinguishable from Allen's by fully orange back. Female and mixed green-and-orange–backed male Rufous distinguished from Allen's by tail feather details (narrow in Allen's, wider and notched in Rufous). Female Rufous distinguished from female Calliope and Broad-tailed by central throat patch, orange rump, and more orange in tail (none in Calliope). CONSERVATION STATUS Near Threatened (Decreasing).

Adult male

Female/immature

Adult male flaring tail

Allen's Hummingbird

Selasphorus sasin

Open woodlands, scrub, chaparral
LENGTH 3.5–4 in. WINGSPAN 4–4.5 in.

Small but stocky hummingbird with sturdy, straight, black bill. Both sexes wear a copper vest and white collar with metallic green back and dark chocolate-colored wings, though female and immature male coloration is a muted, subtle version of the resplendent adult male. Male shows a brilliant copper-marigold gorget, rufous cheeks, and bronze-colored head. Outer tail feathers on both sexes are remarkably narrow, creating a starburst appearance in flight. BEHAVIOR Male puts on a dazzling courtship display, swinging hypnotically from side to side, wings buzzing like a bumblebee. Courtship and mating occur along the Pacific coastline, followed by females moving inland to build nests, incubate eggs, and raise young on their own. They feed on nectar and small insects, frequenting feeders as well. Subspecies *S. sasin sasin* migrates south to overwinter in South Central Mexico, while subspecies *S. sasin sedantarius* remains along the California coast year round. VOCALIZATIONS Both sexes give ticking chip while feeding. Males give three-part call, like a metallic, buzzy version of a Common Yellowthroat's call, to protect territory; females use the same call in response to males' pendulum display. ABUNDANCE Rare in AZ, seen only in migration in areas below Mogollon Rim. LOOKALIKE Rufous Hummingbird. Narrow outer tail feathers help with identification of Allen's while in flight; male Allen's shows more green than orange on back. ALTERNATIVE NAMES Copper-throated Hummingbird, Pumpkin Spice Hummingbird, Coastal Copper Hummingbird. CONSERVATION STATUS Least Concern.

Adult male

Female/immature

Broad-tailed Hummingbird

Selasphorus platycercus

Dry coniferous forests, adjacent habitats
LENGTH 4 in. **WINGSPAN** 5.25 in.

The largest of our *Selasphorus* species. Green overall with a very slight pale eye ring, whitish breast and belly, tail that sticks out distinctly beyond wingtips, and short, straight, stout, dark bill. Male has ruby-red to purplish gorget; thin, pale stripe connecting eye ring to bill and neck; greenish flanks; and dark undertail. Female has green-speckled cheeks and throat, buffy flanks, green rump, and tail with orange base, black margins, and white tips. **BEHAVIOR** Feeds on nectar and catches insects by flycatching and gleaning from vegetation and spiderwebs. Male display involves a quick, high upward flight followed by a long, trilling dive. **VOCALIZATIONS** Chips and chitters similar to Rufous, but higher and clearer. Male wings produce high trill in flight and during display dives, like a small, tinkling bell flying through the forest. **ABUNDANCE** Common breeding resident at high elevations from Northern to Southeast AZ and Central to Western NM. **LOOKALIKES** Female Rufous and Allen's hummingbirds, Calliope Hummingbird. Compared to adult female Rufous and Allen's, adult female Broad-tailed is larger, has less orange in rump and at base of tail, shows buffy (not orange) flanks, and lacks a central throat patch. Compared to Calliope, Broad-tailed is larger with a heavier bill, orange in the tail, a tail that extends well beyond the wingtips at rest, and in adult males, a solid (not streaked) gorget and a darker undertail. **CONSERVATION STATUS** Least Concern.

Adult male

Female/immature

Broad-billed Hummingbird

Cynanthus latirostris

Canyon streams, high-elevation riparian areas; will frequent desert habitat in landscaped areas of parks, cities, suburbs within its range
LENGTH 3.5–4 in. **WINGSPAN** 4.5–5 in.

Small hummingbird with brilliant coloration. Males show various shades of green on back, from emerald to peridot, with a gleaming sapphire gorget; bright orange-red, black-tipped bill; green to blue underparts; and a notched tail with white undertail coverts. Female and immature male show deep chartreuse on back, with a dingy white belly and pronounced white line with downward curve extending from eye toward nape. Immature male shows orange on bill and some stippled blue-green coloration, hinting toward adult plumage. **BEHAVIOR** Feeds on nectar from flowers and feeders, as well as insects. Males will sometimes gather in groups, or leks, to show off their pendulum-like courtship display. Of several subspecies, only subspecies *magicus* occurs in the United States. **VOCALIZATIONS** Call is a repetitive *chiddit chiddit*. Male's song during courtship starts with one high-pitched chip, followed by a drumroll of sorts, with five to seven single-note squeaks, finished by another drumroll. **ABUNDANCE** Primarily a resident of Western and Central Mexico, the northern tip of their range includes a small patch of the Southwest from Tucson through Southeast AZ to the bootheel of NM; often observed in spring/summer as a vagrant throughout both states. **CONSERVATION STATUS** Least Concern.

Adult male

Female/immature

White-eared Hummingbird

Basilinna leucotis

Mountainous, canyon-rich pine-oak and conifer forests
LENGTH 3.5–4 in. WINGSPAN 4.5–5 in.

Heavy bodied and round-headed hummingbird with a broad tail and short, orange, black-tipped bill. Both sexes are dark green-bronze overall with bold white ear stripes and dark tails. Male has dark green flanks; head and throat shimmers an iridescent purple to green. Female's eye stripe accentuated by thick, dark line below eye; flanks adorned with messy rows of green spots. BEHAVIOR More commonly observed from Mexico to Nicaragua. In the United States, almost always observed at feeders. VOCALIZATIONS Rapid, dry chips and rattles. ABUNDANCE Rare summer visitor to extreme Southeast AZ and Southwest NM.

KEY SITES AZ: Huachuca Mountains (Sierra Vista, Miller Canyon, Ash Canyon), Chiricahua Mountains (near Portal), Paton Center for Hummingbirds in Patagonia. LOOKALIKE Female Broad-billed Hummingbird. White-eared distinguished by spotted flanks, shorter bill, and more distinct facial stripes. CONSERVATION STATUS Least Concern.

Violet-crowned Hummingbird

Ramosomyia violiceps

Riparian areas, riparian-adjacent forests and meadows
LENGTH 4–4.5 in. WINGSPAN 5.5–6 in.

Aptly named, look for the unmistakable blue-violet cap; straight, carrot-orange bill with black tip; and clean white throat and chest (this bird lacks the gorget that many hummingbirds display). Back is a bronzy but relatively lackluster green-brown. Immature shows muted, spotty cap. Dimorphism is not present in this species. BEHAVIOR Feeds on nectar and insects. Look for them hovering near tubular blooms such as honeysuckle. VOCALIZATIONS Calls include a variety of raspy chips, sometimes rapidly uttered like a machine gun, other times as single notes. Song is a soft, monotonous series of descending *see-ooo* notes. ABUNDANCE Relatively uncommon, even within range that just barely bumps up into extreme Southeast AZ. KEY SITE Famous feeder visitor at the Paton Center for Hummingbirds in Patagonia, AZ. CONSERVATION STATUS Least Concern.

Berylline Hummingbird

Saucerottia beryllina

Mountainous, canyon-rich oak and pine-oak forests
LENGTH 3.5–4.25 in. WINGSPAN 5.75 in.

Unlike any other North American hummingbird. Named after the mineral beryllium, large hummingbird has a deep green head, nape, shoulders, and throat; paler belly and flanks; and deep purple-brown lower back, rump, tail, and wings. Long, slightly curved bill has pinkish orange center to lower mandible. BEHAVIOR More commonly observed from Mexico to El Salvador. In the United States, almost always observed at feeders. VOCALIZATIONS A variety of short, thin chips, squeaks, and rattles. ABUNDANCE Rare summer visitor to extreme Southeast AZ. KEY SITES AZ: Huachuca Mountains, Madera Canyon in the Santa Rita Mountains, Chiricahua Mountains (near Portal), Paton Center for Hummingbirds in Patagonia. CONSERVATION STATUS Least Concern.

RAILS, CRANES, AND OTHER MARSHBIRDS

Wetlands are special places in the arid Southwest and are becoming more so in the face of climate change, drought, and water overuse. Special places call for special birds, and the rails, gallinules, coots (family *Rallidae*), and cranes (family *Gruidae*) are happy to oblige. When birding in the region, be sure to delight in the antics of the American Coots that adorn nearly any vegetated pond, listen carefully for the rails and gallinules as they call from the densest reeds, and make the time to attend the annual show put on by enormous flocks of wintering Sandhill Cranes.

ABOVE Common Gallinule with young

Ridgway's Rail

Rallus obsoletus

Marshes
LENGTH 12 in. WINGSPAN 20 in.

Distinguished, medium-sized but somewhat stocky rail may appear a simple gray-brown at first glance, but if you're lucky enough to take a closer look, you'll notice a warm, pale cinnamon-colored wash down the neck and breast, leading to delicate black-and-white barring pattern on the flanks; heavily streaked, gray and dark chocolate-colored back; with white-sided undertail coverts. Orange bill is sturdy and long, and legs are dingy orange-brown. BEHAVIOR Secretive marshbird spends most of its life under the cover of heavy marsh vegetation in both fresh and salt water; a special gland enables it to filter salt from seawater. Uses bill to root around in the mud for crustaceans and invertebrates, its preferred prey, but occasionally springs for a seed or two, which makes it technically omnivorous.

VOCALIZATIONS The rail equivalent of a chip, gives a harsh, raspy, one-note *cack* as well as a series of rapid, repeated *keks* that often accelerate into a low rattle. ABUNDANCE Uncommon, with the *yumanensis* subspecies (the only subspecies that occurs in the region) listed as endangered by the US Fish and Wildlife Service. KEY SITES AZ: lower Colorado River wildlife refuges along western border, westernmost stretches of the Gila River. LOOKALIKES Clapper, Virginia, and King rails. Distinguished from Clapper and King by range, from Virginia by size and lack of distinct gray face. ALTERNATIVE NAMES Western Clapper Rail, Salt Rail. CONSERVATION STATUS Near Threatened (Decreasing).

Virginia Rail

Rallus limicola

Freshwater and brackish marshes
LENGTH 8–10.5 in. WINGSPAN 13–15 in.

This richly patterned rail is slightly larger than a Sora. A solid gray face, orange-red bill, and white supercilium top off a pale cinnamon-colored breast and belly, black-and-white barring on flanks, deep cinnamon-colored wing patches, and heavily streaked cocoa-brown and dark chocolate-colored back. Legs match the bill but tend to be masked by a coating of mud. BEHAVIOR Much like Verdin, they build dummy nests, but often give themselves away with their grunting vocalizations. Broad feet are built for moving through mud. VOCALIZATIONS Song is a very squeaky, repetitive *tick-it*, and calls include a stuttery *kik-kik-kik-er*; during breeding season, listen for what can only be described as the calls of small marsh pigs, affectionately grunting in courtship and warding off unwelcome guests from nest sites. ABUNDANCE A common breeder in Northern NM and from Central to Northern AZ, a winter resident across Southern AZ, and a year-round resident across much of NM and along the Rio Grande, Gila, and Colorado rivers. CONSERVATION STATUS Least Concern.

Sora

Porzana carolina

Freshwater wetlands with plentiful emergent vegetation such as cattails, rushes
LENGTH 8.75–9 in. WINGSPAN 13 in.

The diminutive chicken of the marsh. Stocky and short-tailed with triangular yellow bill; black mask; gray face, throat, and breast; brown crown and nape; brown back with black spots and fine white streaks; black-and-white–striped flanks; and long, greenish yellow legs. Juvenile is less patterned, more buffy than gray, and has dull blackish yellow bill. BEHAVIOR Easy to detect by ear at dawn or dusk, Sora spend their days creeping through dense cover, foraging for small invertebrates on the water's surface, and using long toes to pull seeds from wetland plants. Often walks along edges of dense stands, tail flicking frequently. Will occasionally swim. VOCALIZATIONS A descending and then rising *per-WEEP*; a short, loud, and clear *EEP*; and a long, descending whinny that first accelerates and then trails off: *sko-WEE-ee-ee-e-e-e*. ABUNDANCE An uncommon breeding resident from Central to Northern NM and from Central to Northwest AZ. Winters in Western AZ and across the southern portions of both states. Lingers year round between these seasonal ranges and southward along the Rio Grande. Uncommon migrant wherever suitable habitat exists. CONSERVATION STATUS Least Concern.

Adult

Juvenile

Common Gallinule

Gallinula galeata

Freshwater and brackish marshes
LENGTH 12.5–14 in. WINGSPAN 21–24 in.

Medium-sized, plump rail, easily identified by bright, waxy red shield extending up the forehead from its matching, yellow-tipped bill. Body is deep brown-gray on back; deep charcoal-gray on head, neck, belly and flanks; with bright white stripe along the sides and on outer tail feathers. Chickenlike legs and feet a dingy yellow-green, with large feet appearing out of proportion with body. Juvenile lacks red shield and bill. BEHAVIOR Walks slowly atop marsh vegetation and along heavily vegetated shoreline, dining on plants along the way. When swimming, displays a rocking motion, moving its head forward and back, akin to a wooden toy duck pulled by a string. VOCALIZATIONS A verbose rail with a wide variety of cackles, clucks, grunts, whinnies, and clickety clacks that often give away its presence before you see it; most raucous during the breeding season. ABUNDANCE Commonly observed year round in Western and from Central to Southern AZ and from Central to Southwest NM. LOOKALIKES American Coot, Purple Gallinule. Similarities among juveniles only, with Common Gallinule distinguished by darker coloration and bright white stripe down side. CONSERVATION STATUS Least Concern.

Adult

Juvenile

American Coot

Fulica americana

Open water in natural and human-made landscapes; still, shallow water with plentiful emergent and floating vegetation
LENGTH 15–17 in. WINGSPAN 21–24 in.

Very ducklike, but more closely related to the rails. Solid brownish to grayish black overall with thick, medium-length, yellow-green legs; red eyes; and heavy white bill with black ring toward tip and blackish-red, fleshy patch at base. Large feet have fleshy lobes surrounding but not connecting toes. First winter birds lack red patch and black bill ring. Juvenile is gray-brown overall with unmarked grayish bill and dark eyes. BEHAVIOR Ornery waterbird often seen stalking or chasing other coots and fowl. Forages for vegetation and algae, but also some invertebrates, by diving and skimming off water's surface. Will occasionally forage on open shorelines or lawns. Forms large flocks, especially in winter. VOCALIZATIONS A raspy *burrup-burupp-burr-burr*, a high *hi-CUP-hi-CUP*, and short nasal honks. ABUNDANCE Common year round, with greater abundance in winter. CONSERVATION STATUS Least Concern (Decreasing).

Black Rail

Laterallus jamaicensis

Shallow freshwater and brackish marshes
LENGTH 4–6 in. WINGSPAN 8–11 in.

A rail of the tiniest proportions, sporting lovely, solid charcoal-gray plumage on the head and breast, with deep chestnut-brown shoulders deepening to white-speckled, dark chocolate-brown on back and lower belly. Eyes are ruby-red, with jet-black bill that is relatively short for a rail. BEHAVIOR Exceedingly elusive, rarely leaves the protection of marsh vegetation, nesting and hunting for aquatic invertebrates under the cover of reeds and rushes. More often heard than seen, the male is reliably vocal on late spring evenings when seeking a mate. VOCALIZATIONS Song is a rising and falling *kiky-doo* or *kik-kiky-doo*. Calls vary, from a growling, territorial *kurrr* to squealy *eenk* and *kik* calls. ABUNDANCE Very rarely observed, but present year round in Western AZ along the Colorado River. KEY SITES Lower Colorado River wildlife refuges. CONSERVATION STATUS Endangered.

Sandhill Crane

Antigone canadensis

At night, roosts in large groups on shallow lakes, rivers, playas; by day, groups disperse into agricultural fields, wetlands, grasslands LENGTH 41–47 in. WINGSPAN 73–79 in.

Unmistakable; our only commonly occurring crane. Gray overall with rust-colored undertones; long, dark bill; patch of bare, red skin on forehead; light cheek; long neck; bushy tail; and long, dark legs. First winter birds browner overall, with pale bill and no bare forehead patch. BEHAVIOR Sandhill Crane migration is among the greatest wildlife spectacles in North America. Flocks containing tens of thousands of birds soar into key roosting sites at dusk and back out at dawn, always with a raucous cacophony of calls. An omnivore, but feeds primarily on agricultural grain. As winter progresses, observations of cranes practicing elaborate courtship dances become more common. VOCALIZATIONS Resonant, rattling yodel that can travel for miles, especially from large flocks. ABUNDANCE Extremely numerous at key winter roosts and a common winter resident in surrounding areas. KEY SITES AZ: Whitewater Draw Wildlife Area; NM: Bosque del Apache National Wildlife Refuge, Elephant Butte Reservoir. CONSERVATION STATUS Least Concern.

SHOREBIRDS

With their lanky legs and specialized bills perfect for plucking tasty critters from the muck, the stilts and avocets (family *Recurvirostridae*), plovers (family *Charadriidae*), and sandpipers (family *Scolopacidae*) are built for life at the water's edge. Most abundant in the Southwest during migration and in winter, some of these birds—such as the American Avocet and Black-necked Stilt—remain unmistakable, while others fade into muted plumage and create some of the most frustrating identification challenges in the birding world. When you're trying to distinguish the most similar birds in these families, pay special attention to subtle differences in overall shape, size, plumage, bill shape, leg and bill color, foraging style and location, and vocalizations.

ABOVE American Avocet

Black-necked Stilt

Himantopus mexicanus

Mudflats, marshes, flooded fields, ponds, water treatment facilities
LENGTH 13–16 in. WINGSPAN 28–30 in.

Finely dressed, elegant shorebird is reminiscent of a tiny flamingo in stature, with bright coral-pink legs. Clean white belly and throat extend to the face, where they meet contrasting black pattern that extends from the forehead to a broad, thick eyeline accented by a bright white patch of "eyeshadow," and then continues down the nape (black-naped would more aptly describe this bird) and onto satiny black back and wings. Long bill is straight and black, and side profile of head is very round. Juvenile shows pale pink legs and brown plumage. BEHAVIOR Look for them delicately wading along shorelines of mudflats and shallow ponds, using their bills to probe for invertebrates. Rarely observed swimming. Often gathers in small flocks. Known to perform several types of distraction displays to lure predators from nesting areas.
VOCALIZATIONS No song, yet fairly vocal with a variety of high-pitched *yeep* and *kek* calls to sound alarm or communicate with young.
ABUNDANCE Commonly observed in suitable habitat throughout AZ and NM, with breeding, migratory, wintering, or permanent status depending upon the location and elevation.
CONSERVATION STATUS Least Concern.

Breeding adult

Nonbreeding adult

American Avocet

Recurvirostra americana

Shallow, open water with limited vegetation; mudflats; flooded agricultural fields
LENGTH 17–18 in. **WINGSPAN** 29–32 in.

Long-necked wading bird with extremely long, bluish gray legs; white underparts; and boldly patterned, black-and-white–barred wings. Long, thin bill is dark and distinctly upturned. Nonbreeding birds have grayish white head, neck, and breast. Breeding birds develop a rich, cinnamon-rust–colored head, neck, and breast, retaining white only near the base of the bill and around the eyes. **BEHAVIOR** Forages, often in large groups, for small invertebrates and some plant material by lowering head and swaying bill from side to side. Will occasionally plunge bill into water to capture larger prey. **VOCALIZATIONS** Usually silent, but will make high-pitched chirps and gull-like squeals when disturbed. **ABUNDANCE** Breeding resident across much of NM and a winter resident along the Colorado River in AZ. Most abundant and widespread during migration. **CONSERVATION STATUS** Least Concern.

Breeding adult

Nonbreeding adult

Underwing

Black-bellied Plover

Pluvialis squatarola

Open mudflats, sparse agricultural land such as sod farms, recently tilled or harvested fields LENGTH 11.5 in. WINGSPAN 24–29 in.

Striking plover in its far-north breeding grounds, but a frustratingly drab peep usually seen standing far away in an open field when spotted in the Southwest. In breeding plumage, black mask, throat, and belly contrast with white crown, nape, sides of neck, and undertail. Wings are black with white speckling. Nonbreeding birds are grayish brown above with light speckling, white belly, lightly streaked breast, and faint white eyebrows and chin. Black armpits in flight are diagnostic. Larger and heavier billed than other plovers. BEHAVIOR Like other plovers, forages for invertebrates on open ground, running short distances before either capturing prey with quick thrusts of the bill or running on to the next spot. VOCALIZATIONS In flight, gives a drawn out, descending, and plaintive whistle reminiscent of Lesser Goldfinch. Squabbling birds give a similar call, but much harsher and ascending. ABUNDANCE A rare migrant in the Southwest. Is most abundant in Eastern NM, but is still uncommon and local. Increasingly rare farther west. LOOKALIKES Golden-plovers, other nonbreeding plovers. Breeding Black-bellied distinguished from golden-plovers by lack of golden tones on back and clean white undertail. In all plumages, black armpits are diagnostic. CONSERVATION STATUS Least Concern (Decreasing).

Killdeer

Charadrius vociferus

Mudflats, wetlands, lawns, parks, gravel lots, agricultural fields
LENGTH 8–11 in. WINGSPAN 18–19 in.

Slender, elongated plover with tawny coloration on back and wings, cheek, and top of head. Bold, black double breast bands contrast with clean white breast, throat, nape, and belly. Eyes are deep amber, matching the russet-orange rump and tail feathers that are most often visible only in flight. White outer tail feathers with black spots, and white horizontal stripe along wing visible in flight. Juvenile lacks double breast bands, showing only one. BEHAVIOR Distinct sprint-stop, sprint-stop, and head-bobbing behaviors are a few of its many endearing traits, along with its convincingly feigning injury to deter predators from its nests which are built precariously on the ground in open areas. It is not uncommon to find Killdeer nests in gravel driveways and busy parking lots, their eggs resembling small rocks. VOCALIZATIONS Frequently makes namesake *kill-DEEer* call, particularly in flight. Alarm call is a high-pitched, resonant, single-note *tee* that can sometimes escalate into a trill. ABUNDANCE Commonly observed year round throughout the region. LOOKALIKE Semipalmated Plover. Killdeer distinguished by double breast band, longer bill, and amber eyes. CONSERVATION STATUS Least Concern (Decreasing).

Semipalmated Plover

Charadrius semipalmatus

Mudflats, agricultural fields, sewage treatment ponds, marshes, beaches
LENGTH 6.5–8 in. WINGSPAN 17–18.5 in.

Small, round, short-necked, and stubby billed plover with pale cocoa-brown back and wings, white belly, and single black breast band. White forehead patch contrasts with black around the large, round eyes and cheeks. Black-tipped orange bill, orange legs. Juvenile shows brown breast band and facial plumage and pale orange legs. BEHAVIOR Much like Killdeer, displays sprint-stop, sprint-stop behavior, stopping to tug at subterranean worms or pluck insects from the ground. They are long-distance migrants, nesting in the arctic tundra and migrating through the United States to overwinter along North, Central, and South American coastlines, extending as far south as Argentina. VOCALIZATIONS Most commonly an urgent, high-pitched, repetitive, and whistling *chee-REE-up*. Other vocalizations include a rattling alarm call and flight calls reminiscent of gulls. ABUNDANCE Observed during their long-distance migration route along the lower Colorado River corridor in westernmost AZ and in extreme Northwest NM. LOOKALIKE Killdeer. Semipalmated Plover has orange base to bill and lacks Killdeer's double breast band and elongated silhouette. CONSERVATION STATUS Least Concern.

Breeding adult

Nonbreeding adult

Mountain Plover

Charadrius montanus

Nests in shortgrass prairie, semi-desert grassland, disturbed habitats including overgrazed pasture, fallowed or harvested agricultural fields; in winter, playas, mudflats, sod farms, fallow or recently harvested fields LENGTH 8–9 in. WINGSPAN 20–23 in.

Easy to identify in its plainness and upright posture, similar to that of a Killdeer. Unpatterned sandy brown above and white below with dark wingtips, white underwings, and black patch near the base of tail. Back color fades into dingy white face and breast. Breeding birds develop black crown patch and lores. Juvenile darker overall with more heavily scalloped back and wings. BEHAVIOR Like other plovers, forages for invertebrates on open ground with quick stops between short bursts of running. Nests on open ground in a shallow scrape. VOCALIZATIONS Song is a high, coarse, and rattling *gee-gee-gee*, slightly rising and quickening before descending on abrupt final notes. Also gives varied calls including a toadlike *grrrt* and a squeaky *poEEP*. Frequently silent. ABUNDANCE Uncommon and local. A breeding resident in Northeast NM and migrant and wintering resident near Phoenix and along the lower Gila and Colorado rivers. Migrants outside of this range are a regular rarity. LOOKALIKES Other nonbreeding plovers. Mountain Plover distinguished by unpatterned back and plain white underparts. CONSERVATION STATUS Near Threatened (Decreasing).

Breeding adult

Nonbreeding adult

Snowy Plover

Charadrius nivosus

Beaches, shorelines, salt flats
LENGTH 6–7 in. WINGSPAN 13.75–18 in.

Diminutive and delightful, this small floof of a plover sports a black bill, gray legs, and dusky, pale brown back and forehead, separated by a white band at the nape. In breeding season, adult shows black forehead/crown patch, eye patches, and partial collar. Nonbreeding adult lacks black patches and displays a paler gray-brown plumage. BEHAVIOR Despite often completing two brood cycles per breeding season, this species is in decline because of loss of habitat and nesting sites being disturbed by human activity. It nests along beaches and scurries along the shore as if being blown by the wind, stopping to pluck small invertebrates from the sand. VOCALIZATIONS Most common is the *too-WHEET*, often concluded with various rattly chittering and *quirr* calls. ABUNDANCE An uncommon breeding resident in Southeast NM and in limited patches of Southern AZ. KEY SITES AZ: Whitewater Draw Wildlife Area, Willcox Playa; NM: various lakes and reservoirs scattered throughout Southeast and East Central NM. CONSERVATION STATUS Near Threatened (Decreasing).

Upland Sandpiper

Bartramia longicauda

Prairies, grasslands, croplands, high-elevation meadows
LENGTH 11–12.5 in. **WINGSPAN** 18–19 in.

Elegant and long-necked sandpiper with long, yellow legs; straight, yellow bill; and speckled back with golden and dark brown coloration. Brown speckling extends to breast and down flanks, leading to a buffy white belly. In flight, a thin white sliver may be visible on the outermost wing feather. **BEHAVIOR** Nocturnal migrators, they emit flight calls in the darkness of night. Breeding male will perch conspicuously on fence posts calling for a mate or will fly in circular patterns over breeding grounds. They forage through shortgrass prairie, picking up seeds and small invertebrates along the way. **VOCALIZATIONS** Song is a long, whistling, and musical trill, very similar in pitch and somewhat less in cadence to the "quick, THREE beers!" of an Olive-sided Flycatcher. Calls most commonly a repetitive, two- to three-note pipping sound. **ABUNDANCE** A migrant in the Southwest, most commonly encountered in Eastern NM. **LOOKALIKE** Buff-breasted Sandpiper. Upland distinguished by elongated body, darker spotting on breast, and yellow bill. **CONSERVATION STATUS** Least Concern.

Whimbrel

Numenius phaeopus

Mudflats, sparse and flooded agricultural fields
LENGTH 17–18 in. WINGSPAN 32–33 in.

Our miniature Long-billed Curlew. Large, long-necked sandpiper with a long, strongly downward-curving bill. Mottled brown, tan, and white above with dark eyeline and two dark crown stripes. Paler underside with fine dark streaks on face, neck, and breast that become messier and coarser in the flanks. Legs are pale and underwings are brown with dark barring and grayish trailing edge. BEHAVIOR Arctic breeder that winters mainly along the coast; observed as a fleeting visitor in the Southwest. Takes slow, deliberate steps while foraging for crustaceans and other large invertebrates, probing deeply into mud with its long bill. VOCALIZATIONS A loud, high *wip-wip-wip-wip*, varying in speed and quality but always at a steady pitch. Also a whistled, warbling, two-part whinny. ABUNDANCE Uncommon and local spring migrant along the lower Colorado and Gila rivers. Rare elsewhere. KEY SITE Yuma, AZ. LOOKALIKE Long-billed Curlew. Whimbrel is smaller, more drab brown overall, and has a shorter bill and striped face and crown. CONSERVATION STATUS Least Concern (Decreasing).

Long-billed Curlew

Numenius americanus

Prairie, grassland, cropland, mudflats, estuaries, flooded fields
LENGTH 19.5–26.5 in. WINGSPAN 25–35 in.

Unmistakable and accurately named, this sizeable shorebird's plumage is heavily patterned, with back showing various shades of warm brown meeting a washed cinnamon-and-buff belly. The bill is roughly quadruple the length of the head, and the lower mandible is pink at the base, deepening to dark gray/black. Deep black eyes accented by pale buff, almond-shaped eye rings. BEHAVIOR This bird doesn't walk—it struts, head bobbing slightly back and forth with each step. Along the shore, it uses its long bill to unearth crustaceans and invertebrates from the sand; on land, tilting its head from left to right, it seeks worms and other insects. VOCALIZATIONS Distinct, loud *curLEEEU*, though sometimes the first syllable is less pronounced. Often follows this call with a series of twittering *whit-whit-whit* sounds. ABUNDANCE Uncommon migrant throughout much of AZ and NM. Breeds in grassland/prairie habitat in extreme Northeast NM. Can also be observed overwintering in far southeast AZ and near the lower Colorado River. CONSERVATION STATUS Least Concern (Decreasing).

Breeding adult

Nonbreeding adult

Marbled Godwit

Limosa fedoa

Mudflats, flooded agricultural fields, wetland edges
LENGTH 17–19 in. WINGSPAN 28–32 in.

Very large shorebird with long, dark legs and long, slightly upturned, bicolored bill. Upper parts marked with messy brown, russet, buff, black, and white barring. Breeding birds show orange bill with black tip and dark barring on flanks and belly. Nonbreeding birds have pink bill with black tip and plain, unbarred, buffy flanks and belly. In flight, note brownish orange underwings, dark wingtips and wrists, and legs that protrude behind the tail. BEHAVIOR Walks slowly in search of aquatic invertebrates, which it captures by burying its long bill nearly to the hilt. A social forager, but often seen singly in the Southwest. VOCALIZATIONS Calls include a harsh, reedy *huEK-huEK-huEK* and a nasal *hEEhoo-hEEhoo* or *heennk-heennk-heennk*, repeated at varying speeds, a steady pitch, and with clear pauses between notes or phrases. ABUNDANCE Uncommon and irregular migrant across the region, most abundant in Eastern NM and along the lower Colorado River. Has a very limited winter range in extreme Southwest AZ. CONSERVATION STATUS Least Concern (Decreasing).

Breeding adult

Nonbreeding/immature

Stilt Sandpiper

Calidris himantopus

Freshwater wetlands, ponds, flooded agricultural fields, rainwater pools
LENGTH 8–9 in. **WINGSPAN** 16–17 in.

Medium-sized sandpiper with long and slender neck, long legs, and very slightly decurved black bill. Breeding adult shows deep chocolate-brown scalloped pattern on back, a chocolate-and-cream–colored barred belly, and a chestnut-brown cheek and crown separated by a pale, creamy white supercilium. Nonbreeding adult lacks the chestnut facial marking and are a washed-out version of their breeding selves, showing dusky gray-brown rather than dark brown. Breeding adult has olive-yellow legs; nonbreeding/juvenile has bright yellow legs. **BEHAVIOR** Slowly forages in freshwater habitats, much like the dowitcher. Often found in mixed flocks with dowitcher and yellowlegs species. **VOCALIZATIONS** Calls vary from rattly trill to buzzy *trrrit trrrit*. **ABUNDANCE** Uncommonly observed during migration along NM's southern edge. **LOOKALIKES** Short-billed and Long-billed dowitchers, Greater and Lesser yellowlegs, Pectoral Sandpipers. Stilt Sandpiper differentiated by legs extending past tail in flight, bright creamy white supercilium, and slightly decurved bill. **CONSERVATION STATUS** Least Concern.

Breeding adult

Nonbreeding adult

Sanderling

Calidris alba

Mudflats, sandy shorelines
LENGTH 7–8 in. WINGSPAN 13–14 in.

Medium-sized, stocky sandpiper with heavy black bill, black legs, white underside, and striking white wing stripes visible in flight. In seldom observed breeding plumage, has grayish back shingled in white-lined, rufous-and-black feathers and shows fine black streaks and spots and varying amounts of rufous in the face, neck, and breast. In more frequently observed nonbreeding plumage, plain, pale gray above with gray eyeline and blackish shoulder; much plainer than other peeps. Juvenile has blackish wings speckled in white. BEHAVIOR Arctic breeder with primarily coastal breeding range, Sanderlings are always exciting to spot among the usual peeps. Much like an overexcited plover, scuttles quickly along bare ground, stopping to probe very rapidly for invertebrates. VOCALIZATIONS A high *twip*, given singly or in a series. During courtship, males give a series of low croaks and females respond with a complex song, but this display is not observed outside of the breeding season. ABUNDANCE Very uncommon to rare fall migrant. Most common in Eastern NM. LOOKALIKES Other nonbreeding sandpipers. Sanderling distinguished by straight, stout bill; pale and unmarked gray back, neck, and head lacking brownish tones; and solid white underparts and breast. CONSERVATION STATUS Least Concern.

Nonbreeding adult

Breeding adult

Dunlin

Calidris alpina

Coastal estuaries and lagoons, wetlands, agricultural fields, water treatment ponds
LENGTH 6–8.5 in. WINGSPAN 14–15 in.

Small but stocky shorebird with long, black, drooping bill; black legs; and white outer tail feathers. Breeding adult shows black belly patch bordered by white, mottled cinnamon-colored back and crown, and pale face. Nonbreeding adult lacks cinnamon coloration and black belly patch, showing grayish hood leading to gray-brown back and clean white belly. Faint cinnamon-colored wash on cheek. BEHAVIOR Gregarious and often in large flocks, Dunlins forage along shorelines and in shallow water to turn up invertebrates just below the surface. VOCALIZATIONS Flight calls include raspy, chattery chirps and *chit*s. ABUNDANCE Uncommon but regularly observed during migration in Western AZ. LOOKALIKES Sanderling, Western, Semipalmated sandpipers. Dunlin distinguished primarily by drooping bill and slightly larger size. CONSERVATION STATUS Least Concern (Decreasing).

Breeding adult

Nonbreeding adult

Baird's Sandpiper

Calidris bairdii

Dry, open habitats including pastures, sparse grassland, playas; dry edges of wetlands and waterbodies
LENGTH 6–7.5 in. WINGSPAN 14–17 in.

Medium-sized, horizontally postured, and relatively short-legged with very slightly drooping black bill. Long wings extend well beyond tail at rest. White underparts; very dark to black legs; faintly speckled breast; fine streaking on head and neck; clean, unmarked flanks; faint white eyebrows; and a blurry dark line through the eye. Breeding adult is grayish brown above with black spots. Nonbreeding plumage is browner overall with slightly scalloped appearance and faded grayish eyebrow. Juvenile is more strongly scalloped. Most often observed in nonbreeding plumage. BEHAVIOR Forages by picking invertebrates from the ground or gleaning from low-hanging vegetation, often far from water. Probes soil much less frequently than other peeps. VOCALIZATIONS Short, plaintive, and sometimes trilling notes. Also a repeated *pureep-pureep-pureep*. ABUNDANCE Uncommon migrant in Eastern NM and across the southern portions of both states. Most abundant in the fall. LOOKALIKES Semipalmated, Least, Western, Pectoral, White-rumped sandpipers. Baird's distinguished by combination of long wings that extend past tail at rest, dark legs, brownish plumage, and lack of white rump. ALTERNATIVE NAME Black-spotted Grass Sandpiper. CONSERVATION STATUS Least Concern.

Least Sandpiper

Calidris minutilla

Gentle shorelines, estuaries, mudflats
LENGTH 5–6 in. WINGSPAN 10–11 in.

Tiny and plump, with a black, slightly curved bill and characteristic yellow-orange legs showing a greenish tinge. Breeding adult shows brighter legs, a brown-mottled back with rusty speckling throughout, a white belly, and brown speckling down the breast. Nonbreeding adult lacks rusty speckles, showing more subtle variations of dusky brown. Juvenile displays brighter rusty speckles than breeding adult; otherwise difficult to differentiate unless side by side. BEHAVIOR They may be the world's smallest shorebirds, but they are chock-full of spunk and perseverance, migrating thousands of miles per year to their Central and South American wintering grounds. They prefer to gather in small flocks, often with other species, along gentle, muddy shorelines where they can forage for invertebrates. VOCALIZATIONS High-pitched, peeping *CREEP* call, as well as alarm trills. ABUNDANCE Present throughout southern half of NM and AZ during winter. Observed in northern areas of both states during migration. LOOKALIKES Semipalmated, Western, Pectoral sandpipers. Least distinguished by size, yellow legs, thin bill, and in flight, black line running down an otherwise clean, white rump. CONSERVATION STATUS Least Concern (Decreasing).

Breeding adult

Nonbreeding adult

White-rumped Sandpiper

Calidris fuscicollis

Mudflats, edges of wetlands and waterbodies, sod farms, flooded agricultural fields
LENGTH 6–7.5 in. **WINGSPAN** 16–17 in.

Medium-sized sandpiper has a very slightly drooping black bill with a reddish orange base that is often difficult to see. Has white underparts, black legs, speckled to streaked breast and flanks, pale eyebrows, long wings that extend beyond the tail at rest, and white rump visible in flight. Breeding birds are overall brownish gray with rufous tones, with black spots on back and finely streaked head and neck. In more commonly observed nonbreeding plumage, they are paler overall, lack brownish tones, and are almost entirely gray. Juvenile plumage is more brightly patterned and strongly scalloped. **BEHAVIOR** Forages for invertebrate prey by plunging its bill into mud or shallow water, a foraging style very different from the similar Baird's Sandpiper. **VOCALIZATIONS** Very high and sometimes raspy, sparrow-like chips and hummingbird-like creaking. Complex, buzzy song not rendered outside of breeding range. **ABUNDANCE** A rare spring migrant most commonly observed in Eastern NM. Also seen in Southeast AZ, but even more infrequently. **LOOKALIKES** Semipalmated, Least, Western, Pectoral, Baird's sandpipers. White-rumped distinguished from Semipalmated, Least, and Western by longer wings that extend past the tail at rest; from Least and Pectoral by black versus yellow legs; and from Baird's by grayish plumage and white rump. **CONSERVATION STATUS** Least Concern (Decreasing).

Adult

Juvenile

Buff-breasted Sandpiper
Calidris subruficollis

Grasslands, prairies, plowed agricultural fields
LENGTH 8–9 in. WINGSPAN 17–18 in.

Small, dainty, and refined, with a dove-like silhouette. Plumage is a very warm buff wash all over, with mottled back and fine black streaking on the forehead/crown. Legs are yellow, and black bill is thin and short as sandpipers go. Characteristic underwing shows white flashes in flight and in display, with distinct black comma on the wrist. BEHAVIOR More akin to a plover, they move with sprint-stop, sprint-stop motions as they search for insects. Atypical among peeps, these sandpipers form leks for mating displays. VOCALIZATIONS Generally silent, they sometimes make soft chupping sounds in flight. ABUNDANCE Rare, can be observed during migration, most likely in easternmost NM. CONSERVATION STATUS Near Threatened (Decreasing).

Wing display

Pectoral Sandpiper

Calidris melanotos

Sod farms, flooded fields, edges of wetlands and waterbodies; prefers grassy areas over open flats
LENGTH 8.75 in. WINGSPAN 18 in.

If you love the Least Sandpiper but wish it were bigger, this is the peep for you. Medium-sized with often upright posture; slightly drooping, drab orange bill tipped in black; dusky yellow legs; scaly black, gold, white, and brownish gray back pattern; pale eyebrows; and wings that do not extend beyond tail at rest. Dense streaks on breast end abruptly to form crisp border with white underparts. BEHAVIOR Probes in shallow water and mud for mostly insects, but also other invertebrates and occasionally fish. Tends to forage near vegetation rather than in the open. VOCALIZATIONS A harsh, low, squeaky *churt*, often in repetition. ABUNDANCE An uncommon fall migrant most frequently observed in Eastern NM and Southwest AZ. LOOKALIKES Stilt, Baird's, White-rumped sandpipers. Pectoral distinguished from Stilt by shorter, less heavily drooping bill and sharp border between patterned breast and plain underparts; from Baird's and White-rumped by drab yellow versus dark legs. CONSERVATION STATUS Least Concern.

Breeding adult

Nonbreeding adult

Juvenile

Semipalmated Sandpiper

Calidris pusilla

Mudflats, flooded agricultural fields, estuaries, water treatment ponds
LENGTH 6–7 in. **WINGSPAN** 13.5–14.5 in.

Small, plump peep with dark gray-green legs; short, straight, black bill; clean white belly with speckled bib; and wingtips that do not extend beyond tail at rest. Breeding adult shows faint rusty speckles along back, scapulars, and on head, with a cleaner contrast between dark upper parts and white belly compared to nonbreeding birds. Juvenile shows distinct bright white supercilium that fades with adulthood. **BEHAVIOR** Often seen slowly foraging for invertebrates in large flocks along river margins, mudflats, and wet fields. Long-distance migrant. **VOCALIZATIONS** Somewhat frantic, chittery *chirrup* and *cher* calls. **ABUNDANCE** Uncommon, but regularly observed migrant in suitable habitat throughout AZ and NM. **LOOKALIKES** Least and Western sandpipers, Sanderling. Semipalmated distinguished by black legs versus yellow of Least, coloration difference (less rusty than breeding Western and Sanderling, more brown than nonbreeding), and straight, tubular bill. **CONSERVATION STATUS** Near Threatened (Decreasing).

Breeding adult

Nonbreeding adult

Juvenile

Western Sandpiper

Calidris mauri

Flooded agricultural fields, edges of wetlands and waterbodies, mudflats
LENGTH 5.5–6.5 in. WINGSPAN 14 in.

Small and pale overall with long and drooping black bill, black legs, and wings meeting the tail tip at rest. In breeding plumage, finely streaked head; whitish nape with gray to black streaks that continue and break into spots on shoulders, breasts, and flanks; white underparts; and scalloped black, white, and gray wings washed in apricot coloration also seen in crown and behind eye. In more commonly observed nonbreeding plumage, they are very light brownish gray overall with pale face, faintly streaked breast, white underparts, and fine black streaks on the head, neck, nape, and back. Juvenile plumage more heavily patterned and rufous overall. BEHAVIOR Picks and deeply probes with urgency in mud or shallow water in search of invertebrate prey. VOCALIZATIONS Varied sharp chips, *peep*s, and trills. Also a *churt* similar to Pectoral Sandpiper but higher in pitch. ABUNDANCE Common winter resident across most of AZ and southwestern half of NM. Uncommon migrant outside of this range. LOOKALIKES Semipalmated, Least, and White-rumped sandpipers; Sanderling. Western distinguished from Semipalmated by larger head, bulkier body, richer coloration, longer bill, and more heavily streaked flanks; not always distinguishable in nonbreeding plumage. Western distinguished from Least by larger size and black legs; from White-rumped by smaller size, lack of white rump, and wingtips that do not extend beyond the tail at rest; and from Sanderling by larger size, more heavily patterned plumage, and dark streaking. CONSERVATION STATUS Least Concern (Decreasing).

Breeding adult | Nonbreeding adult

Short-billed Dowitcher

Limnodromus griseus

Wetlands, flooded agricultural fields, water treatment ponds, playas
LENGTH 9–12 in. WINGSPAN 18–22 in.

Chunk of a shorebird, shaped like a football in silhouette, with a short neck; long, straight, downturned dark gray bill with dingy yellow-green at the base; and yellow legs. Breeding plumage shows rich variations of head-to-toe mottled colors in cinnamon, deep brown, and buff, with pumpkin-colored wash on the breast. Juvenile shows clean plumage with orange breast and deep cinnamon coloration on wings, in contrast to their pale belly. Nonbreeding adult lacks cinnamon coloration, with mottled gray-brown and dingy, speckled chest and pale barring on flanks. All plumage variations show white upper rump patch in flight. BEHAVIOR During migration, look for them foraging along muddy shorelines, using their bills like plungers, bobbing up and down in a motion akin to a sewing machine needle.

VOCALIZATIONS Calls are a gentle, chirping, almost finchlike *tuu-tuu*, very distinct from the harsher call of the Long-billed. ABUNDANCE Rare. Observed irregularly during migration in suitable habitat along river corridors. LOOKALIKE Long-billed Dowitcher. Short-billed distinguished by vocalization, bulkier silhouette, and straight (not humped) back while foraging. CONSERVATION STATUS Least Concern (Decreasing).

Breeding adult

Nonbreeding adult

Long-billed Dowitcher

Limnodromus scolopaceus

Flooded agricultural fields, shallow edges of wetlands, ponds, rivers, other waterbodies
LENGTH 11.5 in. WINGSPAN 19 in.

Bulky and oval-shaped with short neck; long, straight, slightly downturned dark gray bill with sometimes dingy yellow base; dark cap; pale eyebrows; dark eyeline; dark yellow legs; and white patch above rump visible in flight. Breeding adult is a rich cinnamon to buff color below with dark bars and streaks on flanks and neck. Brownish gray back is mottled with brown, white, black, and buff. Nonbreeding adult is gray overall with thin black streaks on back, dingy breast, pale belly, and pale bars on flanks. Juvenile plumage resembles nonbreeding but with dull, rusty tones. BEHAVIOR Forages in shallow water with repetitive probing and continuous soft vocalizations. VOCALIZATIONS Most commonly a short, harsh, and often repeated *keek*. Also gives varied, harsh, and sometimes trilling *keks*, whistles, and chirps, always higher pitched than Short-billed. ABUNDANCE Common spring and fall migrant across entire region, and winter resident across southern portions of both states. Most frequently observed in Central AZ and along the Gila, Colorado, and Rio Grande rivers. LOOKALIKE Short-billed Dowitcher. Long-billed more common within inland freshwater habitats. Vocalizes while foraging (Short-billed forages silently). Note Long-billed's distinct voice and hunched rather than flat back when feeding. Bill length between the two species overlaps, making for a poor field mark (and name). CONSERVATION STATUS Least Concern.

Wilson's Snipe

Gallinago delicata

Marshy areas with both open and densely vegetated habitat, including flooded agricultural fields, wetlands, ponds, rivers **LENGTH** 10.5–12.5 in. **WINGSPAN** 16–18 in.

Medium-sized, portly shorebird with relatively short, thick legs; short tail; extremely long, straight bill with light base and dark tip; and plumage that walks the line between ornate and cryptic. Rich buff, dark brown, and gray overall with buff-and-brown–striped face; dark cap; three buffy stripes down back; buffy breast marked with thick, dark bars; white underside; and bold dark stripes on flanks. In flight, shows solid dark gray underwing. **BEHAVIOR** Hides in plain sight, slowly and deliberately probing for aquatic invertebrate prey. When approached too closely, explodes into erratic flight before diving abruptly back into the marsh. **VOCALIZATIONS** Varied harsh, heronlike *kek* and squawk. Song, given from a perch, is a high, clear *kipa-kipa-kipa* or *kip-kip-kip*. During swooping courtship displays, wings produce a loud winnowing sound, *hu-hu-hu-hu-hu*, similar in quality to an Eastern Screech-Owl's call but intensifying and rising. **ABUNDANCE** Common migrant and winter resident across the Southwest. Uncommon breeding resident north of Albuquerque in the Carson National Forest and in highest elevations of AZ. **LOOKALIKES** Dowitchers. Wilson's Snipe is shorter legged and striped. **ALTERNATIVE NAME** American Snipe, Winnowing Snipe. **CONSERVATION STATUS** Least Concern.

Breeding adult

Nonbreeding adult/immature

Spotted Sandpiper

Actitis macularius

Rocky shorelines, beaches, shorelines of rivers, ponds, lakes
LENGTH 7–8 in. **WINGSPAN** 14.5–16 in.

A relief to peepwatchers everywhere, identifying this smallish sandpiper in the field is made easy by several diagnostic features. Breeding adult has straight, black-tipped, sunflower-orange bill; pale pink legs; and distinct brown spotting on clean white flanks, belly, and breast. Back is cocoa-brown with dark brown spotting. Nonbreeding and immature birds have yellow legs, dark bill, and lack heavy spotting. Plumage is lovely, with a clean white belly and a faintly patterned back with dusky brown tones that just barely creep onto the upper breast. Dark eyeline and faint white supercilium present on all plumage variations. **BEHAVIOR** Diagnostic butt-bobbing behavior is a dead giveaway, as they bump their tails down and up while foraging along the shoreline. Also look for distinct flight pattern of rapid wingbeats interspersed with graceful glides. **VOCALIZATIONS** As they flush from the shoreline, they almost always vocalize to announce their departure. Calls and song are typically a variation of a stuttery *whe-whe-whe-whEAT*. **ABUNDANCE** Common and widespread, breeding in northern areas of NM and AZ, overwintering in southern areas of both states and into Central and South America **CONSERVATION STATUS** Least Concern.

Breeding adult

Nonbreeding adult/immature

Solitary Sandpiper

Tringa solitaria

Forested wetlands, river margins, drainage ditches, ponds, lakes, flooded agricultural fields
LENGTH 8.5 in. WINGSPAN 22 in.

Satisfyingly distinct for a shorebird. Brown above with a yellowish cast and white spots, white below, with a mottled brown-and-white head, neck, and breast; dark shoulders; white eye rings; dingy yellow-green legs; and medium-length, dark bill, sometimes with olive-yellow base. In flight, shows entirely dark wings and white outer tail feathers marked with dark bars. Nonbreeding birds lighter overall with reduced to no white spotting. Juvenile has white lores and eye rings that together form "spectacles" similar to those of Plumbeous Vireo; compared to adults, has smaller white spots. BEHAVIOR May be found in areas not frequented by other sandpipers. True to its name, usually observed alone, habitually trembling and bobbing its tail while foraging in shallow water for aquatic invertebrates. After being flushed, will often land with wings raised, holding them that way momentarily before settling. VOCALIZATIONS A sharp *weet*, given singly or sometimes in rapid repetition. ABUNDANCE Uncommon but widespread migrant. In spring, restricted to Eastern NM and the southern portions of both states; more widespread in fall. LOOKALIKES Greater, Lesser yellowlegs. Solitary Sandpiper is smaller, with duller legs and clear white spots. CONSERVATION STATUS Least Concern (Decreasing).

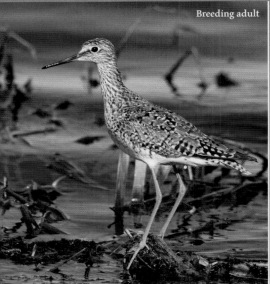
Breeding adult

Nonbreeding adult

Lesser Yellowlegs

Tringa flavipes

Wetlands, flooded fields, tidal flats
LENGTH 9–11 in. WINGSPAN 23–25 in.

You guessed it—this slender shorebird has bright yellow legs, though this is not a particularly unique feature among its peers. The bill is slender, straight, and dark, with faint yellow at the base, roughly equal in length to the head—helpful when differentiating it from the Greater Yellowlegs. Breeding birds are pale gray and tan on the back with white spots; gray streaking on face, neck, and flanks; dark stippling on crown, upper breast, and back; white eye ring; and clean, white belly. Juvenile is often heavily spotted with white on a gray-brown back. Nonbreeding adult and juvenile have smudgy markings on neck rather than distinct streaking. White rump and plain wings in flight. BEHAVIOR Look for their signature high-stepped gait when foraging along shorelines. Will sometimes sway head from side to side while bill tip is submerged, stirring up invertebrate snacks. Flight is buoyant and graceful. VOCALIZATIONS Gentle, whistling single- or double-noted *tuee* or *teeu*. ABUNDANCE An uncommon-to-rare migrant found occasionally in wetland habitats throughout AZ and NM. LOOKALIKES Greater Yellowlegs, Solitary Sandpiper. Lesser distinguished from Greater by lack of heavily speckled flanks during breeding, smudgier breast on nonbreeding/immature, smaller size overall, and shorter, thinner bill. Distinguished from Solitary Sandpiper by brighter legs, faded eye ring, and larger size. CONSERVATION STATUS Least Concern (Decreasing).

Breeding adult

Nonbreeding adult

Adult

Willet

Tringa semipalmata

Mudflats, wetlands, marshes, wet meadows, beaches
LENGTH 13–16 in. **WINGSPAN** 26–27 in.

Large, bulky, pale gray to brownish gray shorebird with long, bluish gray legs; stout neck; black wingtips; white belly; and long, barely upturned bill with light base and black tip. Breeding birds show light brownish gray bars on breast and flanks and dark brown to black mottling on back. Nonbreeding birds have a grayish wash on neck and breast and paler mottling on back. Juvenile resembles nonbreeding birds, but slightly more brownish and with a more intricately patterned back. In flight, note bold white stripe on otherwise black wings. **BEHAVIOR** Probes in mudflats and shallow water, often alone, for aquatic invertebrates. Is more apt to run than take flight when disturbed. **VOCALIZATIONS** Song is a high, clear, whistled, and repeated *pill-will-will-willet*, not heard during migration. Calls include high, repetitive chirps, plaintive and quail-like squawks, and gull-like chatter. **ABUNDANCE** An uncommon to rare spring and fall migrant across the region and an uncommon wintering resident in extreme Southwest AZ near Yuma **LOOKALIKES** Marbled Godwit, Greater and Lesser yellowlegs. Willet distinguished from Marbled Godwit by smaller size, grayish plumage, and shorter, less upturned bill lacking orange to pink base; from yellowlegs by larger size, dark legs, and heavier bill. **CONSERVATION STATUS** Least Concern.

Breeding adult

Nonbreeding adult/immature

Greater Yellowlegs

Tringa melanoleuca

Wetlands, mudflats, flooded agricultural fields, edges of waterbodies including ponds, lakes, rivers; often in deeper water than other shorebirds
LENGTH 12–14 in. **WINGSPAN** 24–28 in.

Elegant shorebird, gray to grayish brown overall, with long, bright yellow legs; white tail, eye ring, and belly; long neck; dainty, round head with fine gray to brown streaks; and a long, barely upturned grayish bill with pale base and dark tip. Breeding birds show dark blackish brown back with white speckling and coarse, dark bars on flanks and breast. Nonbreeding birds paler with dingy gray-green bill base, lightly marked flanks, and smudged grayish breast with pale streaks. Juvenile resembles nonbreeding birds but with grayer bill, more boldly streaked breast, and white throat. In flight, note legs extending well beyond tail. **BEHAVIOR** Walks with high steps through relatively deep water in search of aquatic invertebrates and small fish. Will occasionally run in short bursts to capture prey. **VOCALIZATIONS** A high, squawking *kew-kew-kew* or *kew-kew-kew-kew*, usually three- or four-parted, but sometimes longer. **ABUNDANCE** Common winter visitor in the southern portions of both states and a common migrant farther north. **LOOKALIKES** Lesser Yellowlegs, Stilt Sandpiper. Greater distinguished from Lesser by larger size, more heavily marked flanks, longer bill (usually at least twice the length of head), and longer, usually three- to four-part call; from Stilt Sandpiper by larger size and thinner, slightly upturned grayish bill. **CONSERVATION STATUS** Least Concern.

Wilson's Phalarope

Phalaropus tricolor

Lakes, marshes, ponds, coastal wetlands
LENGTH 8.5–9.5 in. WINGSPAN 15–17 in.

Reversing the usual dynamic of sexual dimorphism, breeding female shows brilliant plumage with a lovely slate-blue crown and back, mahogany wing patches, cinnamon-colored wash on throat, and deep black stripe extending from the eyeline down the neck. Belly is clean white. Legs on breeding adults are dark gray/black, while legs on immature and nonbreeding adults are yellow. Breeding male is a washed-out version of the female, lacking the deep black stripe. Juvenile has dark cap, with distinct and beautiful scaled appearance on back, ranging in color from rufous to gray. Nonbreeding adult has clean white neck and belly, with very slender appearance in comparison to similar species. BEHAVIOR Unique among shorebirds, they are often observed swimming in deep water, heads bobbing back and forth in a mechanical motion. Look for their distinct circling behavior on the water, stirring up prey for easy grabbing. Also rare among shorebirds, they molt during stops along their migration routes. Females court males and defend their mates from predators, eventually leaving the males to raise young as they move on to seek additional breeding opportunities. VOCALIZATIONS Short, nasally, ascending *quit quit* calls, as well as low gurgling calls when roosting during migration. ABUNDANCE Commonly observed during migration in suitable habitat throughout AZ and NM. ALTERNATIVE NAMES Tricolor Phalarope, Cinnamon-throated Phalarope. CONSERVATION STATUS Least Concern.

Breeding male

Breeding female

Nonbreeding adult/immature

Red-necked Phalarope

Phalaropus lobatus

Ponds, wetlands with ample open water; more common in coastal habitats
LENGTH 7–7.75 in. WINGSPAN 13–16 in.

Small, high-floating waterbird with thin neck, short black legs, and thin, pointed black bill. Breeding birds are dark overall with dark gray head and face; small white mark above eye; white throat; red streak from nape, down neck, and into otherwise dark breast; dark back with buffy streaks; white flanks mottled in gray; and white belly and rump. Female more brilliantly colored than male. Nonbreeding male plumage shows streaky gray back; gray nape and partial cap; white face with black ear patch; white neck, throat, and underparts; and slight gray mottling in flanks. Juvenile resembles nonbreeding but darker and more heavily patterned. BEHAVIOR Almost always observed on open water, often with other phalaropes. Swims in tight, rapid circles while feeding on aquatic invertebrates. VOCALIZATIONS Single, high-pitched, and frequently raspy or burbling notes, sometimes combined into long, frenetic phrases. ABUNDANCE Uncommon to rare in migration; most common from Central to Southwest AZ, along the Gila and lower Colorado rivers. LOOKALIKES Wilson's, Red phalaropes. Red-necked distinguished from breeding Wilson's by smaller size, shorter bill, black legs, darker breast, and lack of dark eye and neck stripe; from nonbreeding Wilson's and Red by streaked back. CONSERVATION STATUS Least Concern (Decreasing).

GULLS AND TERNS

Many species of gull and tern (family *Laridae*) are strictly birds of the sea, but others abandon the surf for at least part of the year and head inland. During winter and migration, a few make their way to the arid Southwest where, in their mostly gray, black, and white plumage, they look somewhat out of place, even on the region's largest bodies of fresh water. Though gulls' subtle differences and complicated plumage development can create frustrating identification challenges, and while the similar looking terns can be nearly impossible to get a good look at as they jet around on scythe-like wings, only a handful from this group regularly make their way to the region, sparing Southwestern birders significant struggles.

ABOVE Breeding adult Common Tern

Breeding adult

Nonbreeding adult

Immature

Bonaparte's Gull

Chroicocephalus philadelphia

Lakes, ponds, rivers, coasts
LENGTH 11–12 in. **WINGSPAN** 35.5–39 in.

Small, graceful gull. Breeding adult shows black head with neat white arcs above and below the eyes, solid gray back with black wingtips, black bill, and deep orange-red legs. Nonbreeding adult lacks black head, instead showing a small black patch at the ear and pale pink legs. Immature shows gray-brown mottled plumage on back, with black ear patches on mostly white head. On nonbreeding adult in flight, note solid white border on leading edge of wing, black-tipped primaries, and white tail. Juvenile in flight has black-tipped tail and shows distinct M shape on back. **BEHAVIOR** Agile in flight, resembling the lofty flight of a tern. Gathers in large flocks in the nonbreeding season, plunging into frigid waters to feed on small fish in lakes and rivers. Diet varies with the seasons: they prefer flying insects during breeding and expand to worms and other invertebrates along their migration routes. **VOCALIZATIONS** Most common call during migration is a raspy, grating *keer* or *kek*. **ABUNDANCE** Regularly observed in migration on larger bodies of water throughout AZ and NM. **LOOKALIKES** Black-headed, Little gulls. Bonaparte's distinguished by clean white underwings, black-tipped primaries, and black bill. **ALTERNATIVE NAME** Black-eared Gull. **CONSERVATION STATUS** Least Concern.

Breeding adult

Nonbreeding adult

First winter

Franklin's Gull

Leucophaeus pipixcan

Wetlands, lakes, flooded agricultural fields or grasslands
LENGTH 12.5–14 in. WINGSPAN 34–37 in.

Small and ternlike, with a short, delicate bill. Adult plumage achieved in three years. Breeding adult shows red bill, black head, broken white eye arcs, dark gray back and wings, rosy hue to white breast and belly, and dark red-tinged legs. In flight, white tail, white underwings, and black wingtips terminating in white; show as white spots on black primaries at rest. On upperwing, white band separates black wingtips from gray back and continues along trailing wing edge. Nonbreeding adult has faded black and lightly streaked partial hood, faded eye arcs, and dark bill; may retain some pinkish tones. First summer plumage is similar to nonbreeding adult, but lacks white wingtips and shows limited white band on upperwing. First winter plumage is also similar to nonbreeding adult but with limited white wingtips, more brown in wings, grayish mottling, and partial black band across center of tail tip. BEHAVIOR With bouncy flight, catches insects midair and nabs fish off water's surface. VOCALIZATIONS A high, nasal *keh*, similar to Gila Woodpecker and often rapidly repeated. ABUNDANCE An uncommon to rare migrant, most abundant from Eastern NM to Central AZ. LOOKALIKE Bonaparte's Gull. Franklin's larger and less common and distinguished in breeding plumage by red bill, white wingtips, dark red-tinged legs, and more white around the eyes; in all other plumages, Franklin's distinguished by dark hood versus Bonaparte's dark ear patch. ALTERNATIVE NAME Rosy Gull. CONSERVATION STATUS Least Concern.

Breeding adult

Nonbreeding adult

First winter

Ring-billed Gull

Larus delawarensis

Human-built habitats such as garbage dumps, parks, parking lots, plowed agricultural fields; breeds along inland fresh waterbodies (making "seagull" a real misnomer)
LENGTH 17–21 in. WINGSPAN 41–46 in.

Breeding and nonbreeding adults show a black band on a short, yellow bill, with yellow legs and pale eyes. Breeding adult has clean white head, body, and tail and light gray wings with black outer primaries accented by small white spots and, visible in flight, a bold white spot near the wingtip. Nonbreeding adults show tan streaking—sometimes heavy, sometimes light—but are otherwise similar to breeding in appearance. Juvenile shows heavy gray-tan mottling on the entire body and has a dark bill with pink base, dark wings, and a distinct dark tail band visible in flight. First and second winter plumage shows mottling beginning to fade and black band on bill becoming more prominent. Legs are pink during first two years. BEHAVIOR These omnivorous opportunists are likely the gulls trying to steal your fries at a ball game, waiting outside your car in a big box store parking lot, or circling around a freshly plowed field, taking turns diving for insects and worms. They are acrobats of the sky, strong in flight and often gathering in large flocks, by the thousands in some areas. They forage on the wing, on the ground, and afloat in fresh water. VOCALIZATIONS Varied, boisterous squealing and high-pitched *mews*. ABUNDANCE Common in suitable habitat across the region during migration. Winters in Eastern, Central, and Southern NM and in Western AZ along the Colorado River. CONSERVATION STATUS Least Concern.

Breeding adult

California Gull

Larus californicus

Flooded agricultural fields, rivers, waterbodies within nearly any relatively open habitat type, often in areas with plentiful food-dropping humans
LENGTH 19–21 in. WINGSPAN 52–54 in.

Medium-sized gull with a relatively thin bill and long, pointed wings. Adult plumage achieved in four years. In breeding plumage, shows gray back; white head, neck, breast, belly, and tail; yellow-green legs; yellow bill with red and black spots near the tip of lower mandible; and dark eyes with red orbital rings. Wings have black tips terminating in white, showing as white spots on black primaries at rest. Nonbreeding adults are similar but with heavy brown streaks on head and neck and with yellow eyes. First winter birds are mottled brown overall, with pale face; pale, black-tipped bill; grayish pink legs; and black wingtips. First summer birds are similar but are lighter overall, with more white in face and breast and more gray in mantle. Second winter birds show fully gray mantle, bluish gray bill and legs, and heavy brown mottling. Second summer birds retain fully black wingtips and brown mottling, but develop yellow bill. Third winter plumage is similar to nonbreeding adult but with some black in tail and limited white wingtips. Juvenile is brown overall, with dark bill with light base, dull pink legs, and a black tail. BEHAVIOR The furthest thing from a picky eater, will dive for fish, sift through garbage, chase insects, gulp down small mammals, scavenge for carrion, mooch off agricultural waste, or accept a handout. VOCALIZATIONS Enthusiastic and

(continued)

Nonbreeding adults

First winter

California Gull, *continued*

varied squeals, laughs, and cries. Pitch falls between the higher pitched Ring-billed Gull and lower pitched Herring Gull. **ABUNDANCE** An uncommon to rare migrant across both states. **LOOKALIKES** Ring-billed, Herring gulls. California distinguished from Ring-billed by larger bill and overall size; in breeding plumage by dark eyes and a red spot on bill; and in juvenile plumage by browner coloration and pinkish legs. Distinguished from Herring by smaller size, smaller bill; in breeding plumage by greenish yellow legs and dark eyes; and in juvenile plumage by duller legs and pale base to dark bill. **CONSERVATION STATUS** Least Concern (Decreasing).

Nonbreeding adult

Breeding adult

Second winter

Herring Gull

Larus argentatus

Coasts, reservoirs, lakes, rivers, plowed fields, open garbage dumps
LENGTH 22–26 in. **WINGSPAN** 53–58 in.

Large, bulky, broad-winged gull most recognizable by size, as plumage variations pose an identification challenge in their early years. Adult has pink legs, yellow eyes, yellow bill (brighter on breeding adult), pale gray back, black on outer primaries, and white border on the wings. Breeding adult shows red spot on lower mandible and clean white head, body, and tail. Nonbreeding shows pale but distinct tan streaking on head and body. Juvenile is white and tan overall with dark bill, primaries, and tail band; pink legs help with identification. First, second, and third year transition from heavy to subtle mottled appearance, and dark bill fades to pale yellow with a dark band before taking on full yellow bill in adulthood. **BEHAVIOR** Competitive, boisterous scavengers, these sizable gulls barrel along coastlines, hover over parks and fields, and circle boats, searching for their next meal, which they are happy to steal if they can't catch it themselves. **VOCALIZATIONS** Herring Gulls have quite a lexicon, with calls taking on multiple meanings in a range of contexts. Most commonly, a long, drawn-out *mew* or laughlike alarm call, both high-pitched, clear, and sonorous but not as frantic as the cacophony of Ring-billed Gulls. **ABUNDANCE** Overwinters along the lower Colorado River and from Central to Northeast NM. Uncommon but more widespread in migration. **LOOKALIKES** California, Western gulls. Breeding adult Herring distinguished from California by pale yellow eyes and pink versus yellow legs; from Western in all plumages by thinner versus bulbous bill. **CONSERVATION STATUS** Least Concern.

Breeding adult

Juvenile

Least Tern

Sternula antillarum

Migrates along river corridors, frequently stopping at lakes and open wetlands along their route; nests in sparse, sandy areas near water
LENGTH 9 in. WINGSPAN 19–21 in.

Smallest tern in North America. Slightly smaller than a Lesser Nighthawk, with long and scythe-like wings, lean body, short tail and legs, and a long but slender and sharply pointed bill. Breeding adult has gray back; white underparts; a yellow-orange, black-tipped bill; black cap and eyeline; white forehead; dark outer primaries; and yellow-orange legs. Nonbreeding adult shows faded cap and dark bill and lack dark outer primaries and black eyeline. Juvenile has smudgy brown cap, dark line from behind eye and into neck, a lightly patterned brownish gray back, more extensively dark primaries, dingy orange legs, and a slight orange hue to dark bill. BEHAVIOR A strong flyer, forages for small fish by diving near shore with darting flight and swift wingbeats. VOCALIZATIONS Most commonly, a very high, squeaky *ki-deek* or *KEE-dika*, often strung together and repeated with urgency. ABUNDANCE A regular migrant near Carlsbad, NM, and an uncommon to rare (but increasingly common) migrant across Eastern NM. Breeds regularly in small numbers in Southeast NM. Limited breeding records in AZ. KEY SITES NM: Bitter Lake National Wildlife Refuge, Bottomless Lakes State Park. LOOKALIKES Other Southwestern terns. Least Tern distinguished by size, proportions, and yellow bill and legs. CONSERVATION STATUS Least Concern (Decreasing).

Breeding adult

Caspian Tern

Hydroprogne caspia

Always near open water on shorelines, rivers, lakes, beaches, islands
LENGTH 19–21 in. **WINGSPAN** 49–51 in.

The world's largest tern is graceful, easily recognized by both its heft and its thick, gray-tipped, bright red bill, like long-nosed pliers jutting from their faces. Breeding adult sports a neat black cap extending to just below the eyeline, a clean white body, and pale gray back and underwings. Legs are dark gray/black. Nonbreeding adults and juveniles lack black crown, instead showing scruffy gray plumage in its place. **BEHAVIOR** Cruises the sky with gliding, smooth wingbeats, circling over open water and plunging in like a torpedo for fish. Nests on flat, open land near open water. **VOCALIZATIONS** Raspy, nasally calls and screams. **ABUNDANCE** Commonly observed in westernmost areas of AZ during migration as they make their way to Central and South American wintering grounds. **CONSERVATION STATUS** Least Concern.

Nonbreeding adult

Breeding adult

Breeding adult

Nonbreeding adult

Juvenile

Black Tern

Chlidonias niger

Wetlands, rivers, ponds, lakes, flooded agricultural fields
LENGTH 9.5–14 in. **WINGSPAN** 22–24 in.

Small, dark seabird with short legs; long, tapered wings; pointed, slender black bill; light gray underwings; and a shallowly forked tail. Breeding adults show black head, mantle, breast, and belly; gray wings, back, and tail; white rump; and dark legs. Nonbreeding adult has a black ear patch and partial crown on otherwise white head; grayish white underparts; gray back, wings, and tail; and dark legs. Juvenile is similar to nonbreeding adult, but darker with brownish tones on the back, a more scalloped appearance, and pinkish orange legs. On nonbreeding adults and juveniles in flight, note the dark spot on side under leading wing edge. **BEHAVIOR** Forages for insects midair during slow, bouncy, swallowlike flight, only rarely diving for small fish like other terns. **VOCALIZATIONS** A raspy, truncated *kik*, often repeated.

ABUNDANCE Uncommon to rare migrant with greatest abundance from Central to Eastern NM. **LOOKALIKES** Other terns. Black Tern is similar in nonbreeding plumage but is always darker than similar species. **CONSERVATION STATUS** Least Concern (Decreasing).

Breeding adult

Nonbreeding adult

Breeding adult

Common Tern

Sterna hirundo

Breeds on both fresh and sea water; overwinters primarily in saltwater areas; prefers rocky islands, beaches near open water
LENGTH 12–15 in. **WINGSPAN** 29.5–31.5 in.

Full of dramatic, sharp angles, the Common Tern resembles origami with its long and pointed wings, sharply pointed bill, and deeply forked tail. Breeding adult is clean white on the throat and tail, with a gray wash over the belly, pale gray back, dark wingtips, and a black cap that extends just over the eyes and back to the nape. Black-tipped, deep orange bill matches bright orange legs. Nonbreeding adults and immature terns lack full black cap, showing a white forehead and patchy ash-gray crown, distinct charcoal-gray carpal stripe on the wing, and black on the legs and bill. **BEHAVIOR** Gregarious birds, graceful in flight, floating over the water with distinct, rowing pattern, diving for small fish near the surface. **VOCALIZATIONS** Very vocal, ranging from nasally, high-pitched, kazoo-like squeals to grating *keer* calls, and many raspy trills and growls in between. If you are near a flock, you'll likely hear the din before you see them. **ABUNDANCE** Uncommon but regularly observed in suitable habitats along the lower Colorado River corridor and less frequently in Southeast AZ at water treatment facilities and other wetland habitats. **CONSERVATION STATUS** Least Concern.

Breeding adult

Nonbreeding adult

Forster's Tern

Sterna forsteri

Wetlands, rivers, ponds, lakes, flooded agricultural fields; often uses smaller waterbodies than typical of other terns
LENGTH 13–14 in. WINGSPAN 31 in.

Small, trim seabird with short, orange legs; long, tapered wings; a long, heavy, pointed bill; silvery gray underwings; and a long, deeply forked white tail with points that, in breeding plumage, extend beyond the wingtips at rest. Breeding adult has a bright orange-red, black-tipped bill; black cap extending to just below the eye; white face, breast, and underparts; and light gray wings and back. Nonbreeding adult has a black bill, a dark black spot around and behind the eye, gray primaries, and a white face, crown, and nape. Rufous-orange–washed juvenile quickly fades to resemble nonbreeding adult but with a lightly scalloped pattern on brownish gray back. BEHAVIOR Flies low and slow over relatively shallow water before making acrobatic, spiraling dives to capture fish. Will also catch insects midair or from the water's surface. VOCALIZATIONS Common calls include a high, raspy *KEER*; a short, dry *chuck*; and a rattling *eehh*, often strung together into more complex phrases. Lower in pitch than Common and Arctic terns. ABUNDANCE An uncommon migrant across the Southwest. LOOKALIKE Common Tern. Forster's has thicker bill, longer tail, less black in underwings, and in juveniles and nonbreeding adults, black ear patches instead of a cap. ALTERNATIVE NAME Swallow-tailed Tern. CONSERVATION STATUS Least Concern.

CORMORANTS, LOONS, AND PELICANS

Although there are only a handful of natural lakes in the Southwest, the landscape is dotted with human-made reservoirs, golf courses, urban ponds, and constructed wetlands. To change the environment even more, people have introduced nonnative fish to these built habitats to create sport fishing opportunities, to manage vegetation and insects, and purely by accident. This tinkering has had mixed consequences for humans and native aquatic ecosystems, but birds that eat fish—including cormorants (family *Phalacrocoracidae*), pelicans (family *Pelecanidae*), and loons (family *Gaviidae*)—have been happy to take advantage.

ABOVE American White Pelicans

Breeding adult

Nonbreeding adult/immature

Common Loon

Gavia immer

Wetlands, ponds, lakes, rivers, coastal areas near shoreline
LENGTH 26–35 in. **WINGSPAN** 41–52 in.

Breeding adult is strikingly beautiful, with inky black head and bill, deep red eyes, thin black-and-white pinstripes down the throat and upper breast divided by a thick black collar, and a black-and-white–speckled and checkerboard-patterned back. Nonbreeding adult and juvenile are far less showy, with pale blue-gray bills, faded striped throats, gray plumage on the backs, and white bellies. Bill is hefty but comes to a fine point, perfect for grasping fish. **BEHAVIOR** Rarely seen on land, usually spotted floating in the water, with much of their bodies submerged. Graceful and silent as they dive beneath the surface for fish. Their odd foot placement, toward the back of the body, is ideal for swimming but not for walking and can strand them if they are forced to land during an extreme weather event. In flight, they use quick, shallow wingbeats and a direct flight pattern—no bobbing, no weaving. **VOCALIZATIONS** Haunting and melancholy, their sonorous yodels, hoots, and wails are emblematic of the foggy, still lakes and ponds of Northeast forests; in the Southwest, they can be heard during migration. **ABUNDANCE** Uncommon, but regularly observed in migration throughout the region. **CONSERVATION STATUS** Least Concern.

Breeding adult

Nonbreeding adult

Nonbreeding adult

Neotropic Cormorant

Nannopterum brasilianum

Lakes, ponds, rivers
LENGTH 23–25 in. WINGSPAN 40–41 in.

Slender, lanky waterbird with long neck and tail. Dark purple-black plumage shows a scaly appearance and greenish sheen in the right light. Long, straight grayish to orange-yellow bill ends in a downturned hook, and chin is deep sunflower-gold, outlined in white. Feet are large and webbed. During breeding season, white whiskery tufts near the ears. Juvenile shows dusky brown coloration on head, neck, and chest. BEHAVIOR Excellent fishers, they can be observed swimming with necks extended like periscopes out of the water, diving for relatively long periods of time while hunting and eating prey. Unlike other cormorants, Neotropics will sometimes dive for fish from the air. Colonies nest in trees along the water's edge. Can be seen basking on the shore or perched in trees, with wings extended to dry wet feathers. VOCALIZATIONS Hoglike or bullfrog-like croaks and grunts. ABUNDANCE Commonly observed year round throughout Central and Southeast AZ, with sightings increasing in recent years. Also seen along Rio Grande corridor in NM and along the lower Colorado River in AZ. Numbers increasing as human-made waterbodies increase opportunities for suitable habitat. LOOKALIKE Double-crested Cormorant. Neotropic distinguished by smaller body, longer tail, and white outline on yellow chin that's absent on Double-crested. Bare patch of skin extends above the eye on breeding Double-crested, below the eye on Neotropic. Juvenile Neotropic distinguished by size and darker chest. CONSERVATION STATUS Least Concern.

Breeding adult

Nonbreeding adult

Nonbreeding adults

Double-crested Cormorant

Nannopterum auritum

Wetlands, lakes, rivers, ponds
LENGTH 28–35 in. **WINGSPAN** 45–48 in.

Large, chunky, and low-floating waterbird with a long, sinuous neck; short legs; short tail; turquoise eyes; and thin, sharply hooked bill, with broad patch of bare facial skin from base to back of eye. Adult is black with brownish iridescence in scalloped wing feathers and has yellow-orange facial skin and a dark gray to orange bill. Juvenile is lighter and browner overall, especially in neck and breast. Breeding adult develops deep orange-red facial skin and a spindly white or black crest behind each eye. **BEHAVIOR** Dives from water's surface and swims to capture fish and other aquatic animals. Frequently seen drying off with outspread wings or flying in sloppy V-shaped formations. Roosts and nests in large colonies, usually in dead water-adjacent or submerged trees. **VOCALIZATIONS** Deep piglike or froglike croaks. **ABUNDANCE** Expanding in range and abundance in response to human development. A year-round resident near Phoenix; along the Colorado, Rio Grande, and Pecos rivers; and at other scattered locations with waterbodies and plentiful fish, where it can also be found during winter and in migration. Breeds mostly from the central to northern portion of the region. **LOOKALIKE** Neotropic Cormorant. Double-crested is larger, heavier bodied, and shorter tailed; lacks white border at base of bill; and has more extensive facial skin. In flight, Double-crested's tail much shorter than neck compared to Neotropic, which has neck and tail of similar lengths. **CONSERVATION STATUS** Least Concern.

Breeding adult

Nonbreeding adult

Breeding adult

American White Pelican

Pelecanus erythrorhynchos

Coastal waters, rivers, bays, estuaries, lakes, reservoirs
LENGTH 50–65 in. WINGSPAN 95–114 in.

Massive birds showing bright, clean white plumage on entire body, with exception of black flight feathers visible only when extended. Large, long, straight bill is bright yellow-orange, matching the legs. During breeding, small protrusion extends from upper mandible, like a small, incomplete horn. Immature shows dusky gray patches on head, neck, and back. BEHAVIOR Cartoonish and prehistoric in appearance, these large, gangly birds are contradictingly graceful and regal in flight, often soaring in V-shaped formations with slow and mechanical wingbeats—seemingly effortless, despite their size. Unlike Brown Pelicans, they do not plunge into water like a torpedo; rather they dabble like a duck, sweeping fish into their bill pouches. VOCALIZATIONS Typically silent, though grunts are used during copulation and by young at nesting sites. Chicks are known to vocalize from inside their eggs when conditions become uncomfortably warm or cold. ABUNDANCE Commonly observed during migration throughout AZ and NM. Overwinters from Central to Southwest AZ on large waterbodies and along the Gila, lower Colorado, and other sizeable rivers. CONSERVATION STATUS Least Concern.

HERONS AND OTHER WADING BIRDS

Like dinosaurs with long legs and necks, serious faces, and intimidating bills built to satisfy their carnivorous desires, the bitterns, herons (family *Ardeidae*), and ibises (family *Threskiornithidae*) make their Triassic origins obvious. Despite their lack of hooked bills and sharp talons and seemingly contrary to their sinuous and gangly appearance, these birds are formidable predators, with invertebrates, fish, amphibians, reptiles, and even small mammals on the menu. When birding in the Southwest, watch for herons hunting in the shallows and along the edges of open water, look twice for bitterns lurking in the densest patches of reeds, and scan flooded agricultural fields for foraging flocks of ibises.

ABOVE Adult Great Egret

American Bittern

Botaurus lentiginosus

Freshwater marshes, wetlands
LENGTH 24–33 in. WINGSPAN 34–38 in.

Bulky, medium-sized heron with heavy, buffy-rufous streaking on creamy white underside. Back is warm brown. Posture is often hunched, making the thick neck appear short. Yellow bill is straight, long, and comes to a sharp, knifelike point. Eyes and legs are yellow. In flight, underwings show dark brown that contrasts with warm plumage on body. BEHAVIOR Extremely secretive. Look for them hiding among the reeds with necks held high or crouched in waiting as they seek prey. Will sometimes sway back and forth with neck extended, giving its best windblown cattail impression. Flight pattern is more erratic and hurried than the gliding, smooth wingbeats of other heron species. VOCALIZATIONS Very low-frequency, glugging *punk-er-lunk* or *oomp-a-chunk* breeding call sounds like motor oil being poured from a jug. When flushed, gives a harsh, grumpy croaking call similar to that of many heron species. Most vocal during breeding season. ABUNDANCE Rare and local in winter and breeding season. Rare but regularly observed throughout migration in suitable marsh habitat. LOOKALIKE Immature Black-crowned Night-Heron. American Bittern distinguished by longer, slimmer bill; longer neck; and warmer coloration. CONSERVATION STATUS Least Concern (Decreasing).

Least Bittern

Ixobrychus exilis

Ponds, lakes, slow-moving rivers, wetlands, always with plentiful and dense emergent vegetation
LENGTH 11–14 in. WINGSPAN 16–18 in.

Tiny, heavily camouflaged, and incredibly skulky heron. Buffy straw-brown overall with long, yellow legs and toes; long, stabbing, yellow bill with sometimes dark upper surface; long, white-striped, and frequently curled neck; and often hunched posture. Adult male shows blackish back with two white stripes. Female and juvenile have brown but still darkish backs with fainter buffy stripes. BEHAVIOR More often heard than seen. Stays perfectly still when hunting for fish and other aquatic life. Surveys the water intently, usually from within dense wetland vegetation where its small size enables it to perch by grasping cattails and rushes. VOCALIZATIONS An apelike *oh-oh-oh-oh-oh*; a loud, sharp, *kek-kek-kek-kek*; and a short, often squeaky *ERT*.

ABUNDANCE Uncommon and local year round in AZ along the lower Colorado and Gila rivers and in NM along the lower Rio Grande and Pecos rivers. Occasionally found using isolated patches of suitable habitat in the southern portions of both states; rarely found farther north. LOOKALIKE Green Heron. Least Bittern distinguished by smaller size, lighter plumage, and behavior. CONSERVATION STATUS Least Concern.

 Adult

 Immature

Great Blue Heron

Ardea herodias

Marshes, wetlands, rivers, lakes, coastal areas, estuaries, water treatment facilities
LENGTH 23–33 in. **WINGSPAN** 36–39 in.

These iconic denizens of shorelines are tall, lanky, and, yes, a lovely steel-blue color. Adult is gray-blue overall with accents of chestnut and white, with a long, serpentine neck; long, grayish-orange legs; and a thick, long, orange bill designed to stab prey. Bold black head stripe above the eyes, and wispy plumes that extend from the crown, back, and chest. Immature is muted gray-blue and dusky brown overall, with gray legs and upper mandible. Great White subspecies (*A. herodias occidentalis*) occurs only in Florida. **BEHAVIOR** Superbly stealthy hunters, Great Blues stand statuesque and motionless or cruise shorelines with calculated, slow steps, striking their prey (fish, reptiles, amphibians, and small mammals are all fair game) with astonishing speed and accuracy. In flight, they glide with slow, deep wingbeats, legs trailing behind and neck tucked, creating a signature silhouette. **VOCALIZATIONS** Drawn out, high-pitched croak of an alarm call will scare you out of your wits if it catches you off guard. Males and females are very vocal in the breeding season, giving raspy, croaking *reh roh* and *grawk* calls. They also communicate for various purposes by clicking and clapping their bills. **ABUNDANCE** Abundant year round in suitable habitat throughout AZ and NM. **CONSERVATION STATUS** Least Concern.

Breeding adult

Nonbreeding adult

Breeding adult

Great Egret

Ardea alba

Wetlands, ponds, lakes, mudflats, slow-moving rivers and backwaters, agricultural fields; nests colonially in large trees near water
LENGTH 37–41 in. WINGSPAN 51–57 in.

Very large, long-legged, long-necked, and all-white heron. Legs and feet are black. Long, sharp bill is yellow-orange. In breeding plumage, develops a partially dark upper mandible, mint-green lores, and long, wispy plumes that extend from the back and lay over the tail at rest. BEHAVIOR Forages by walking slowly in shallow water or on adjacent land, bolting with surprising speed to stab or snatch passing prey. A powerful and opportunistic predator, will take everything from fish and other aquatic life to reptiles and small mammals. VOCALIZATIONS Low, rattling croaks. ABUNDANCE Common year-round resident at significant wetlands and along major rivers, including the Colorado, Gila, Rio Grande, and Pecos rivers; most numerous in the Phoenix metropolitan area and west along the Gila and Colorado rivers. Scattered appearances farther north, mostly in June and July. Frequents golf courses, urban ponds, and other human-made wetlands. Most abundant in winter. LOOKALIKES Snowy, Western Cattle egrets. Great Egret distinguished from Snowy by larger size, black feet, and yellow-orange bill. Great is significantly larger and longer necked than Western Cattle. CONSERVATION STATUS Least Concern.

Breeding adult

Nonbreeding adult

Breeding adult

Snowy Egret

Egretta thula

Wetlands, mudflats, beaches, agricultural fields, marshes, coastal areas
LENGTH 22–26.5 in. WINGSPAN 37–40 in.

Medium-sized, delicate heron with pure white plumage overall, black legs, and diagnostic yellow feet. Bill is black with bright yellow skin patch at base. Breeding plumage shows ornate, filament-like feathers, which once led them to be exploited to near extinction by the fashion industry. BEHAVIOR Though they employ many of the same stalking behaviors as the Great Blue, Green, and other heron species, the Snowy tends to patrol the waters on the move, wading through shallow water and chasing down prey. Male and female take turns incubating eggs in breeding season, nesting in trees. VOCALIZATIONS Typically silent with the exception of breeding season, when female and male give grating, reprimanding squawks. ABUNDANCE A common breeding resident at significant wetlands and along major rivers including the Colorado, Gila, Rio Grande, and Pecos rivers; a common migrant throughout the region. LOOKALIKES Great, Western Cattle egrets. Snowy distinguished by black bill and yellow feet on black legs. CONSERVATION STATUS Least Concern.

 Breeding adult

 Nonbreeding adult

Western Cattle Egret

Bubulcus ibis

Short grasslands, agricultural fields, other sparse upland habitats, typically near open water
LENGTH 18–22 in. WINGSPAN 38 in.

Quite stocky and stout-necked for a heron, more similar in shape to a night-heron than to their all-white relatives. Completely white with dark legs and sharp, relatively short, yellow bill. Breeding birds develop bright orange legs and bill and peachy tan, wispy plumes on the crown, breast, and lower back. BEHAVIOR Forages in groups on land mostly for insects, but will also take other small prey when opportunity arises. Forages much less frequently in water than other herons. Known to forage alongside horses, cattle, and farm equipment that kick up insects. Nests in trees or wetlands, frequently with other heron species. VOCALIZATIONS Calls are varied but all repetitive, nasal, and ducklike—*kuh-kuh-kuh, kuh-ow-kuh-ow, kuh-wa-wa-waa*.

ABUNDANCE Uncommon, year-round resident along the lower Colorado and Gila rivers west of Phoenix and a breeding resident in Central and Southeast AZ and in NM along the Rio Grande and Pecos rivers. An uncommon migrant elsewhere. LOOKALIKES Great, Snowy egrets. Western Cattle much smaller and thicker necked than Great; lacks yellow feet and, in adults, black bill of Snowy. CONSERVATION STATUS Least Concern.

Adult

Juvenile

Green Heron

Butorides virescens

Ponds, lakes, rivers, wetlands, typically with dense emergent vegetation, waterside trees
LENGTH 16–18 in. WINGSPAN 2–27 in.

Small, dark, chunky, and short-legged heron with a short neck often held curled, giving this bird a football-shaped appearance. Breeding adult has long yellow-and-black bill; yellow lores; blackish cap that is often raised; plum-brown nape, neck, and breast; white streaks in the throat and neck; grayish underparts; mottled iridescent green to purple back; and yellow-orange legs. Juvenile is lighter overall with white-streaked, cinnamon-colored head, neck, and breast; dingy, gray-green back; and lighter bill and legs. BEHAVIOR Hunts for fish and other aquatic animals from the water's edge, usually on the ground, on floating vegetation, or from a woody perch. Rarely forages with legs submerged in water like other herons. VOCALIZATIONS Most common call is a sharp, truncated *KEOW*. Also a repeated *keh-keh-keh*. ABUNDANCE Uncommon and local. Year-round resident in Southern and extreme Western AZ and from Central to Southwest NM. A breeding resident from Central to Northwest AZ and in NM along the Rio Grande. A migrant across much of the region. LOOKALIKE Least Bittern. Green Heron distinguished by larger size, darker plumage, and behavior. CONSERVATION STATUS Least Concern (Decreasing).

Adult

Juvenile

Black-crowned Night-Heron

Nycticorax nycticorax

Wetlands, marshes, estuaries, rivers, lakes, reservoirs
LENGTH 22.5–26.5 in. WINGSPAN 45–47 in.

Stocky, squat heron with uncharacteristically bulky features—thick neck and bill, short legs, and plump body. Breeding adult shows namesake black crown extended to black back, pale gray-white belly, and gray wings. Legs are yellow, bill is black, eyes are large and red. Juvenile is brown overall with white spots on back and heavily streaked underside. Immature plumage is a blend of juvenile and adult plumages, with charcoal-gray cap, gray-brown back, and faint streaking on underside. BEHAVIOR Situationally gregarious, they tend to roost and nest together but go their separate ways to hunt, waiting patiently at water's edge, primarily at dusk and into the night. During the day, look for them perching inconspicuously in high tree limbs or among thick vegetation. They retain their compact appearance in flight, with rounded wings and neck tucked in. VOCALIZATIONS Most commonly gives a grating, nasally *twok* in flight or when flushed. Also uses a variety of squawks and croaks. ABUNDANCE Common and abundant, breeding throughout much of Northern AZ and NM; observed year round through much of the southern parts of both states. CONSERVATION STATUS Least Concern (Decreasing).

Breeding adult

Nonbreeding adult

White-faced Ibis

Plegadis chihi

Shallow, sparse wetlands; flooded agricultural fields
LENGTH 19–23 in. WINGSPAN 36 in.

Somewhere between a shorebird and a heron. Gangly yet graceful with a long, thick, gray, downward-curving bill; reddish facial skin; red eyes; long neck; and long, dark legs. Best viewed in good light—often appears dark overall but iridesces green, gold, purple, and blue. Breeding adult shows a white border around deeper red facial skin, reddish legs, and rich purple-brown head, neck, nape, and underparts. Nonbreeding adult and juvenile are duller overall, lacking the white facial border and with subtle streaking in the head and neck. BEHAVIOR Forages for invertebrates and other small prey by probing in mud or water, sweeping its bill from side to side, or by directly grabbing unsuspecting critters. Most often observed in large flocks that can number in the hundreds. VOCALIZATIONS Relatively quiet, infrequently giving Mallard-like laughs and grunts. ABUNDANCE Locally common. Lingers year round along the Colorado and Gila rivers in AZ, breeds in SE NM and along the Pecos and Rio Grande rivers, and occurs as a migrant throughout the rest of the region. LOOKALIKE Glossy Ibis (regionally rare). White-faced distinguished by facial pattern and, at close range, eye color, (red in White-faced, dark in Glossy). Occasional hybrids may be seen. CONSERVATION STATUS Least Concern.

VULTURES, HAWKS, AND OTHER RAPTORS

It's a scene many over-adventurous Southwestern birders know all too well: you're meandering through the desert, parched, and hopefully heading toward your vehicle, when the long-winged silhouette of a Turkey Vulture temporarily blocks the sun, both of you wondering if you'll manage to find your way. Along with the falcons, the American vultures (family *Cathartidae*), the Osprey (family *Pandionidae*), and the kites, eagles, and hawks (family *Accipitridae*) are collectively referred to as raptors and are set apart from other birds by their sharply hooked bills and, in all but the vultures, sharp talons adorning the toes. Though some species breed in the Southwest or are year-round residents, the real show happens during winter and migration, when they pass through the region in incredible numbers.

ABOVE Black Vultures

Adult / Immature

Adult

California Condor

Gymnogyps californianus

Nests on mountain cliffs, canyon walls
LENGTH 46–54 in. **WINGSPAN** 105–110 in.

North America's largest bird, with a wingspan of nearly 10 feet. Adult coloration is black, with bright white patches under the wings visible in flight, and white legs. Primary feathers have long, fingerlike appearance. Cowl of feathers around neck accentuates peachy pink bald head and heavy orange-yellow bill. Immature condor shows gray head and lacks bright white underwing plumage. **BEHAVIOR** Scavenges for carrion while soaring above coastlines and beaches, canyons, forests, meadows. Young take up to eight years to reach maturity. They are social creatures, gathering together both to roost and scavenge. **VOCALIZATIONS** Typically silent. Communicates with grunts and snorts at the nest. **ABUNDANCE** Rare and local. **KEY SITES** AZ: Historic Navajo Bridge, Marble Canyon, Grand Canyon area. **LOOKALIKES** Turkey Vulture, Golden Eagle. California Condor distinguished from Turkey Vulture by darker bill and pinkish orange head, from Golden Eagle by bald head and smaller bill, and from both by larger size and combination of white underwing coverts and black flight feathers. **CONSERVATION STATUS** Critically Endangered.

Black Vulture

Coragyps atratus

Open, low-elevation habitats including grasslands, deserts, usually not far from more heavily forested areas; frequent visitor to urban landscapes **LENGTH** 24–27 in. **WINGSPAN** 54–59 in.

Large raptor with wide, round-tipped wings held flat in flight; rounded tail; featherless, gray, and wrinkled head; narrow, grayish, and sharply hooked bill; and short gray legs. Entirely black except for silvery gray hands (flight feathers) visible in flight. **BEHAVIOR** Forages from the sky for carrion, often in flocks with other species. Glides for shorter durations than Turkey Vulture, with snappier wingbeats and never rocking from side to side. Occasionally takes live prey including small birds and mammals. Roosts in large groups in tall trees or on human-made infrastructures such as power lines and bridges. **VOCALIZATIONS** Mostly silent; sometimes soft, raspy hisses and barks. **ABUNDANCE** An uncommon year-round resident from Central to Southern AZ. **LOOKALIKE** Turkey Vulture. Black Vulture distinguished by lack of red head, narrower and more sharply hooked bill, and underwing pattern. Also easily distinguished by shape: Black Vulture is similar in shape to Black Hawk, while Turkey Vulture is similar in shape to Zone-tailed Hawk. **CONSERVATION STATUS** Least Concern.

Turkey Vulture

Cathartes aura

Roadsides, countryside, open areas; often frequents urban environments
LENGTH 26–33 in. **WINGSPAN** 66–72 in.

Stately vulture with signature bald red head, sharply hooked bill with white at the tip, and black-brown plumage overall, with the exception of gray flight feathers all along the trailing edge of the wing. Legs are pale gray-white. Primaries have fingerlike appearance when extended. Juvenile has gray head. **BEHAVIOR** Look for them teetering back and forth in their dihedral (V-shaped) flight posture as they ride the thermals on a sunny day, often in funnel-shaped flock formations called kettles, sometimes with other raptor species. They circle the skies in search of carrion, using their keen sense of smell more so than sight. You can often spot them along roadsides snacking on roadkill, tearing flesh with their sharply hooked bills, or roosted in trees with a small flock, often basking with wings extended. They will defecate on their legs to keep cool in hot weather and are known to vomit on or in the direction of intruders as a defense mechanism. **VOCALIZATIONS** Lacking the anatomical ability to make "song," they vocalize through low-pitched, throaty hissing. **ABUNDANCE** Common and abundant throughout the region in spring and summer with only a handful of birds lingering through the cooler months. **LOOKALIKES** Black Vulture, Zone-tailed Hawk. Turkey Vulture distinguished from Zone-tailed Hawk by lack of banded tail, yellow bill, and yellow legs; from Black Hawk by narrower, longer wings and flight style; from both by fleshy red head. **CONSERVATION STATUS** Least Concern.

Pair with juveniles at nest

Adult male

Osprey

Pandion haliaetus

Nearly any open body of water; nests over or near water within dead trees or atop poles LENGTH 21–23 in. WINGSPAN 60–70 in.

Extremely large but slender with long, narrow wings that have fingerlike flight feathers extending from the tips and a characteristic bend midway along the leading edge (the wrist). Brownish black above and white below with yellow eyes and a white cap, dark eyeline, and white throat that together form a striking facial pattern. Dark necklace across female's breast almost entirely absent in male. Juvenile shows reddish orange eyes and buffy tones. In flight, mostly white below with mottled underwings and dark wrist patches. BEHAVIOR Foraging style made for cinema—soaring on frequent wingbeats above water, sometimes stalling briefly before plunging downward with feet thrust forward, splashing into the water before emerging with a fish in its talons. While in the air, will orient fish face-forward for ease of flight. VOCALIZATIONS Short, repetitive, whistly, and sometimes squawking high-pitched chirps. ABUNDANCE Uncommon and local. Breeds primarily at higher elevations; in AZ along the Mogollon Rim to the White Mountains and in Northern NM. In winter, most common along the Colorado River and near Phoenix in AZ. Can be seen in migration across the entire region. CONSERVATION STATUS Least Concern.

Adult  Juvenile

White-tailed Kite

Elanus leucurus

Grasslands, open woodlands, marshes, fields, meadows
LENGTH 13–15 in. **WINGSPAN** 40–43 in.

Adult

Distinctive, small raptor with solid white tail and body, black shoulder patches, gray back and wings, and yellow legs. Severe-looking white face has large red eyes with "smoky eye" rings of dark gray plumage, sharply hooked black bill with yellow at the base. Juvenile shows ruddy wash on breast and crown. **BEHAVIOR** Look for them hovering over open spaces, head and tail angled toward the ground, scanning for prey. Nests in treetops, preferring live oak. **VOCALIZATIONS** Sweet, soft, whistling *teep*. **ABUNDANCE** Uncommon, local resident in extreme Southeast AZ and in the bootheel of NM. **KEY SITES** Foothills of the Huachuca Mountains, Galiuro Mountains, and Chiricahua Mountains of Southeast AZ; Animas Mountains and Valley of extreme Southwest NM. **CONSERVATION STATUS**

Adult

Adult

Juvenile

Golden Eagle

Aquila chrysaetos

Open habitat including grasslands, deserts, woodlands, chaparral, agricultural fields; nests on nearby rugged and often remote cliffs FEMALE LENGTH 30–33 in. WINGSPAN 78–86 in. MALE LENGTH 28–31 in. WINGSPAN 73–82 in.

Among the continent's largest birds and a reason to look up occasionally from the ground and beyond trees and into the open sky. In flight, note relatively small head, long tail, and long, straight wings held in a slight V shape. Adult is rich dark brown and shows golden head and mantle in good light. Juvenile has white base to dark-tipped tail and white patches near wingtips in flight. Bill is large, strongly hooked, and dull yellow-gray. Legs are fully feathered. BEHAVIOR Soars at extremely high altitudes with few flaps, foraging for mostly mammalian prey and carrion. VOCALIZATIONS Varied, gentle, and whistling strings of high-pitched notes. Not as forceful or squeaky as Bald Eagle vocalizations. ABUNDANCE Uncommon year round throughout much of the Southwest. A migrant and winter visitor to extreme Southwest AZ and Southeast NM. LOOKA-LIKE Bald Eagle. Golden has smaller head and fully feathered legs, flies with wings held in a slight V shape rather than flat, and lacks white head and tail. Juvenile Golden distinguished by white patches near wingtips, distinct white base to dark-tipped tail, and lack of white mottling in underparts, back, nape, and underwings. CONSERVATION STATUS Least Concern.

Adult male

Adult female
Adult female

Northern Harrier

Circus hudsonius

Grasslands, agricultural fields, marshes, meadows
LENGTH 18–20 in. WINGSPAN 41–46 in.

Medium-sized, owl-like raptor with distinct long-tailed, flat-faced silhouette in flight. Female shows brown coloration overall with whitish, streaked underside. Male is gray (the "gray ghost") with pale undersides. Both sexes have characteristic white rump; black-banded tail; flat, owl-like face; and hooked bill. In flight, they often have a dihedral (V-shaped) posture. BEHAVIOR Look for harriers cruising low over fields and meadows, scanning for rodents and other prey. Stiff facial plumage funnels sound to their ears, much as in owls. Male typically has multiple mates, up to five at one time, and provides food to mates and young while females tend to nests. VOCALIZATIONS Typically silent, but may give fast, repetitive, high-pitched *kee* when courting or mobbed by other birds in flight. ABUNDANCE Common and abundant in winter throughout AZ and NM. CONSERVATION STATUS Least Concern (Decreasing).

Adult

Immature

Adult

Sharp-shinned Hawk

Accipiter striatus

Nests in dense conifer forests within mixed-deciduous or pure conifer woodlands; in migration and winter, nearly any wooded habitat including suburban areas
FEMALE LENGTH 11.5–13.5 in. **WINGSPAN** 23–25 in.
MALE LENGTH 9.5–10.5 in. **WINGSPAN** 21–22 in.

Our tiniest accipiter. Small-headed, heavy bodied, and bug-eyed, with rounded wings and long tail showing black-and-gray bands and, when not fanned, obvious sharp corners. Adult is slate-gray above with a dark hood (cap and nape); red eyes; heavy rufous-orange bars on the cheeks, throat, breast, and underwing coverts; flight feathers with black-and-white bands; white undertail coverts; and long, thin yellow legs. Juvenile shows brown instead of gray, replaces the dark hood with a brown head and pale eyebrows, and has yellow eyes, a messier underwing pattern, and thick, vertical brown streaks on the breast, belly, and flanks. When perched, middle toe is noticeably longer than the rest. **BEHAVIOR** Forages for birds on short, crisp wingbeats in dense vegetation. Tends to take hidden perches within trees and shrubs. **VOCALIZATIONS** A repetitive, high-pitched *ki-ki-ki-ki-ki*. **ABUNDANCE** Uncommon but widespread. A year-round breeding resident at high elevations from Northwest to Southeast AZ and from Northern to Southwest NM. A winter visitor at lower elevations. **LOOKALIKE** Cooper's Hawk. Sharp-shinned distinguished by quicker wingbeats; small, rounded head situated behind leading wing edges in flight; forward-set eyes; dark nape; square-cornered tail; and behavior. **CONSERVATION STATUS** Least Concern.

Adult | Immature

Immature

Cooper's Hawk

Accipiter cooperii

Wooded areas, including dense forests, urban parks, backyards
FEMALE LENGTH 16–18 in. **WINGSPAN** 30–35 in.
MALE LENGTH 14–15.5 in. **WINGSPAN** 24–30 in.

Medium-sized, long-bodied accipiter with long tail, rounded wings, and boxy head topped with a dark gray cap. Adult shows slate-blue back and pale, rufous-barred underside, including wings. Tail shows rounded edges and is banded in charcoal-gray with white terminal band. Yellow legs; yellow, hooked bill with black at tip; and amber eyes. Female significantly larger than male. Juvenile shows heavy brown streaking down underside, with brown plumage on back. **BEHAVIOR** Impressive aerialists, they weave through trees with relative ease to ambush prey—primarily small birds—in wooded areas and backyard feeders. Flight pattern is a distinctive pump-pump-pump-glide. **VOCALIZATIONS** Most common call is during breeding at nest sites. A repetitive, high-pitched, laughlike one-note *cack*. **ABUNDANCE** Common and abundant. **LOOKALIKE** Sharp-shinned Hawk. Cooper's distinguished by rounded versus square-edged tail, longer neck that extends squared (versus rounded) head in front of leading wing edges in flight, and a dark cap versus the hood of the Sharp-shinned. **ALTERNATIVE NAMES** Square-headed Hawk, Aerial Ambush Hawk, Gray-capped Hawk, Big Blue Darter. **CONSERVATION STATUS** Least Concern.

American Goshawk

Accipiter atricapillus

Large expanses of old-growth coniferous forest, often with many openings created by riparian corridors, meadows, aspen groves; in winter, heavily wooded sites at lower elevations **FEMALE LENGTH** 23–26 in. **WINGSPAN** 43–46 in. **MALE LENGTH** 21–24 in. **WINGSPAN** 39–41 in.

Largest and most powerful of the North American accipiters—what an overconfident Sharp-shinned Hawk might see if it looked in a mirror. Chunky and round-winged for an accipiter, with long secondaries that give wings a bulging, trailing edge in flight. Adult is dark gray above and light gray–barred below, with a black cap and eyeline, bold white eyebrows, red eyes, yellow bill base and feet, white rump, and a long, light-and-dark–banded tail. Juvenile is brown and buff with buffy eyebrows, yellow eyes, and thick, dark, messy bars down the throat, breast, and belly. **BEHAVIOR** Watches from perch before snatching bird or mammalian prey from dense understory brush. Extremely territorial around nest sites. **VOCALIZATIONS** Usually silent away from nest. Calls at nest include high, repetitive barking and sharp squeals. **ABUNDANCE** Uncommon to rare. A year-round breeding resident at high elevations from Northern to Southeast AZ and from Northern to Southwest NM. A winter visitor at lower elevations. **CONSERVATION STATUS** Least Concern.

Bald Eagle

Haliaeetus leucocephalus

Lakes, rivers, reservoirs, coastal areas
FEMALE LENGTH 32–38 in. WINGSPAN 78–92 in.
MALE LENGTH 28–35 in. WINGSPAN 70–85 in.

A ubiquitous national emblem, these heavy bodied raptors have an awe-inspiringly wide wingspan with long, fingerlike primaries. Adult shows white-plumed head (bald is a misnomer, from the term "piebald," meaning white-headed); heavy, hooked yellow bill; lemon-yellow eyes; chocolate-brown plumage on body; solid white tail; and yellow feet and lower legs. Legs are heavily feathered, like brown trousers. Female significantly larger than male. Juvenile shows mottled brown and white overall, with brown head and dark terminal band on tail. BEHAVIOR Often soars at extreme heights on warm, sunny days, with flat, broad wings that rarely flap. May be observed perched in trees or on the ground near prey. Both hunters and scavengers, they are opportunists and will eat mammals, fish, birds, even garbage. They are social and have been observed engaging in play with objects ranging from sticks to trash. Often mates for life. VOCALIZATIONS High-pitched series of chirping whistles, unexpectedly soft for a sizable bird. ABUNDANCE Common year round along the Salt and Verde rivers in AZ and at several high-elevation lakes above the Mogollon Rim. Winters near water across much of AZ and NM, including along the Rio Grande, Pecos, and Gila rivers. Rare in extreme southern portions of both states. KEY SITES Central AZ: Salt River/Verde River corridor (nearly two-thirds of the state's Bald Eagle population nests here); NM: Gila Cliff Dwellings National Monument. CONSERVATION STATUS Least Concern.

Mississippi Kite

Ictinia mississippiensis

Open areas with patches of tall trees; most frequently spotted in grasslands near rivers, in agricultural fields, suburban parks **LENGTH** 13–15 in. **WINGSPAN** 30–32 in.

Lean and relatively small with long, pointed wings; long, square tail; round head; red eyes; and small but sharply hooked bill. Adult is gray overall, lightest in the nearly white head and darkest in the wings, with black tail, black wingtips, black eye patches, and white secondaries visible in flight. Juvenile shows brown streaks in the breast, belly, and underwing coverts and has banded black-and-gray tail. **BEHAVIOR** Spends most of its time in graceful, bouncy flight, foraging for insects that it catches in its talons, often eating on the wing. **VOCALIZATIONS** A high, whistly, and descending *WEH-peeeew*. **ABUNDANCE** Uncommon to rare and very local, with breeding and migratory records concentrated in Southeast AZ and near Albuquerque, below Las Cruces along the Rio Grande, and in Eastern NM. **KEY SITES** AZ: San Pedro River near Winkelman and Saint David; NM: Pecos River near Santa Rosa and Roswell. **LOOKALIKE** White-tailed Kite. Mississippi Kite gray instead of bright white overall and lacks black shoulder patches. **CONSERVATION STATUS** Least Concern.

Adult

Juvenile

Common Black Hawk

Buteogallus anthracinus

Wooded streams, rivers
LENGTH 17–22 in. WINGSPAN 44–47 in.

Bulky, dark charcoal-gray hawk with hooked, black-tipped yellow bill; yellow legs and feet; wide white tail band; and narrow white tail tip. Juvenile is dramatically mottled in deep gray/black/brown and warm buff, with much paler, barely yellow bill. Wings are broad and rounded in flight. BEHAVIOR Often observed perched near streambanks and rivers, their preferred hunting grounds. They will wade—much like shorebirds but lacking their stealth—into water to round up their prey. VOCALIZATIONS Squeaky, whistling *kwee*. ABUNDANCE Uncommon. Breeds along rivers and streams at mid-elevations from North Central to Southeast AZ and Southwest NM. LOOKALIKE Zone-tailed Hawk. Common Black Hawk distinguished by broader wings with rounded edges and fan-shaped tail, versus pointed wings with fingerlike primaries and longer, straighter tail of the Zone-tailed. White stripe higher on tail than on Zone-tailed. Voice and behavior can be helpful clues in differentiating between these two species. CONSERVATION STATUS Least Concern (Decreasing).

Adult | Juvenile

Adult

Harris's Hawk

Parabuteo unicinctus

Relatively open, low-elevation desert and semi-desert with plentiful mesquite, saguaro, or other tall, columnar cacti; regular visitor to urban areas, where tasty urban birds and tall perches in the form of light posts and electrical towers abound
FEMALE LENGTH 18–24 in. **WINGSPAN** 40–48 in.
MALE LENGTH 18–24 in. **WINGSPAN** 38–45 in.

An iconic sight atop a saguaro cactus. Medium-sized with long, yellow legs; large yellow bill; long tail; and obvious white rump. Adult is coffee-brown with cinnamon-colored shoulders, upper leg feathers, and underwing coverts; tail has a crisp, white base, wide black band, and white tip; with dark gray flight feathers. Juvenile shows coarse brown streaking from the lower breast through the belly and has gray and indistinctly banded tail and flight feathers. **BEHAVIOR** Matriarchal, spending much of the year in packs of multiple males and a dominant female. Hunts cooperatively, watching from high, obvious perches, sometimes stacked atop one another for a better view, before executing coordinated, multibird attacks on mammalian prey. Will also forage independently for reptiles, insects, birds, and occasionally carrion. **VOCALIZATIONS** Extremely raspy screams and squawks. **ABUNDANCE** An uncommon year-round resident. Most abundant in extreme Western, Southern, and from Central to Southeast AZ, but also in Southeast NM and less commonly from Alamogordo southwest to Rodeo. **ALTERNATIVE NAMES** Matriarch Hawk, Cactus Hawk. **CONSERVATION STATUS** Least Concern (Decreasing).

Gray Hawk

Buteo plagiatus

Wooded riparian areas, thornscrub, forest edges; nests in willow and mesquite stands along streams, rivers
LENGTH 19–22 in. WINGSPAN 32–35 in.

Lovely, small to medium-sized buteo, pale gray overall with fine gray-and-white–barred underside. Tail shows wide black-and-white banding, with white undertail coverts. Bill is hooked, black at the tip and yellow at the base; legs are yellow. Juvenile is heavily patterned and brown on back, with creamy white belly splotched with dark brown and finely barred tail. BEHAVIOR Perches and soars relatively low to the ground searching for prey, preferably lizards. Though a buteo, its flight pattern follows the pump-pump-glide characteristic of accipiters. Courtship displays consist of impressive aerial acrobatics. VOCALIZATIONS Melancholy, three-note, high-pitched whistle during breeding season, to claim territory and communicate with mates. One-note alarm call used by both sexes year round. ABUNDANCE Common in suitable habitat within restricted Southeast AZ range. CONSERVATION STATUS Least Concern.

Adult

Immature

Adult

Red-shouldered Hawk

Buteo lineatus

Open habitats with scattered deciduous trees; river corridors, wetlands, agricultural fields, parks
FEMALE LENGTH 19–24 in. **WINGSPAN** 40–42 in.
MALE LENGTH 17–23 in. **WINGSPAN** 37–40 in.

Medium-sized and reminiscent of a Gray Hawk in full color, with long, broad wings and long, banded tail. Adult rusty-orange overall; solid in the head, breast, and shoulders; densely barred on belly and underwing coverts; and with coarse brown streaks in nape and upper back. Black-and-white barring adorns upperwing coverts, flight feathers, and tail. Juvenile more brownish overall, more white below with coarse brownish orange bars and gray-banded tail. In flight, note semi-transparent patch near wingtips. **BEHAVIOR** Hunts from tall perch or while circling overhead, taking a wide variety of prey including small mammals, birds, amphibians, and reptiles (especially snakes). Flight style is similar to that of an accipiter, with quick, snappy wingbeats and short glides. **VOCALIZATIONS** Extremely vocal; a forceful *keeAH*, similar to Red-tailed Hawk but shorter, clearer, and rapidly repeated. **ABUNDANCE** Uncommon to rare with extremely limited Southwestern range. A year-round and wintering resident at scattered sites near Phoenix. Observed rarely in winter and migration along the lower Colorado and Gila rivers. **KEY SITES** AZ: Hassayampa River Preserve (year round) and Rio Salado Habitat Restoration Area (winter). **LOOKALIKE** Broad-winged Hawk. Red-shouldered distinguished by brighter orange coloration, black barring in wings, thinner barring below, and see-through patch near wingtips. **CONSERVATION STATUS** Least Concern.

Adult (light morph)

Adult (light morph)

Broad-winged Hawk

Buteo platypterus

Dense forest
LENGTH 14–17 in. WINGSPAN 32–38 in.

Relatively small but girthy buteo, with distinctively broad wings that come to a clear point at the wingtips; otherwise rounded in appearance. Light morph adult has chunky, ruddy brown head and back; pale, heavily barred underparts; broadly banded black-and-white tail; and brown border on pale underwings. Dark morph is deep cocoa-brown overall with distinctly banded tail and pale border on underwings. Juvenile shows lighter, dusky brown coloration with heavy streaking on sides of belly and narrow banding on tail. BEHAVIOR Hunts small mammals from perches in deep forest or cruising along forest breaks. Gathers in remarkably large flocks for migration, often with other species. VOCALIZATION Very high-pitched, nasally, two-syllable *PIK-eee* whistle. ABUNDANCE Rarely observed during migration in Eastern NM, even less commonly from Central to Southeast AZ. LOOKALIKE Red-shouldered Hawk. Broad-winged distinguished by brown versus reddish barring on chest and brown streaking on belly. CONSERVATION STATUS Least Concern.

Adult (light morph)

Adult (dark morph)

Adult (light morph)

Swainson's Hawk

Buteo swainsoni

Grasslands, prairies, sagebrush
LENGTH 19–22 in. WINGSPAN 46–54 in.

Medium-large buteo with slender body and long wings. Plumage varies among individuals, though female tends to have brown head versus the gray head of the male. Back is brown overall, belly is clean and pale, tail is narrowly banded, and breast is rufous-brown. In flight, look for dihedral (V-shaped) flight posture, much like a Turkey Vulture, and a creamy white upper border on dark espresso-shaded flight feathers. Dark and red morphs are relatively rare. BEHAVIOR Will often perch relatively low to the ground, on fence posts or snags, seeking grasshoppers and other grassland treats. Gregarious outside of the breeding season. Will often nest in a lone tree in a sea of grass. VOCALIZATIONS Harsh, labored, whistling shriek, similar to Red-tailed Hawk. ABUNDANCE Common and abundant throughout breeding season and migration in preferred habitat across the region. LOOKALIKE Red-tailed Hawk. Swainson's distinguished by brown hood and bib, lack of distinct belly band, and darker flight feathers. ALTERNATIVE NAMES Brown-hooded Hawk, Grassland Hawk, Grasshopper Hawk. CONSERVATION STATUS Least Concern.

Adult

Immature

Zone-tailed Hawk

Buteo albonotatus

Nests along permanent rivers, streams within cottonwood/willow forests; forages nearby over open habitats including deserts, grasslands, chaparral, sparse woodlands
LENGTH 18–22 in. WINGSPAN 47–55 in.

Easy to miss among kettles of similarly shaped Turkey Vultures. All black to brownish black with long, narrow wings; yellow legs; a black-tipped, yellow bill; gray skin between bill and eyes; gray flight feathers with thin black bands; black trailing wing edges; and a banded tail showing black and white below and gray and white above. Juvenile shows more thinly banded gray-and-black tail and lacks black trailing wing edge. BEHAVIOR Forages for mostly mammals and birds from the air, hiding in plain sight with a similar flight style to and often alongside Turkey Vultures. Extremely vocal near nest sites. VOCALIZATIONS Screams similar to Red-tailed Hawk, but less raspy and lower pitched. ABUNDANCE Uncommon and somewhat local summer resident from Northwest AZ to Southwest NM. Also farther east along the Rio Grande as far north as Santa Fe, in the Lincoln National Forest, and around Carlsbad Caverns National Park. Rare in winter. LOOKALIKES Common Black Hawk, Turkey Vulture. Zone-tailed distinguished in flight by longer, narrower wings and, when perched, by shorter legs and gray (not yellow) skin between the bill and eye. In flight, can be confused with Turkey Vulture, but note Zone-tailed's larger, feathered head and banded tail. CONSERVATION STATUS Least Concern.

Red-tailed Hawk

Buteo jamaicensis

Open habitat types, including deserts, grasslands, chaparral, agricultural fields, urban and suburban areas, roadsides, sparse forests **FEMALE LENGTH** 20–26 in. **WINGSPAN** 45–52 in. **MALE LENGTH** 17–22 in. **WINGSPAN** 44–51 in.

A bulky buteo with diverse plumage variation. Look for brown plumage on back and head, with pale belly and brown-streaked belly band. Also shows two pale arced stripes along the shoulders when perched, like the straps of a backpack. Namesake rufous tail shows from above, paler on the underside. Underwings are typically pale with dark border and a diagnostic, smudgy dark brown bar extending from the shoulder toward the wrist, accentuated by a pronounced comma shape at the wrist. Dark morph shows deeper, warmer brown color overall. Juvenile often shows banding on tail, lacking reddish coloration. Rare Harlan's form shows white tail with dark band at the tip. **BEHAVIOR** Incredibly common roadside hawks. You're likely to spot them on utility poles, posts, and snags along just about any highway on the continent. This ubiquitous bird soars in broad circles over the landscape in search of mammals—primarily rodents, though they will occasionally snack on birds and reptiles. Their wingbeats are heavy, and when diving for prey, they move at a calculated, controlled pace before crushing their prey with their talons. Known to hunt in pairs. **VOCALIZATIONS** Iconic raspy, high-pitched, extended scream is a true claim to fame. You're likely to hear the majestic call of the Red-tailed Hawk performing voice-over work for Bald Eagles and other birds of prey on the big screen. **ABUNDANCE** Common and abundant. **CONSERVATION STATUS** Least Concern.

Adult females (light morph [left] and dark morph)

Adult female/immature (light morph)

Adult male (light morph)

Rough-legged Hawk

Buteo lagopus

Open habitats; grasslands, deserts, agricultural fields
LENGTH 19–21 in. WINGSPAN 52–54 in.

Large hawk with feathered legs and, for a buteo, a relatively small bill, long tail, and long, narrow wings. Exhibits incredible (and incredibly frustrating) plumage variation, with dark underwing edges, dark wrist patches, and a dark tip to an otherwise light tail being among the only reliable field marks across variants (faded in juvenile). Typical (light) female and juvenile have a whitish head, nape, throat, and breast with brown streaking most heavily concentrated in the breast; solid dark bellies; white rump; and, in flight, light underwings with dark wrist patches. Males are variable but similar, with thin, dark undertail bands and bellies that may be completely dark, densely barred, or almost completely white. Breast streaking may become so dense as to create a bibbed appearance. Dark morph is almost entirely dark brown with pale flight feathers and a white-banded and dark-tipped tail. Often flies with wingtips held back, creating an M-shaped silhouette. BEHAVIOR Hunts mostly for mammals from the air or a perch. Frequently hovers with help of strong headwinds. Often perches on thinner branches than do similar hawks. VOCALIZATIONS Piercing and relatively clear screams. ABUNDANCE An uncommon winter visitor across the northern portions of both states. Rare farther south. CONSERVATION STATUS Least Concern.

Adult (light morph)

Adult (dark morph)

Adult (light morph)

Ferruginous Hawk

Buteo regalis

Open country, open deserts, prairies/grasslands
LENGTH 22–27 in. **WINGSPAN** 53–56 in.

The largest buteo in North America and similar to an eagle in appearance, with heavily feathered legs, large head, and long, broad wings that come to a distinct narrow tip. More common light morph is creamy white underneath, with finely barred rufous (or ferruginous) legs that create a V-shaped pattern in flight. Back/upper parts are a lovely blend of rufous, brown, and gray. Uncommon dark morph shows deep chestnut-brown overall, but with white tail and flight feathers. **BEHAVIOR** Hunts by soaring and hovering or from the ground, feeding primarily on small mammals. Will use old owl and hawk nests as foundations for the large, bulky branches they use to complete nest construction. **VOCALIZATIONS** Hoarse, long, high-pitched screech. **ABUNDANCE** Common resident in northern half of NM and AZ, overwintering throughout suitable habitat in southern half. **CONSERVATION STATUS** Least Concern.

OWLS

Those without a spark for birds overlook much of what the avian world has to offer, but few are not captivated by an owl sighting. With their expressive eyes, swiveling heads, intimidating talons, and soft feathers that grant them silent flight, the typical owls (family *Strigidae*) and barn owls (family *Tytonidae*) capture the attention and imaginations of birders and nonbirders alike. Though some of the Southwest's fourteen species of owls are easy to find, such as the Great Horned Owl that haunts even dense urban spaces, spotting a habitat specialist like the Whiskered Screech-Owl or Boreal Owl requires intentional treks to unique ecosystems on the region's edges, and crossing paths with an infrequent visitor like the Short-eared Owl will take both persistence and a bit of luck.

ABOVE Juvenile Great Horned Owls in nest

Barn Owl

Tyto alba

Open areas, including marshes, grasslands, agricultural fields, parks
LENGTH 13–16.5 in. **WINGSPAN** 39–49 in.

Medium-sized owl with distinctive heart-shaped face; long, rounded wings; short tail; and top-heavy body that results in a bobbing flight pattern. Plumage is peach, tan, and gray on back, with peachy buff belly, pale underwings, and white face outlined in deep cinnamon-brown. Female shows deeper coloration and pronounced spotting on chest. Can appear white in flight. **BEHAVIOR** Nocturnal. Cruises low over open areas, listening for prey. Nests in cavities and abandoned human-made structures, primarily barns and silos. **VOCALIZATIONS** Piercing, shrill scream, like something out of a horror movie. Long hissing call is given when alarmed or protecting nest site. **ABUNDANCE** Common and abundant. **CONSERVATION STATUS** Least Concern.

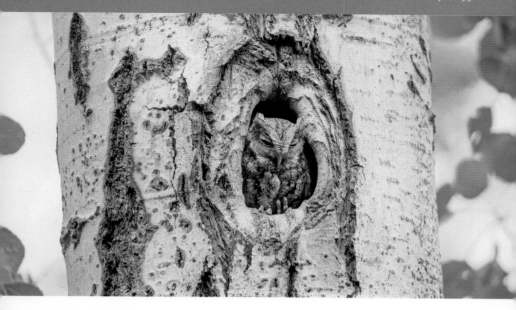

Flammulated Owl

Psiloscops flammeolus

Ponderosa pine and pine-oak woodlands
LENGTH 6–6.75 in. WINGSPAN 16 in.

Tiny owl, smaller than a European Starling and similar to a screech-owl in shape, with cryptic, barklike plumage; short ear tufts often held flat; and completely dark eyes. Brown, gray, white, and varying amounts of buff to rufous with dark streaks down the belly, this owl comes in gray, red, and intermediate morphs. In all morphs, shows a flame-colored stripe from the shoulder into the back (on the side above the wing in flight), the trait that earned the species its name. Fledglings are gray-and-white floofs with barred underparts, similar to screech-owls but with dark eyes. BEHAVIOR Strictly nocturnal and expertly camouflaged, this tiny owl is more often heard than seen. Hunts flying insects from a perch by night and vanishes into cavities or against tree bark by day. VOCALIZATIONS A single, short, and low-pitched *hoot* given repetitively with several seconds between notes. Also a slightly faster but also repetitive *hoodoo-hoo, hoodoo-hoo* and varied screeches and cries. Heard only at night and notoriously difficult to locate by ear. ABUNDANCE Uncommon breeding resident at high elevations in Northwest and Central to Eastern AZ and from Central to Western NM. LOOKALIKES Screech-owls. Flammulated Owl distinguished by smaller size, namesake flame-colored back streak, and dark eyes. CONSERVATION STATUS Least Concern (Decreasing).

Whiskered Screech-Owl

Megascops trichopsis

Montane pine-oak woodlands; frequents sycamore-lined streams
LENGTH 6.5–8 in. WINGSPAN 16–18 in.

Small, stocky owl with ear tufts and mottled gray-brown plumage resembling tree bark, providing remarkable camouflage when perched. Deep golden eyes; whiskers extending from facial area, barely perceptible in the field. BEHAVIOR Nocturnal. Cavity nesters preferring oak and sycamore. Feeds almost exclusively on arthropods such as moths, katydids, centipedes, and scorpions. VOCALIZATIONS Slow, evenly paced *toots*, as well as a series of *toots* that sound like Morse code. ABUNDANCE Resident in high-elevation areas of far Southeast AZ. KEY SITES Chiricahua and Huachuca mountains of Southeast AZ. LOOKALIKE Western Screech-Owl. Whiskered tends to occupy higher elevation habitats and gives a slow, evenly paced call versus the trill of the Western. Though extremely difficult to discern in the field, Whiskered is smaller, with whiskers emerging from facial disk, and has a paler bill surrounded by longer and denser bristles. CONSERVATION STATUS Least Concern.

Western Screech-Owl

Megascops kennicottii

Wooded habitats ranging from low desert to about 6000 feet, including saguaro forests; mesquite, riparian, and pine-oak woodlands; suburban areas
LENGTH 8–9 in. WINGSPAN 22–24 in.

Small and boxy with short tail, frequently raised ear tufts, a dark border around face, dark bill, dark belly streaks, and yellow eyes. Usually light gray overall but may show brownish tones. Fledgling is gray, heavily barred, and fluffy with a dark-bordered face. BEHAVIOR Nocturnal and crepuscular (dusk-active). Hunts from a perch for rodents and large insects, but also other invertebrates, bats, birds, reptiles, amphibians, and even fish. Usually snatches prey from the ground, but occasionally takes animals in flight. Roosts in cavities during the daylight hours, often peering out at passersby. VOCALIZATIONS Most frequently, a bouncing ball-type series of short, whistly, descending, and accelerating *hoots*. Often sung by a pair as a duet, with the female's voice notably higher. Also a series of short barks, *wep-wep-wep-weper-weper*. Other sounds include short whinnies and bill snaps. ABUNDANCE A common year-round resident throughout much of the region. Is mostly absent from Eastern NM. LOOKALIKES Flammulated Owl, Eastern and Whiskered screech-owls. Western Screech-Owl larger than Flammulated, with light eyes; best differentiated from Eastern and Whiskered by voice and range (which overlaps slightly with Whiskered at mid-elevations). CONSERVATION STATUS Least Concern (Decreasing).

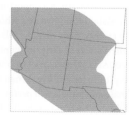

Great Horned Owl

Bubo virginianus

Adapted to a wide variety of habitat types, including low-desert saguaro forests, conifer and deciduous montane forests, parks, urban areas, suburbs
FEMALE LENGTH 20–24.5 in. **WINGSPAN** 48–56 in.
MALE LENGTH 18–23 in. **WINGSPAN** 42–52 in.

Unmistakable, quintessential owl in silhouette, with long, catlike ear tufts and plump, oblong body shape. In flight, head appears squat and wings rounded. Female slightly larger than male; otherwise plumage is similar, mottled gray-brown overall with clean white throat patch, reddish facial disc, and golden eyes. **BEHAVIOR** An opportunist and prolific predator, dining on everything from scorpions to rabbits, even smaller owls. Nocturnal, but often active at early dusk and beyond dawn. **VOCALIZATIONS** Low, deep sequence of *hoot*s, typically four or five. When in breeding pairs, male and female will perform a duet, with the female's *hoot*s at notably higher pitch. Many other vocalizations, from screams to coos and hisses, are made by young and adults at or defending the nest. **ABUNDANCE** Common and widespread resident. **CONSERVATION STATUS** Least Concern.

Northern Pygmy-Owl

Glaucidium gnoma

Relatively open pine and pine-oak woodlands, typically with interspersed patches of chaparral, meadow, or riparian forest
LENGTH 6–7 in. WINGSPAN 12–15 in.

Tiny and long-tailed with a very round head, lacking ear tufts. Adult is gray to brownish overall with light eyes; white eyebrows; white speckling on the head and flanks; larger white spots on the wings; white central breast and belly with crisp, dark streaks; banded tail often held to one side; and, from behind, two white-bordered black "eye" spots at the base of the neck. Fledgling resembles adult but fluffier overall and with more gray. BEHAVIOR Diurnal (day-active), foraging for songbirds with bouncy flight, like a woodpecker. More likely to perch on steeply diagonal branches than are other small owls. VOCALIZATIONS A series of short, steady pitched, and fluty *hoots* given both singly (*hoot-hoot-hoot*) or as pairs (*hoothoot-hoothoot-hoothoot*). Pace varies, but is usually delivered more slowly than that of Northern Saw-whet Owl. ABUNDANCE Uncommon year round across Northern and Eastern AZ and from Central to Western NM. LOOKA-LIKES Ferruginous Pygmy-Owl, Boreal Owl, Northern Saw-whet Owl. Northern Pygmy-Owl distinguished from Ferruginous by lack of rufous bands in tail and range (there is little overlap). Distinguished from Boreal and Northern Saw-whet owls by smaller size, longer tail, and darker face with a less obvious facial disc. CONSERVATION STATUS Least Concern (Decreasing).

Ferruginous Pygmy-Owl

Glaucidium brasilianum

Saguaro forests, mesquite bosques
LENGTH 6–7 in. WINGSPAN 14–16 in.

Neatly dressed tiny owl with a plump body and small, squat head. Ferruginous plumage on head and back, white belly with heavy streaking, and matching bright yellow bill, eyes, and feet. Tail is relatively long and rufous-barred. BEHAVIOR Diurnal (day-active) species but a crepuscular (dawn and dusk) hunter, nesting in cavities vacated by woodpeckers or other vacant spaces in snags and large cacti such as saguaro. Will vocalize throughout the day. VOCALIZATIONS Monotone, repetitive series of evenly spaced *hoots*, with the sound quality of a children's recorder instrument. ABUNDANCE Uncommon resident from South Central to Southeast AZ, population in decline. LOOKALIKE Northern Pygmy-Owl. Ferruginous distinguished by rufous bands in tail and range. CONSERVATION STATUS Least Concern (Decreasing).

Elf Owl

Micrathene whitneyi

Deserts, saguaro forests, pine-oak-juniper woodlands, riparian woodlands, thornscrub
LENGTH 4.5–5.5 in. WINGSPAN 12–13 in.

Bluebird-sized owl—the world's smallest owl—with lovely gray, brown, white, and rufous mottled coloration on body; bright gold eyes; and a black-bordered facial disc with white eyebrows. BEHAVIOR A true night owl, most active when the sun is fully set, awakening to dine on insects. Nests in abandoned cavities in trees and large cacti, but will make use of human-made structures such as nest boxes. This fierce little package will join forces with fellow elves to mob predators. Also known to collect worm-sized threadsnakes, bringing them to their nest sites to dine on parasites. VOCALIZATIONS Male vocalizes at night, with a series of high-pitched but soft notes, ascending to descending; both sexes give soft, whistling *mew* calls. Alarm call is a barklike *chee-rr*. ABUNDANCE Common summer resident from Central to Southern AZ and in Southwest NM. KEY SITE Saguaro National Park, AZ. CONSERVATION STATUS Least Concern (Decreasing).

Burrowing Owl

Athene cunicularia

Sparse and relatively treeless habitats, including grasslands, vacant lots, golf courses, airports, parks; often near irrigated agricultural fields
LENGTH 8–9.5 in. WINGSPAN 21 in.

Small and potato-shaped, with uniquely long legs, short tail, a bold white unibrow and chin, and bright yellow eyes. Adult is sandy to buffy brown with whitish spots above and white with messy brown spots and bars below. Juvenile is long-legged and awkward, with fluffy buff-and-gray plumage. BEHAVIOR Diurnal (day-active), gregarious, and the only owl to roost and nest underground. Does not dig burrows, but instead uses erosive features, holes dug by small mammals, drainage pipes, or artificial burrows built by conservationists. Forages for a wide variety of small prey including invertebrates and small mammals, often from the ground or a very low perch; rarely ventures far from the ground. VOCALIZATIONS Displaying male gives a loud, clear, quail-like, and repeated *hoo-hoowoo-hoo-hoowoo*. When disturbed, produces a shrill, screaming chatter, particularly near active burrows. Also a variety of other calls, including a very convincing rattlesnake-like shriek from juveniles in burrows. ABUNDANCE Uncommon, local, and frequently displaced; mostly a summer resident in the northern two-thirds of the region but sticks around all year farther south. Often overlooked by developers, Burrowing Owls can be displaced or trapped in their burrows as habitat is converted for human uses. CONSERVATION STATUS Least Concern (Decreasing).

Spotted Owl

Strix occidentalis

Mature pine, pine-oak, or mixed conifer woodlands within narrow canyons or rugged, forested mountains
LENGTH 17–19 in. **WINGSPAN** 39–40 in.

A large, chunky owl with dark eyes, short tail, round head lacking ear tufts, and pale, dark-bordered facial disc. Adult is brown and buff overall with white mottling above, large white spots on the chest and belly, and pale crescents between the eyes that form an X, centered above the orange-yellow bill. Fledglings are tan overall with pale gray barring and greenish yellow bills. The Southwestern subspecies, the Mexican Spotted Owl (*S. occidentalis lucida*), is paler overall compared to it relatives farther north. **BEHAVIOR** Nocturnal and dusk-active. Mostly hunts for small mammals by swooping down from a perch on silent wings. Hunting, which usually involves visiting multiple haunts, starts just before sundown and continues until just before dawn. **VOCALIZATIONS** Gives a broad array of vocalizations composed of clear and resonant whistles, hoots, and barks. Most commonly heard vocalizations are a three-part, hooting contact call (*wuh-HOOHOO-hoo*) and a drawn out, descending and then sharply rising cry (*eerrrrr-WIP*). **ABUNDANCE** Uncommon and local at high elevations with disjointed populations from Northwest to Southeast AZ and from Central to Western NM. Mexican Spotted Owl is listed as Threatened by the US Fish and Wildlife Service and is declining in the face of threats including logging and climate change–fueled wildfires. **CONSERVATION STATUS** Near Threatened (Decreasing).

Long-eared Owl

Asio otus

Edge habitat, requiring both open forage and dense forest cover for roosting/nesting
LENGTH 13–16 in. WINGSPAN 36–40 in.

Medium-sized owl with notably long ear tufts, giving it an almost comical look of perpetual surprise. Plumage is finely mottled in various shades of brown, including buff, dusky, and dark browns. Facial disc is a muted pumpkin color, with black-and-orange ear tufts, and distinct white lines running vertically from bill to ear tufts. BEHAVIOR Roosts inconspicuously during the day, typically adjacent to the trunk of a tree in densely vegetated habitat. Communal winter roosts may be abandoned if disturbed by overzealous onlookers. Hunts in late-night darkness by cruising low over open range. VOCALIZATIONS Very vocal species. Strange dog- or monkeylike barking and squealing, as well as low, deep, breathy *hoot*s. ABUNDANCE An uncommon year-round resident across much of the region and a winter visitor to Southwest AZ and Southeast NM. CONSERVATION STATUS Least Concern (Decreasing).

Short-eared Owl

Asio flammeus

Open habitats including grasslands, wetlands, agricultural fields
LENGTH 14–17 in. **WINGSPAN** 34–40 in.

Medium-sized owl with a round head, short tail, and ear tufts so tiny that they often go unseen. Adult is brown to grayish brown, with white and buff spots above and pale with dark brown streaking below. Facial disc is pale, and yellow eyes are surrounded by dark patches. In flight, note heavily patterned wings showing barred wingtips, a dark patch on the underwing just beyond the wrist, and a large, pale patch at the base of the primaries on the upperwing. Fledglings are fluffy and gray, with large black eye patches. **BEHAVIOR** Mostly nocturnal and crepuscular (dawn/dusk active), but also forages during the day. Mostly hunts for mammals with low, harrier-like flight. Frequently seen perching very low or on the ground. **VOCALIZATIONS** Song is a series of short, low, monotone and rapidly repeated *ho-ho-ho-ho* sounds. Also gives raspy barks, screams, and a catlike meow. Wing display produces a sharp rattle. **ABUNDANCE** Rare. Most frequently observed during migration and winter in Southwest and Southeast AZ and in Southwest and Northeast NM. **LOOKALIKE** Northern Harrier. Short-eared Owl differentiated by larger head, shorter tail, and shorter, more rounded wings. **CONSERVATION STATUS** Least Concern (Decreasing).

Boreal Owl

Aegolius funereus

Subalpine spruce-fir forests
LENGTH 8–11 in. WINGSPAN 21–24.5 in.

Small owl with blunt, short tail and blocky, tuftless head. White spots contrast with brown back, and heavy, warm-brown streaks adorn creamy white belly. Dark brown crown is speckled in white, and brown-bordered, grayish facial disk contrasts with bright yellow eyes. BEHAVIOR Spends the day roosting in trees, typically adjacent to the trunk, selecting a new roosting tree daily. Hunts at night by perching in tree branches, awaiting prey to ambush. VOCALIZATIONS Most common call is rapid series of ten to twelve *hoot*s. ABUNDANCE Rare. Resident population at key sites in North Central NM. KEY SITES NM: San Juan, Sangre de Cristo, and Jemez mountains. CONSERVATION STATUS Least Concern.

Northern Saw-whet Owl

Aegolius acadicus

Mature conifer, pine-oak, mixed-conifer forests; frequently but not always near riparian corridors or other wet habitat **LENGTH** 7–8 in. **WINGSPAN** 17–19 in.

Tiny but robust with large, tuftless head; yellow eyes; dark bill; and prominent facial disc. Adult is brown to buffy brown with white spots above and white with thick brown streaks below. Head has thin white streaks, and pale facial disc shows varying amounts of white, gray, brown, and buff, always with a white V shape between the eyes. Fledglings are brown above and cinnamon-colored below and show a bolder version of the white V shape seen in adults. **BEHAVIOR** Strictly nocturnal and more often heard than seen. Mostly hunts for small rodents from a perch at night. During the day, is extremely difficult to find within dense tree cover. **VOCALIZATIONS** Most commonly a flute-like *toot-toot-toot* repeated rapidly with less than a second between notes; heard only at night. This song's likeness to the sound of a saw being sharpened on a whetstone earned this owl its name. Also a shrill, rising scream and varied whistles and barks. **ABUNDANCE** Uncommon. Year-round breeding resident at high elevations from South Central to Northwest NM and from Northern to Southeast AZ. **CONSERVATION STATUS** Least Concern (Decreasing).

TROGONS

Hiking on a humid summer morning in the forested mountains of Southeast Arizona or Southwest New Mexico can make a person wonder if they've wandered off course and landed somewhere south of the international border. Hearing a trogon (family *Trogonidae*) calling from somewhere off in the distance only strengthens those suspicions, but don't be fooled. Despite being predominantly birds of tropical forests, two of these resplendent mountain recluses make their way to the Southwest, and, being entirely unique from all other birds in the United States, they make it clear that this region is like no other in the country.

ABOVE   Female Elegant Trogon

Adult male

Adult female

Elegant Trogon

Trogon elegans

Montane riparian corridors
LENGTH 11–14 in. WINGSPAN 12–15 in.

With colors evoking a Christmas tree ornament, this ornate, colorful, charismatic bird is heavy bodied, with a long tail, short neck, and squat, round head. Male plumage shows a deep, copper-glinted green on the back and head, with lipstick-red belly and a white band across the breast. Primaries are an ashy gray with black tips, and tail is bright white with delicate black barring on underside. Bill is crayon-yellow; eyes are large, round, and black. Female appears similar to male, but with gray head and upper parts, vertical white patch extending from behind the eye, and rosy lower belly. Juvenile also shows white vertical line behind eye and mottled green/gray/copper plumage overall. Wings appear short and round in flight. BEHAVIOR Charismatic birds, they skulk through riparian oak and sycamore canopy feasting upon fruit and awaiting insects. Look for them dining on mistletoe berries when in season. Selects vacated woodpecker cavities as nest sites, primarily Acorn Woodpeckers and Northern Flickers. They perch upright, inconspicuously awaiting prey before ambush. VOCALIZATIONS Distinct barking *koink koink koink* call is most common. Also a rapid series of harsh *kek-kek-kek*s. ABUNDANCE Uncommon but regularly observed throughout the breeding season. KEY SITES AZ: Madera Canyon (Santa Rita Mountains), Cave Creek Canyon (Chiricahua Mountains), other Southeast AZ Sky Islands. CONSERVATION STATUS Least Concern (Decreasing).

Adult male Adult female

Eared Quetzal

Euptilotis neoxenus

Streamside pine-oak or conifer forests within montane canyons
LENGTH 13–14 in. WINGSPAN 24 in.

Medium-sized and pigeon-shaped but with a long tail, often hunched posture, and frequently hidden plumes above the eyes. Adult and juvenile are dark iridescent green above with small, gray bill; dark blue in the uppertail; and a mostly white undertail with a black base often showing bold white spots on a black background. Adult male shows nearly black head, green breast, and bright red belly and undertail coverts. Adult female shows grayish head and breast with less extensive red on belly. Juvenile similar to female but with buffy green head and breast and even less extensive red below. BEHAVIOR Gleans insects and fruit from vegetation, often while hovering. Seemingly lazy, lingering in the same area and spending much of its time perched. A cavity nester, will take advantage of abandoned woodpecker holes. VOCALIZATIONS Song is a series of quickly repeated, subtly trilling *toot*s. Calls are similar to those of blackbirds and include a short, ascending squeal and harsh, abrupt chatter. ABUNDANCE Almost exclusively found in Mexico, an extremely rare visitor to Southeast AZ's Chiricahua, Huachuca, and Santa Rita mountains. Occurs even more rarely outside of this range, with one record each from Central AZ's Superstition Mountains and Southwest NM's Pinos Altos Mountains. LOOKALIKE Elegant Trogon. Eared Quetzal distinguished by larger size, dark bill, plumage, overall shape, and voice. CONSERVATION STATUS Least Concern (Decreasing).

KINGFISHERS

Crowned with spiky crests and built to function as living fishing spears, the kingfishers (family *Alcedinidae*) are an aptly named bunch. Their unique form isn't only suited for fishing, though. They also put their chisel-like bill, compact body, and fused, shovel-like toes to work excavating impressively deep nesting burrows along the water's edge. In the Southwest, look for the widespread Belted Kingfisher on bare branches above water, surveying for fish preplunge, and watch closely for the much less conspicuous Green Kingfisher as it lurks along vegetated shorelines within its limited Southeast Arizona range.

ABOVE Male Green Kingfisher

Adult male

Adult female

Belted Kingfisher

Megaceryle alcyon

Lakes, ponds, wetlands, riparian areas
LENGTH 11–14 in. WINGSPAN 18–23 in.

Stocky, ragged crested, heavy billed bird with short legs and squared, long tail. Male is slate-blue on the back and head, with a clean white belly and collar, and slate-blue breast band. The female wears the pants in this species—or in this case, the belt—with a deep rufous belly band in addition to the blue breast band. Juveniles of both sexes show rusty patches on the belly. BEHAVIOR Perches on snags, rocks, and open branches along wetland or lakeside shorelines, seeking fish. Will also hover above the water surveying for prey. When a potential meal is spotted, the kingfisher performs an impressive plunge, often fully submerging itself in the water. Easily startled from their perches and very vocal. Burrow nesters, typically digging a nest in soft soils on the water's edge. VOCALIZATIONS Jarring, distinctive rattle given in flight and when startled on/from perch. ABUNDANCE Common resident throughout Northern NM and in the White Mountain region of AZ. Common in suitable habitat throughout the region in winter. CONSERVATION STATUS Least Concern.

Adult male

Adult female

Green Kingfisher

Chloroceryle americana

Heavily vegetated areas with relatively clear water, including rivers, creeks, lakes, ponds LENGTH 8–9 in. WINGSPAN 11–13 in.

Small, stocky, large-headed, and short-legged with a slight crest and a thick, pointed bill that is longer than the head. Both sexes are deep forest-green with white speckling above, darkest in the wings, and white below with white collars and outer tail feathers. Male shows a bright rufous breast, while the female shows two ragged green bands. BEHAVIOR Watches from a low perch over water before diving in headfirst for small fish. Pumps tail up and down frequently and often chooses an inconspicuous perch obscured by vegetation. VOCALIZATIONS Dry clicks, delivered either in evenly spaced pairs or in longer, more jumbled phrases. Sound quality often described as reminiscent of two small stones being hit together. Also a harsh, buzzy *jeeer* in flight. ABUNDANCE A regular rarity in extreme Southeast AZ. KEY SITES AZ: Juan Bautista de Anza National Historic Trail from Tubac to Tumacacori-Carmen, Patagonia Lake State Park, San Pedro Riparian National Conservation Area. LOOKALIKE Belted Kingfisher. Green Kingfisher distinguished by green color, smaller size, shorter crest, and proportionally larger bill. CONSERVATION STATUS Least Concern (Decreasing).

WOODPECKERS

With their short tails and parrotlike feet that make perching vertically a breeze, their long tongues that can pull morsels from even the deepest cavities, their sturdy bills perfect for hammering away at bark, and their reinforced skulls that enable them to withstand their high-impact lifestyle, woodpeckers (family *Picidae*) seem to be built merely as avian pickaxes. These birds would likely disagree with that assertion, though, with representatives such as the Gilded Flicker being more inclined to probe into soil and Lewis's Woodpecker preferring to catch insects in midair instead of enduring the jackhammering tactics of their compatriots. With a woodpecker for nearly every habitat, watch for their undulating flight and listen for their laughlike calls from the lowest deserts to the highest forests.

ABOVE Gila Woodpecker

Lewis's Woodpecker

Melanerpes lewis

Breeds in pine, oak, pinyon-juniper woodlands, preferring burned or otherwise open areas with standing dead trees; rather nomadic outside breeding season, in saguaro forests, parks, agricultural areas

LENGTH 9.5–11 in. **WINGSPAN** 19–20.5 in.

Distinct, oily green back and head may appear blackish when cast in shadow, with rosy pink belly, red face, and silvery gray collar/upper breast that looks like it's dusted in frost. **BEHAVIOR** Unlike other woodpecker species, will catch insects on the wing like a flycatcher. Lacks buoyant flight pattern, instead emulating the slow and steady flight pattern of a crow. Cavity nester, borrowing existing cavities from other woodpecker species. **VOCALIZATIONS** Relatively quiet, with a variety of chatters and peeps. During breeding, male will give a repetitive, rolling *churr* note. **ABUNDANCE** Uncommon with year-round populations from the central to northern portions of both states. More widespread but nomadic and equally uncommon in winter. **ALTERNATIVE NAMES** Flycatching Woodpecker, Pink-bellied Woodpecker, Green-backed Woodpecker. **CONSERVATION STATUS** Least Concern (Decreasing).

Adult

Juvenile

Red-headed Woodpecker

Melanerpes erythrocephalus

Sparse woodlands with scattered trees; forest and riparian edges, orchards, parks
LENGTH 7.5–9.25 in. WINGSPAN 16.5–17 in.

Medium-sized and strikingly colored with a round head, bulky body, short tail, and long, spikelike bill. Adding brilliant contrast to its namesake bright red head, adult is white below and black above, with white secondary flight feathers showing as an entirely white lower back when perched. Juvenile has dingy gray-brown head and breast showing varying amounts of red with age, white underparts with an ashy rub, and a back pattern similar to adult but with messy rows of dark spots across the prominent white patch. BEHAVIOR Highly omnivorous, feeding on a broad range of foods including fruit, seeds, insects, and occasionally other birds' eggs. Hunting tactics are equally diverse and include excavating insects from wood, flycatching, and foraging on the ground like a flicker. Maintains food caches, a behavior shared by only a small handful of woodpecker species. VOCALIZATIONS Most commonly a harsh, somewhat buzzy, and often repeated *KWEER*. Also dry rattles, excited yipping, and harsh scolding. Drumming is rolling, soft, slow, and gentle. ABUNDANCE An uncommon summer resident in Northeast NM. Rare elsewhere. CONSERVATION STATUS Least Concern (Decreasing).

Adult male

Adult female

Acorn Woodpecker

Melanerpes formicivorus

Oak woodlands, slopes, canyons, preferring diversity of oak species to ensure successful forage; common in urban areas where oaks are plentiful
LENGTH 7–9 in. **WINGSPAN** 14–17 in.

Often and aptly described as clownlike, with bright, pale eyes set off against a black cheek; creamy white face; and neon-red cap, with black patch at the base of thick, black bill. Back is black, with black streaks on white belly. Female plumage differs just slightly, with red cap covering only the back of the head, leaving a black headband; on male, cap extends forward, meeting white plumage on face. White wing patches and rump are distinct in flight. **BEHAVIOR** Gregarious and vocal, they clamor about in oak woodlands collecting acorns, which they hoard by jamming them into specially made holes in nearby telephone poles and tree trunks. Cooperative breeders and relatively unbothered by human activity. **VOCALIZATIONS** Squeaky, gravelly variations of a two-note *WA-kuh WA-kuh* call (like a dog's squeaky toy). **ABUNDANCE** Common and abundant wherever oaks abound. **CONSERVATION STATUS** Least Concern.

Adult male

Adult female

Gila Woodpecker

Melanerpes uropygialis

Sonoran Desert habitats including saguaro forests, mesquite-lined washes, brushy scrub; common in developed areas with native desert landscaping
LENGTH 9–9.5 in. WINGSPAN 16–16.5 in.

Medium-sized and round-headed with a short tail and heavy, pointed bill. Back, wings, rump, and tail show messy black-and-white bars. Head, neck, nape, and underparts are uniform tan with slight yellow wash in flanks and lower belly. In flight, note obvious white patches near wingtips. Adult male shows restricted red crown atop the head. Juvenile resembles female. BEHAVIOR Forages for insects by pounding into tall, columnar cacti and the trunks or larger branches of trees. Along with Gilded Flicker, is responsible for the nesting cavities commonly seen in saguaro cacti. A frequent backyard visitor and culprit of mysteriously draining hummingbird feeders. VOCALIZATIONS A repeated *henk-henk-henk*, similar in pitch and quality to a dog squeaky toy. Also a high, rolling whinny. Drumming is long, forceful, and evenly paced. ABUNDANCE Common at low elevations from Central to Southern AZ. Less commonly strays farther north into AZ or west into extreme Southwest NM. LOOKALIKE Gilded Flicker. Gila Woodpecker distinguished by smaller size, shorter bill, black-and-white (not black-and-tan) back, lack of golden underwings, white underwing patches, and tan, unpatterned underparts. CONSERVATION STATUS Least Concern.

Adult male

Adult female

Williamson's Sapsucker

Sphyrapicus thyroideus

Breeds primarily in ponderosa pine, also in mixed forests containing Douglas fir and aspen; in winter, in lower elevations and broader range of forested habitats
LENGTH 8.5–9.5 in. **WINGSPAN** 17 in.

The largest of our sapsucker species and highly dimorphic. Adult male is shiny black overall with red throat patch, two white stripes through the face, yellow belly, black-and-white–barred flanks, white rump, and sharply contrasting white wing patches. Adult female has grayish to buffy brown head, black breast, yellow belly, white rump, and black-and-white–barred back, flanks, rump, and tail. Juvenile resembles adult but without the black breast in females and lacking red and yellow coloration in both sexes. **BEHAVIOR** Performs a fluttery flight display near nest cavity. Often uses the same nest tree year after year, but tends to drill a new cavity. **VOCALIZATIONS** A raspy, cracking, hawklike scream; varied trills and laughs; and drumming that starts with a speedy rattle before slowing and quieting into distantly and evenly spaced short bursts. **ABUNDANCE** An uncommon breeding resident at high elevations in Northeast AZ and Northwest NM. More widespread but still uncommon in winter from North Central to Southeast AZ and from North Central to Southwest NM. Can be observed year round where breeding and wintering ranges overlap. **ALTERNATIVE NAME** Black-breasted Sapsucker. **CONSERVATION STATUS** Least Concern.

Adult male

Adult female

Yellow-bellied Sapsucker

Sphyrapicus varius

Mixed hardwood and conifer forests
LENGTH 7–9 in. WINGSPAN 13–16 in.

Both sexes have striking black-and-white faces, bright red foreheads, and white facial stripes curving downward toward chest. Black chest patch meets pale wash of yellow on belly, though the yellow coloration is not always present (and can be a dingy white), with faded black barring on flanks. Vertical white wing patches along side of wing when perched. Black-and-white back, heavily spotted and barred. Fully black-bordered throat patch is white in female and red in male. BEHAVIOR Leaves signature shallow, gridlike forage/drilling pattern on tree bark. Creeps up and down tree trunks, using stiff tail feathers for balance. VOCALIZATIONS Whining, nasally *mew*. Territorial call is a gull-like, raspy squeal. ABUNDANCE Uncommon East Coast species, though regularly observed during migration throughout Eastern NM, where it abuts its traditional wintering grounds in Texas and Mexico. LOOKALIKE Red-naped Sapsucker. Yellow-bellied distinguished by absence of small red patch on nape, full black border around red throat patch, and messier back pattern. CONSERVATION STATUS Least Concern (Decreasing).

Adult male

Adult female

Red-naped Sapsucker

Sphyrapicus nuchalis

Breeds primarily in aspen, also in mixed forests of ponderosa pine and Douglas fir; in winter, inhabits lower elevations in wooded habitats including riparian forests, parks, agricultural windbreaks
LENGTH 7.5–8.5 in. WINGSPAN 16–17 in.

Medium-sized, mostly black and white, with a subtly peaked head and heavy, pointed bill. Adult shows crisp, white wing patches when perched; two rows of white bars down back; yellow-washed, grayish underparts; a red crown, throat, ear patch, and nape; and a black stripe through eye bordered by two white stripes, the lower curving down neck and fading to dingy yellow against black breast. Female distinguished by partially white throat patch and sometimes white nape. Juvenile is brownish version of adult, but with no red feathers. BEHAVIOR Courtship includes pointing bill skyward to display red throat. Often reuses nest cavities. VOCALIZATIONS A loud, squeaky *wyahh-wyahh*; harsh chatter; and repeated, boisterous squawking. Drum is a short burst followed by erratically spaced single and double taps. ABUNDANCE Our most common sapsucker. Found in summer at high elevations in Northeast AZ and Central to Northwest NM, more widely in winter across Western AZ and the southern half of both states, and year round where breeding and wintering ranges overlap. Rare outside of migration in extreme Eastern NM. LOOKALIKE Yellow-bellied Sapsucker. Red-naped distinguished by red nape, back pattern, incomplete (male) or extremely thin (female) black border around throat patch, and range. CONSERVATION STATUS Least Concern (Decreasing).

Adult male

Adult female

American Three-toed Woodpecker

Picoides dorsalis

Burnt, beetle-ridden, or otherwise disturbed stands of mature spruce-fir forest
LENGTH 8.5 in. **WINGSPAN** 15 in.

A small black-and-white woodpecker with a relatively dainty bill. In all plumages, black with varying amounts of messy white barring above and white with black barring below; white throat; white stripe below the eye; thin, white eyebrow; and white speckling on forehead and wings. Male and juvenile of both sexes show small, yellow crown spot. **BEHAVIOR** Forages for mostly insects, sometimes in pairs, by peeling and pecking away loose bark from dead and dying conifers, specializing in trees damaged by beetles and wildfire. **VOCALIZATIONS** A very high *PIK*, a squeaky *kwi-kwi-kwi-kwi*, and short, descending whinnies. Drumming is short, accelerating, and fading, reminiscent of a wooden ball bouncing to a buzzy stop. **ABUNDANCE** Uncommon and inconspicuous at high elevations from Northwest AZ, south through Flagstaff and eastward along the Mogollon Rim to the White Mountains, and then northeastward to the Southern Rockies north of Santa Fe. Also in the Lincoln National Forest, NM. **KEY SITES** AZ: Grand Canyon National Park; NM: Santa Fe National Forest. **LOOKALIKES** Downy, Hairy woodpeckers. American Three-toed distinguished by black-barred underparts, white speckling in less sharply patterned head, and white barring rather than a white patch on the back. **CONSERVATION STATUS** Least Concern.

Adult male

Adult female

Downy Woodpecker

Dryobates pubescens

Wooded areas including deciduous and mixed forests, parks, orchards, urban spaces
LENGTH 5.5–7 in. WINGSPAN 9.5–12 in.

Sparrow-sized, short-billed woodpecker with typical straight-backed posture, uses stiff tail feathers to balance along tree trunks and branches. Clean, white underneath with heavy black-and-white pattern on head and body. Face is black with bold white stripes and puff of white at base of bill; wide white stripe down center of black back. White spots on wings and black spots on outer tail feathers. Male shows small red patch on back of head. Western individuals show lighter white spotting on wings than those in the Eastern United States. BEHAVIOR Active and spritely woodpecker, creeping along even the smallest branches looking for insects. Frequent visitor to backyard suet feeders. Flight pattern is buoyant and bobbing, typical of most woodpecker species. VOCALIZATIONS Shrill, rapid, slightly descending whinny, as well as single-note PIK. Rapid drumming during breeding season; otherwise foraging drum is slow, light, and irregular. ABUNDANCE Resident in high-elevation deciduous forest and open woodland areas across much of NM and in Northeast AZ. LOOKALIKE Hairy Woodpecker. Downy distinguished by smaller size, shorter bill, spotted outer tail feathers, voice, and foraging style. CONSERVATION STATUS Least Concern.

Adult male

Adult female

Ladder-backed Woodpecker

Dryobates scalaris

Arid, brushy habitats from low desert to pinyon-juniper woodlands
LENGTH 6.25–7 in. WINGSPAN 13 in.

The Downy Woodpecker of the desert. Small woodpecker with a blocky head and relatively short, pointed bill. Both sexes mostly black and creamy white with black-and-white–barred back, white underparts with black bars and spots on the flanks, white throat, black stripe down the center of the white nape, and black malar and eye stripes that connect near the back of an otherwise white face. Male shows a red cap; female cap is black. Juvenile shows a partial red cap near the rear of head. BEHAVIOR Hunts for insects through gentle tapping, bark flaking, and gleaning from vegetation. Instead of foraging on thick trunks, takes to small plants and the smaller branches of trees, frequently displaying upside down and nuthatch-like posture. VOCALIZATIONS A sharp *PIK*, sometimes repeated and often followed by a descending, rattling whinny. Drum is an extremely rapid, short buzz. ABUNDANCE Common year round across Central to Western AZ, Central to Eastern NM, and the southern portions of both states. LOOKALIKES Arizona, Downy, Hairy woodpeckers. Ladder-backed distinguished from Arizona by black rather than brown coloration. Does not overlap in range with Downy or Hairy and lacks these species' bold white back patches. CONSERVATION STATUS Least Concern.

Adult male

Adult female

Hairy Woodpecker

Dryobates villosus

Varied forests including conifer, mixed-conifer, pine-oak, deciduous, riparian; also wooded urban landscapes
LENGTH 7–10 in. WINGSPAN 13–16 in.

Medium-sized with a slight crest and a stout, sharp bill that is about as long as the head. Both sexes black-and-white overall, sometimes with a brownish cast. Sports a black back with a bold white central patch, rows of white spots on the wings in varying quantities, grayish white underparts, and white and usually unmarked outer tail feathers. Black cap and nape, black malar stripe, and thick black eyeline are bordered below by a white stripe that continues into nape and above by white eyebrow that continues into crown. Male shows a small red patch on rear of head. Juvenile shows a usually red, but rarely yellow, patch toward the front. BEHAVIOR Forages for insects along the trunks and thicker branches of trees, on fallen logs, or on the ground. VOCALIZATIONS A short, high *PIK*, higher in pitch than Downy Woodpecker and repeated more slowly, and a high, rattling whinny, steadier and more even-pitched than Downy. Extremely fast drumming given in one-second-long, widely spaced bursts. ABUNDANCE Common year round at high elevations across Northern and Eastern AZ and all but extreme Eastern NM. LOOKALIKE Downy Woodpecker. Hairy distinguished by larger size, longer bill, unspotted outer tail feathers, voice, and foraging style. CONSERVATION STATUS Least Concern.

Adult male

Adult female

Arizona Woodpecker

Dryobates arizonae

Montane pine-oak woodlands
LENGTH 7–8 in. WINGSPAN 13–14.5 in.

Small woodpecker, set apart from other species by its brown coloration. Both sexes show solid brown back—a shade of cool brown that closely resembles the bark of an oak—and white belly, heavily spotted and barred with brown. Head is brown with bold white patch from eye to nape, centered by brown cheek patch. Male shows red patch on back of head. **BEHAVIOR** Forages low in the canopy, starting at the base and spiraling upward, searching for insects. Cavity nester, beginning excavation in winter to prepare for breeding season. Often nests in walnut, but can be found in oak, sycamore, or even agave. **VOCALIZATIONS** Relatively quiet in comparison to its woodpecker peers, though both sexes have a *PIK* call and a squeaky, high-pitched *WEE-kah*. Male gives rapid, descending rattle. **ABUNDANCE** Uncommon in the Madrean Sky Island Region of Southeast AZ and Southwest NM. **KEY SITE** Feeders at Santa Rita Lodge in Madera Canyon, just south of Tucson, AZ. **CONSERVATION STATUS** Least Concern (Decreasing).

Adult male (red-shafted)

Adult female (red-shafted)

Northern Flicker

Colaptes auratus

Open mid-elevation and montane woodlands, parks, urban and suburban areas, edge habitat
LENGTH 11–12.5 in. WINGSPAN 17–20 in.

Sizeable but sleek woodpecker with lovely tawny plumage on back, delicately spotted and barred with black; buffy belly with black spots that appear as if painted on; black crescent at neckline. White rump patch is distinct in flight, only sometimes visible when perched or foraging. Rosy red underwing plumage flashes in flight (in Eastern United States, this shows yellow). Male shows lipstick-red mustache stripe (black in Eastern United States). BEHAVIOR Often observed foraging on ground at the base of trees; known to use their long, probing tongues to feed on subterranean ants. Flight pattern is an exaggerated, buoyant bobbing, much like other woodpeckers. VOCALIZATIONS Song is a piercing, extended *kik-kik-kik-kik*. Also a loud, single-note *keer* when flushed or in flight; an excited *wikah-wikah-wikah*; and a short, turkeylike whinny. Drumming is loud, evenly spaced, and rapid. ABUNDANCE Common year round resident at middle and upper elevations and winter resident regionwide. LOOKALIKE Gilded Flicker. Northern distinguished by red underwings versus gold, slightly larger size, and gray versus tan nape. CONSERVATION STATUS Least Concern (Decreasing).

Adult male approaching nest cavity

Adult female

Gilded Flicker

Colaptes chrysoides

Saguaro forests throughout the Sonoran Desert; less commonly in developed areas with plentiful saguaro
LENGTH 10–12 in. WINGSPAN 18–20 in.

The largest and longest billed of the woodpeckers with which it commonly shares habitat. Adult is tan with black bars above and whitish tan with black spots below, and has a warm camel-colored crown and nape; a gray face, neck, and throat; and a large black spot on the breast. Male shows bright red malar. In flight, note obvious white rump and golden underwings and undertail. BEHAVIOR Forages mostly on the ground, especially for large seed-harvesting ants that are common in Sonoran Desert uplands. Excavates large nesting cavities in saguaro cacti. VOCALIZATIONS Song is a loud *kik-kik-kik-kik*. Also a high, clear *keer*; an excited *wikah-wikah-wikah*; and a short, turkeylike whinny. Drumming is short and buzzy. Voice is higher and thinner than Northern Flicker. ABUNDANCE Common year round from Central to Southwest AZ, as far north as Kingman along the western border and diminishing along the boundary of the Chihuahuan Desert southeast of Tucson. LOOKALIKES Northern Flicker, Gila Woodpecker. Gilded distinguished from Northern by tan nape, golden underwings and undertail, and slightly squeakier voice; ranges overlap slightly, most extensively in winter. Gilded distinguished from Gila Woodpecker by larger size, longer bill, and black-and-tan (not black-and-white) back. CONSERVATION STATUS Least Concern (Decreasing).

FALCONS

It may come as a surprise to learn that falcons (family *Falconidae*) are more closely related to parrots and woodpeckers than to their fellow raptors, considering their hooked bills, sharp talons, and taste for flesh. Maybe even more surprising is that despite being powerful predators that make even hawks seem calm and approachable, falcons have proven to be among the birds most affected by human influences on the landscape. If you're lucky enough to spot one of these high-energy predators in the Southwest, be sure to thank those who took quick action to bring the Peregrine Falcon back from the brink of extinction through Endangered Species Act protections and who hold onto hope for the Aplomado Falcon, a critically endangered bird that, while once abundant in New Mexico, is now so uncommon that it wasn't included in this section.

ABOVE American Kestrel

Adult

Juvenile

Adult

Crested Caracara

Caracara plancus

Open deserts, grasslands, scrub
LENGTH 20–23 in. WINGSPAN 48–49 in.

Large, striking falcon, hawklike in appearance, with silky black plumage above and below and white face, throat, and neck bordered by black barring around the collar. Crest is black, contrasting with a bright, vulture-like, deep pink face fading to a heavy, marigold-colored bill with a sharp, pale blue, hooked tip. Legs are deep yellow. Juvenile plumage is gray-brown overall, with dingy white on neck, pinkish face, and gray legs. White wingtips visible in flight. BEHAVIOR Look for these sentinels perched conspicuously on high, open branches and posts, looking over open land for their next meal. Caracaras hold their wings flat in flight and often cruise low to the ground like a Northern Harrier. Can be seen in mixed flocks, often with vultures, and will feed on carrion as well as live prey. They are opportunistic omnivores, often spotted foraging along the ground. Pairs create roughly constructed nests in the canopy, or even in large cactus species, using collected materials, and will build atop the previous year's nest. VOCALIZATIONS Typically silent, but will sometimes give a rattle when alarmed or during breeding season that has a mechanical, insectlike sound quality. ABUNDANCE Uncommon but regularly observed resident of Central Southern AZ. CONSERVATION STATUS Least Concern.

Adult male

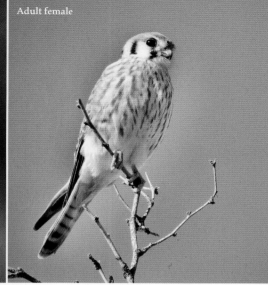
Adult female

American Kestrel

Falco sparverius

Open areas with plentiful perches from which to hunt and cavities in which to nest, including deserts, grasslands, open forests, agricultural fields, parks, suburban landscapes **FEMALE LENGTH** 10–12 in. **WINGSPAN** 21–24 in. **MALE LENGTH** 9–11 in. **WINGSPAN** 20–22 in.

North America's smallest falcon, dove-sized with a small, round head; short, hooked bill; long tail; and long, pointed wings. Both sexes are rufous with black bars or spots above and pale below, with a pale face, small black nape patch, dark wingtips, black bar in front of and behind eye, gray cap with sometimes rufous center, and rufous tail. Male is brighter overall, with slate-blue wings, black spots on buffy belly, and black tail tip. Female is paler, showing lighter belly with faint rufous streaks or bars, wings that match back in color, lighter back pattern, banded uppertail, and less distinct, dark tail tip. **BEHAVIOR** Searches from a perch or hovers in stationary flight for varied small prey including invertebrates, reptiles, mammals, and birds. Usually takes prey from the ground, but will occasionally take birds or bats in flight. Habitually pumps tail when perched. **VOCALIZATIONS** A sharp *kee-kee-kee*, usually with fewer than six repeated notes. **ABUNDANCE** Common year round across both states. **LOOKALIKE** Merlin. American Kestrel is smaller and paler overall, sports a rufous tail, and has a more contrasting facial pattern. **CONSERVATION STATUS** Least Concern.

Merlin

Falco columbarius

Open forests, grasslands; adapted to urban and suburban environments
LENGTH 9–12 in. WINGSPAN 21–26 in.

Small falcon with a stocky body and head. Heavy brown streaking on pale underside, and fairly solid head and back. Pale eyebrow on both sexes, though female is typically creamy white, with paler brown plumage overall versus gray/slate plumage on most males. Legs and bill are yellow, and bill ends in a sharp, hooked, black tip. Wings are sharply pointed and are heavily and uniformly patterned underneath. Tail feathers are banded. Southwest most commonly sees Taiga and Prairie forms. The dark Pacific Northwest (black) form is extremely rare. BEHAVIOR Look for speedy wingbeats and forceful flight—not fluttery or gliding like a kestrel. A fierce and effective predator, primarily targeting birds on the wing. VOCALIZATIONS High-pitched, chattering, whinnying scream, lasting several seconds. ABUNDANCE Uncommon but widespread; overwinters throughout AZ and NM. LOOKALIKES American Kestrel, Prairie Falcon. Merlin distinguished by lack of pronounced mustache stripe, relatively solid gray-brown back, and heavily streaked underparts. CONSERVATION STATUS Least Concern.

Adult

Immature

Peregrine Falcon

Falco peregrinus

Open habitat including deserts, sparse forests, urban and suburban areas; hunts near open water and rests on high ledges, natural or human-made
FEMALE LENGTH 18–21 in. **WINGSPAN** 39–43 in.
MALE LENGTH 15–18 in. **WINGSPAN** 35–39 in.

The region's largest regularly occurring falcon. Similar in size to a Harris's Hawk, with a squared head, long tail, short and hooked bill, and long, pointed wings. With longer wings and shorter tails than other falcons, wingtips reach tail tip when perched. Adult is slate-gray above and white with dense black barring below, with yellow bill, yellow feet, and a black cap and vertical eye stripe that give a helmeted look. Juvenile is brownish overall, with coarse brown streaking below and a less extensive but still dark eye stripe. **BEHAVIOR** The fastest bird in flight, scans from the air or a perch before taking out avian prey during dives that can exceed 180 miles per hour. **VOCALIZATIONS** A fast, raspy *rek-rek-rek* or a whining, elongated, and ascending *reehk-reehk-reehk*. **ABUNDANCE** Uncommon and local. Breeds across much of AZ and less frequently in Northern and Southwest NM. Mostly a passing migrant throughout the remainder of NM and a migrant or winter visitor in Southwest AZ. **LOOKALIKE** Juvenile Prairie Falcon. Peregrine distinguished by darker coloration, more contrasting facial pattern, and lack of dark armpits in flight. **CONSERVATION STATUS** Least Concern.

Adult

Juvenile

Prairie Falcon

Falco mexicanus

Deserts, grasslands, prairies, meadows, agricultural fields; frequently near cliffs or rocky outcrops
LENGTH 15–19 in. **WINGSPAN** 36–44 in.

Large falcon with brown plumage on back, heavily patterned, with deep brown and creamy white underwing and diagnostic dark armpit patches. Belly is creamy white with fine brown streaking that shows heavier on juvenile. Brown head with white cheek and eyebrow and pronounced brown mustache stripe. Bill and legs are yellow, paler in juveniles. Pointed wings and long tail in flight. **BEHAVIOR** As aggressive as they are beautiful, these fierce falcons show little mercy when defending their territories from competitors. They prey primarily on small mammals, stealthily surveying open prairie from just above the grassline. Look for their nests on steep bluffs and cliff ledges. **VOCALIZATIONS** Highly variable, gull-like, hoarse but squeaky *kik-kik-kik* call. **ABUNDANCE** Uncommon but widespread resident. **CONSERVATION STATUS** Least Concern.

PARROTS

The Southwest has a complicated history with parrots. The Thick-billed Parrot, a New World holotropical parrot (family *Psittacidae*), was once the only representative of this order in the region. However, excessive hunting and the degradation of its preferred montane forest habitat led to its extirpation in the mid-1900s, and because of that, the species is not included in this book. Taking its place as the order's sole naturalized representative is an Old World parrot (family *Psittaculidae*), the Rosy-faced Lovebird. Unfit for survival in natural settings, these parrots make a living in and around Phoenix, where the palm-rich and heavily irrigated landscapes in ways resemble their native African oases. Though it is generally assumed that their limited, suburban range precludes any negative impact they could have on native species, the future of this exotic is uncertain in this climatically changing and increasingly modified region.

ABOVE Adult Rosy-faced Lovebird

Rosy-faced Lovebird

Agapornis roseicollis

Areas with palms and saguaro cacti in parks, golf courses, human-made water features; native to woodlands of southwestern Africa LENGTH 6–7 in. WINGSPAN 9–10 in.

Small, round-headed parrot with pale, grassy green plumage overall; a pop of electric-blue on the rump; and a peachy coral face with pale bill. Eyes are dark with thin, pale eye rings. BEHAVIOR Gregarious, as parrots are, gathering in small, cacophonous flocks and displaying endearing courtship behaviors. Their name is inspired by their coupled sleep position, where they perch side by side and tuck their faces toward each other, beak to beak, forming the shape of a heart. Diets consist primarily of seeds and berries. Cavity nesters, often engaging in communal nesting. VOCALIZATIONS High-pitched squeaks and squeals. ABUNDANCE Locally common. Small introduced/naturalized population in Phoenix area has grown to roughly 2000 birds. KEY SITES Kiwanis Park in Tempe, Gilbert Water Ranch. CONSERVATION STATUS Least Concern (Introduced).

TYRANT FLYCATCHERS AND BECARD

Wherever there are flying insects zipping about, there will be tyrant flycatchers (family *Tyranniadae*). A highly speciated family, they occupy nearly all habitat types and have diverse life histories, but their shared quest for insects has led to birds that are alert and highly energetic, with wide, pointed bills; stiff tails; and exceptionally agile flight. Some, most notoriously those in the *Empidonax* genus, can be infuriatingly similar and nearly impossible to identify to species in the field, while others, such as the Vermilion Flycatcher, are unmistakable. Once mislabeled as a tyrant flycatcher, the Rose-throated Becard, a bird that enters the United States only in extreme Southeast Arizona and Southern Texas, is now placed with the Tityras (family *Tityridae*).

ABOVE Dusky-capped Flycatcher

Adult male

Adult female

Rose-throated Becard

Pachyramphus aglaiae

Healthy, mature riparian forests with plentiful cottonwoods and sycamores
LENGTH 7–7.5 in. **WINGSPAN** 11–12 in.

Medium sized and resembling a flycatcher, with a long tail, rounded wings, and a dark, heavy, vireo-like bill. Male is gray overall, darker above, with a black cap and a pale reddish pink central throat patch. Female is lighter overall with rufous-brown wings and tail, pale gray cap and back, and buffy white face, nape, throat, and underparts. **BEHAVIOR** Gleans insects and fruit from vegetation, often associating with other species doing the same. Typically more than a foot long, oblong, and dangling in the riparian canopy, the giant, globular nest of this bird is often the most obvious indicator of its presence. **VOCALIZATIONS** A high, whistled, rising and then descending *weet-soo* and a squeaky rattle often followed by a descending whistle. **ABUNDANCE** More abundant in Mexico, a rare summer visitor to Southeast AZ. **KEY SITES** AZ: Santa Cruz River from Tubac to Tumacacori-Carmen, Sonoita Creek State Natural Area through the town of Patagonia. **CONSERVATION STATUS** Least Concern.

Northern Beardless-Tyrannulet

Camptostoma imberbe

Riparian woodlands, mesquite bosques
LENGTH 4–5.5 in. WINGSPAN 7–8 in.

Tiny, slender flycatcher is gray overall with a slight hint of dull olive-green, darker gray wings, and two dingy white/buffy wingbars. Head is gray, with a subtle crest that becomes more pronounced when the bird is perturbed. Short, blunt-tipped bill is pinkish orange, with black on upper mandible. Juvenile shows rufous-colored wingbars; otherwise similar to adult. BEHAVIOR Look for the Tyrannulet's stereotype-busting behavior—hopping about low- and midcanopy like a warbler, rather than flying out from its perch. Frequently flicks tail as it forages for insects. VOCALIZATIONS Sweet, clear whistle, *pee pee pee*, sometimes up to nine *pee* notes in one song. Also gives a slurred trill call, as well as a slightly more melancholy, single *pee* note. ABUNDANCE Uncommon, short-distance migrant; breeds in limited riparian habitat in Southeast AZ. KEY SITES AZ: Paton Center for Hummingbirds, Sonoita Creek Preserve in Patagonia. CONSERVATION STATUS Least Concern (Decreasing).

Dusky-capped Flycatcher

Myiarchus tuberculifer

Riparian and pine-oak forests sheltered in montane canyons
LENGTH 6.25–7.25 in. **WINGSPAN** 10–11 in.

Slender bodied, long and sharp billed, long tailed, and with a proportionately large and slightly crested head, the Dusky-capped Flycatcher is a brighter, miniaturized version of the Ash-throated Flycatcher. Grayish brown above with a gray throat and breast, pastel yellow belly and undertail coverts, rufous-edged primary and secondary flight feathers, two very faint wingbars, and sometimes obvious rufous tail edges. From below, very limited rufous color in tail extends to tip. **BEHAVIOR** Inconspicuous; forages like a phoebe, scanning intently from a perch before swooping out to snap up insects in flight, often returning to the same or a very nearby perch. Nests in cavities excavated by woodpeckers. **VOCALIZATIONS** With varied trills, yelps, and mournful whistles, voice is similar to other *Myiarchus* flycatchers in composition, but in quality is more like a pewee. Dawn call is a rolling and repeated *whip-weeer-wedeer*. Common call is a one-second long, repeated, and descending whistle. **ABUNDANCE** In summer, common but local in Southwest NM and from Southeast to North Central AZ. Inhabits forests typically with denser tree cover and at higher elevations than Ash-throated Flycatcher, though there is overlap. **LOOKALIKE** Ash-throated Flycatcher. Dusky-capped is smaller, more boldly colored, and has a daintier bill, less rufous in the undertail, and no dark tail tip. All flycatchers in this genus best distinguished by voice. **CONSERVATION STATUS** Least Concern (Decreasing).

Ash-throated Flycatcher

Myiarchus cinerascens

Scrub, mixed woodlands, desert riparian
LENGTH 7–8.5 in. WINGSPAN 11.5–13 in.

Large, peak-headed flycatcher with long, slender body and tail. Dusky gray-brown overall, with pale lemon-yellow belly fading to pale gray-white breast and throat. Wings are accented by cinnamon-colored touches on the primaries, matched by cinnamon tones down the center of the underside of gray-brown bordered and tipped tail. Distinct white wingbars. Dark, slender, straight bill. BEHAVIOR Very vocal, cavity-nesting (secondary, repurposing existing cavities) arid forest and desert dweller, feeding primarily on arthropods, which also provide moisture in an otherwise severely dry and often hot habitat. Most active in the early morning, before the heat of the day sets in. VOCALIZATIONS *Ka-brit*, *ka-breer*, and *whit* calls, as well as soft, whirring, trilly dawn song by male. ABUNDANCE Common breeder throughout region; resident in Southwest AZ. LOOKALIKE Brown-crested Flycatcher. Ash-throated best distinguished by call (*ka-breer*, which is similar to, but softer, less resonate, and less forceful than Brown-crested call), but also distinguished by smaller bill, more subtle transition from yellow to pale gray on belly, and less rufous coloration on tail. CONSERVATION STATUS Least Concern.

Brown-crested Flycatcher

Myiarchus tyrannulus

Riparian woodlands, saguaro forests, oaky slopes, heavily forested urban areas
LENGTH 8–9 in. WINGSPAN 12–13 in.

Large, peak-headed flycatcher with a thick, heavy bill. Plumage is gray-brown overall with hints of olive-green, a pale gray-white throat and chest, and a lemon-yellow belly. Embellishments include primary feathers edged in rufous, two pale wingbars, and rufous highlights on tail. BEHAVIOR Secondary cavity nesters, relying heavily on woodpecker excavations. Uses an array of nest materials from its fellow desert fauna, including fur, skin shed from lizards and snakes, and feathers, to line the nest. Perches conspicuously atop snags and high in the canopy, flitting out to snap up prey and returning to its perch. Often heard before seen. VOCALIZATIONS A broad lexicon of calls including a trilled whistle, a dry *whit* or *whip-purr*, a far-carrying and Cassin's Kingbird-like *brrEEER*, and a single note, clear, loud *wip*. Morning song is a clear, resounding *WHIT* followed by a descending trilly *wee-doo*. ABUNDANCE Common breeding resident from Western AZ to Southwest NM. LOOKALIKE Ash-throated Flycatcher. Brown-crested is easiest to distinguish by vocalization, but also by its big ol' bill, bulkier size (though this is tough to determine unless they are side by side), clean demarcation of yellow to gray on chest, and more rufous in tail. CONSERVATION STATUS Least Concern.

Sulphur-bellied Flycatcher

Myiodynastes luteiventris

Sycamore-dominated riparian woodlands sheltered in montane canyons in pine and pine-oak forests
LENGTH 8.5 in. **WINGSPAN** 14–15 in.

Bulky and distinctly striped flycatcher with a large and rounded head, bulky body, relatively short tail, and long, heavy bill. Both sexes are a mix of blackish to olive-brown with streaky white highlights above, yellow and white with brown streaking below, two bold white stripes across the face, and vividly rufous tail. **BEHAVIOR** Forages from a perch in the high canopy, darting out to catch insects in flight and rarely venturing into lower canopy levels. Makes small cup nests in cavities excavated by woodpeckers. **VOCALIZATIONS** Most commonly, very high-pitched and extremely squeaky squeals: *peee-weeer, WIT-weeer, weeEEER*. Also varied chips and trills. **ABUNDANCE** An uncommon summer resident in Southeast AZ. Sometimes, though rarely, strays farther east into Southwest NM or farther north into East Central AZ. **CONSERVATION STATUS** Least Concern.

Tropical Kingbird

Tyrannus melancholicus

Open, edge habitat; prefers riparian-adjacent areas
LENGTH 8–9 in. WINGSPAN 14–15 in.

Large, substantial flycatcher, with thick, heavy bill; broad wings with pointed tips; and a clearly notched tail. Head is true gray with charcoal-gray ear patch, fading to green-gray on the back, with deep gray primaries and tail feathers. Tail feathers are completely brown underneath. Belly is bright, saturated yellow, reaching all the way up to a white throat. BEHAVIOR In true kingbird style, perches on utility lines, fences, and poles awaiting the next insect meal; flits out to snap it up and returns to its perch to dine. Diet consists primarily of insects but is supplemented by berries and other fruits. Female builds nest, a shallow bowl constructed of twigs and other plant fibers, lined with soft materials such as moss. Fearlessly and aggressively protects the nest site from intruders. VOCALIZATIONS Singsong, chittering trills. ABUNDANCE Ubiquitous throughout Central and South America, but uncommon in this northernmost part of its range. Breeds in extreme Southeast AZ. KEY SITES Southern stretches of the Santa Cruz River, Southeast AZ; sometimes appearing at Sweetwater Wetlands in Tucson, as well as southern stretches of the San Pedro River. LOOKALIKE Couch's Kingbird. Tropical is nearly identical and best distinguished by vocalization: they like to talk, and their trilling calls are distinctly different from the clear whistles and *peer*s of Couch's. Couch's also has a shorter bill, and Tropical has distinct notch in tail. CONSERVATION STATUS Least Concern.

Cassin's Kingbird

Tyrannus vociferans

Sparse habitats with scattered trees, including pine-oak woodlands, oak-dominated grasslands, pinyon-juniper and pine forests, parks, agricultural areas
LENGTH 9 in. **WINGSPAN** 16 in.

A large flycatcher with a round head, slender body, stout bill, and long tail. Both sexes have a dark gray head, neck, and breast; a sharply contrasting white malar and chin; bright yellow belly; olive-gray to brown wings and back; and a dark tail with sometimes pale tip. **BEHAVIOR** Perches high and in the open, watching for insects to pluck, mostly from the sky but also from the ground or vegetation. **VOCALIZATIONS** Most commonly, a loud, raspy, and far-carrying *brrEEER* or *whi-KEER*. Also a more complex song heard at dawn, *whi-ki-ki kurrUP*, and a rapid, squeaky *kideer-kideer-kideer*. **ABUNDANCE** A common breeding resident across all but Southwest AZ and in all of NM, except for in the mountains north of Santa Fe and along the state's eastern edge. Lingers year round in low numbers in and south of Tucson. **LOOKALIKES** Western, Tropical kingbirds. Cassin's distinguished from Western by darker head, brighter belly, sharper border to white malar and throat, and lack of white tail edges. Also ranges into higher elevations than does Western. Cassin's blackish, unforked tail differentiates it from Tropical. **ALTERNATIVE NAME** Charcoal-headed Kingbird. **CONSERVATION STATUS** Least Concern.

Thick-billed Kingbird

Tyrannus crassirostris

Riparian woodlands with oak, sycamores, cottonwoods
LENGTH 8.5–9.5 in. WINGSPAN 14–16 in.

Named for its most prominent feature, the notably thick bill on this otherwise typical tyrant flycatcher easily sets it apart from similar species. Plumage is deep gray-brown with an olive-green tinge on head and back and lemon-yellow wash on belly, fading to dingy white on the chest and throat. Long tail is dark brown/black and just slightly notched. Lacks wingbars. BEHAVIOR In kingbird fashion, perches conspicuously on wire, open branches, and poles, and flies out to snatch flying insects before returning to its perch to snack. Little is known of breeding and nesting behaviors, diets, or population trends, as this species has not been heavily studied/documented. VOCALIZATIONS Tinny, loud, sputtering, and squeaky *ba-REET* and *pitterEER*. ABUNDANCE Locally uncommon. Small breeding populations occur annually within limited range in Southeast AZ and the Bootheel of NM. KEY SITES In AZ, fairly reliable during breeding season at California Gulch in the Atascosa Highlands (difficult to reach without a high-clearance vehicle) or the more accessible Sonoita Creek State Natural Area near Patagonia. CONSERVATION STATUS Least Concern.

Western Kingbird

Tyrannus verticalis

Open habitats with few trees, including deserts, grasslands, agricultural areas, sparse woodlands, parks
LENGTH 8–9 in. WINGSPAN 15–16 in.

Large, round-headed flycatcher with a long tail, slim physique, and stout bill. Both sexes show light gray head, neck, and breast; faded white malar and chin; dusty yellow belly; greenish gray back; dark wings; and a dark tail with reliably white outer edges. BEHAVIOR Hunts for flying insects from conspicuous low to high perches, or takes them from the ground during quick dives from low, hovering flight. Frequently observed chasing larger birds, even raptors, out of its territory with fighter jet–like acrobatics. VOCALIZATIONS Dawn song is a high, stuttering series of chips that escalates into a squeaky outburst: *ki-ki-kidik-ki-SQUEE-SQUEEOO*. Other sounds include high squeaks, chips, and rattles. ABUNDANCE A common breeding resident across both states. LOOKALIKES Cassin's, Tropical kingbirds. Western distinguished from Cassin's by lighter head, paler belly, more diffuse white in malar and throat, and white tail edging; Western found at lower elevations. Western distinguished from Tropical by duller yellow coloration usually ending lower on the breast, blackish undertail, and unforked tail with white edges. CONSERVATION STATUS Least Concern (Decreasing).

Eastern Kingbird

Tyrannus tyrannus

Meadows, fields, open edge habitat near wetlands/water
LENGTH 7.5–9.5 in. **WINGSPAN** 13.5–15 in.

Neatly dressed denizen of prairie and meadow, with deep black on head fading to deep charcoal-gray overall, accented by a neatly white-tipped, squared tail; clean white belly; and pale gray wash on white breast and throat. Subtle, delicate white piping along edges of coverts and primaries. Bill is relatively short; tail is long. Medium-bodied. Small strip of red, orange, or yellow running vertically along crown (the feature for which kingbirds are named) is seldom visible in the field but a treat to see on rare occasions. **BEHAVIOR** Perches low in wait of its next ambush, in and along edges of meadows and wetland shorelines. Dines primarily on a wide variety of insects, but diet shifts to berries in its South American wintering range. Male performs a delightful show of acrobatics in courtship. **VOCALIZATIONS** Rapid, frantic, metallic, high-pitched *d-d-d-d-dzeet dzeets*; also single note, buzzy *zeet* and *kit-tee* calls. **ABUNDANCE** Uncommon to rare breeding resident in Northeast NM. **KEY SITES** NM: regularly but not necessarily reliably observed at Valle de Oro and Maxwell national wildlife refuges. **CONSERVATION STATUS** Least Concern (Decreasing).

Adult male

Adult female

Scissor-tailed Flycatcher

Tyrannus forficatus

Primarily sparse grasslands of the Southern Great Plains; other open habitats such as agricultural areas, other grassland types, parks **FEMALE LENGTH** 10 in. **WINGSPAN** 15 in. **MALE LENGTH** 10–15 in. **WINGSPAN** 15 in.

Large flycatcher with a round head, slender body, stout bill, and an outlandishly long and deeply forked tail. Adult has a grayish white head, back, and breast; blackish wings; and a salmon-washed belly and underwing coverts. Juvenile is similar but with paler wings and a pale yellow wash below. Black-and-white tail is nearly the length of the body in juvenile, longer than the body in adult female, and even longer in adult male. **BEHAVIOR** Takes insects in flight from a plainly visible perch, usually staying relatively close to the ground. Like its kingbird relatives, actively chases intruders out of its territory on agile, acrobatic wings. **VOCALIZATIONS** Similar to Western Kingbird but slightly lower in pitch. Dawn song is a stuttering and repeated *pip-pidip-pip-pida-pida-purEEP*. Other sounds include sharp squeaks, raspy rattles, and a humming, in-flight wing display. **ABUNDANCE** Uncommon summer resident in the southeastern corner of NM and along the state's eastern edge. Rare farther west. **CONSERVATION STATUS** Least Concern (Decreasing).

Olive-sided Flycatcher

Contopus cooperi

Breeds in open montane coniferous forest, burned out areas, edge habitat near conifer forest; during migration, a variety of habitats and elevations, including urban
LENGTH 7–8 in. WINGSPAN 12–14 in.

Stout-chested, stocky flycatcher—the largest of the pewees. Ashy gray-brown overall. Belly is white down the center, often with a pale yellow tinge, and namesake olive-green sides, though greenish tinge is barely perceptible unless awash in the right light. It appears as if it were donning a little gray vest. Throat is pale; bill is bulky relative to overall size. BEHAVIOR Preferring a high vantage point, this bird posts upright in the tallest branches of both live and dead trees. Will fiercely defend its territory against intruders. Long-distance migrant, venturing roughly 7000 miles per year to and from its Central and South American wintering grounds. VOCALIZATIONS A triumphantly whistled command of "quick, THREE beers!" ABUNDANCE Uncommon but regularly observed in breeding and migration range. Notable population decline is thought to result from wintering habitat loss. CONSERVATION STATUS Near Threatened (Decreasing).

Greater Pewee

Contopus pertinax

Relatively open montane pine and pine-oak forests; also riparian woodlands that line canyons within these habitats
LENGTH 7 in. **WINGSPAN** 13 in.

Medium-sized flycatcher, large for a pewee, with a crested head, long tail, and a stout, pointed bill that is black above and bright orange below. Adult is dark overall, brownish gray above, paler gray below with a yellow wash, and with two very faint wingbars. Juvenile has slightly more distinct, buffy wingbars. **BEHAVIOR** Hunts for flying insects with short flights, usually from a very high perch atop a pine. Most easily located by ear. **VOCALIZATIONS** Song is a high, slow, whistled, and repeated *hosay-aREE* or *hosay-hosay-hosay-aREEaa*, often remembered by the mnemonic "José Maria!" Also a classically pewee and frequently repeated *WHIT*. **ABUNDANCE** Uncommon breeding resident at high elevations in Southwest NM, Southeast AZ, and along the Colorado Plateau transition zone from Flagstaff to the Gila National Forest. **LOOKALIKES** Olive-sided Flycatcher, Western Wood-Pewee. Compared to Olive-sided Flycatcher, Greater Pewee wears a less distinct vest and has a bright orange lower mandible. Greater Peewee is larger than Western Wood-Pewee, with a significantly larger bill and less distinct wingbars. **CONSERVATION STATUS** Least Concern (Decreasing).

Western Wood-Pewee

Contopus sordidulus

Open coniferous woodlands, riparian woodlands, edge habitat, open understory
LENGTH 5.5–6.5 in. WINGSPAN 10–11 in.

Medium-sized, peak-headed, and long-winged dusky gray-brown flycatcher. Appears as if wearing a neat, gray vest or cape buttoned only at the top, revealing a small wedge of pale white and/or yellow on the belly. Head is gray-brown and void of facial markings. Two pale white wingbars. BEHAVIOR Perches openly and upright, scanning side to side for flying insects to ambush. Not a tail-pumper like many other flycatchers. Female builds an intricate, inconspicuous nest upon tree branches, securing plant fibers with spiderweb and finishing with moss, lichen, and other materials. VOCALIZATIONS Sings a buzzy little serenade for which it is named, *pee-wee* (and slight variations on that theme), as well as a rattly *zeew*. ABUNDANCE Common breeding-season resident. LOOKALIKES Olive-sided, Willow flycatchers. Western Wood-Pewee distinguished by long primaries, subtle wingbars, and lack of eye ring, as well as vocalization. CONSERVATION STATUS Least Concern (Decreasing).

Willow Flycatcher

Empidonax traillii

Low-lying, brushy, humid, insect-rich areas along rivers and creeks, within floodplains, near seeps or springs, often with dense willow or tamarisk
LENGTH 5.5–6.5 in. WINGSPAN 7.5–9.5 in.

Small, dainty flycatcher with a slightly peaked head, relatively short wingtip projections, and a long, broad tail. Short, wide, and pointed bill is black above and pinkish orange below. Both sexes are grayish to olive-brown above and grayish white, sometimes with a hint of yellow, below. Shows two faint brown wingbars; a dingy white throat; a smudged, grayish breast; and a very thin, barely there eye ring. BEHAVIOR Calls frequently from exposed perches within dense vegetation. Perpetually on the move when foraging for insects to catch on the wing, flitting between plants rather than favoring a single perch. VOCALIZATIONS Song is a short, buzzy *FITZ* or *fitz-BEW*. Most common call is a dry *WHIT*. ABUNDANCE An uncommon migrant across both states and an uncommon breeding resident from Central to Southwest NM, from North Central to Southern AZ, and less commonly along the western edge of the state. The Southwestern subspecies, *E. traillii extimus*, is listed as Endangered under the Endangered Species Act. LOOKALIKES Other *Empidonax* flycatchers. Willow Flycatcher is extremely similar but best differentiated by voice, habitat, short wingtips, minimal or absent eye ring, and drab brownish coloration. CONSERVATION STATUS Least Concern (Decreasing).

Least Flycatcher

Empidonax minimus

Deciduous and mixed woodlands and forests, open woodlands, edge habitat
LENGTH 4.5–5.5 in. WINGSPAN 7–8 in.

Small, olive-gray empid with bold, pale wingbars and eye ring; subtle wash of dusky gray on breast; and very faint wash of pale yellow on belly. Look for fairly short primaries and a rounded head, often lacking the peak-headed shape of other *Empidonax* species. BEHAVIOR The common theme among flycatchers, perches conspicuously awaiting the next unsuspecting insect. A feisty floof, it will fiercely defend its territory from unwelcome visitors. VOCALIZATIONS Easily distinguished from other empids by its telltale *chi-BECK chi-BECK* call. ABUNDANCE Uncommon; regularly observed throughout migration along eastern boundary of NM. LOOKALIKE Willow Flycatcher. Least distinguished by smaller size and bold eye ring, versus faint eye ring of Willow. CONSERVATION STATUS Least Concern (Decreasing).

Hammond's Flycatcher

Empidonax hammondii

During breeding and winter, almost exclusively in old-growth forests, frequently among ponderosa pine, Douglas fir, and aspen; during migration, nearly any habitat type including low deserts, high forests, urban parks LENGTH 5–6 in. WINGSPAN 8.75 in.

Miniscule flycatcher with a tiny and nearly all-dark bill, slight crest, and long wingtip projections that make their notched tail appear relatively short. Greenish gray above and gray with a splash of faded yellow and a dingy vest below, gray head, two dingy white wingbars, and a white eye ring that is pointed behind the eye. BEHAVIOR Watches from a perch for insects to pluck from the sky or glean from vegetation. Frequently flicks tail and wings simultaneously in kinglet-like fashion. VOCALIZATIONS Song a three-part *chitip-churet-chitter*; notes sometimes disjointed or out of order. Notes always buzzy and distinctly two-syllabled. Other calls include a short, dry *whit* and, during breeding, squeaky trills. ABUNDANCE A common migrant across both states, an uncommon breeding resident in the Southern Rockies north of Santa Fe, and an even less common winter visitor to Southeast AZ and Southwest NM. LOOKALIKES Other *Empidonax* flycatchers. Though maddeningly similar, Hammond's best distinguished by very long primaries, seemingly short and notched tail, wing and tail flicking behavior, habitat, and voice. ALTERNATIVE NAMES Western Old-Growth Flycatcher, Long-winged Dusky Flycatcher. CONSERVATION STATUS Least Concern.

Gray Flycatcher

Empidonax wrightii

Open woodlands, oak and juniper woodlands, chaparral, mountain foothills; winters in streamside mesquite bosques
LENGTH 5–6 in. WINGSPAN 8–9 in.

Small, short-winged, relatively nondescript empid. Dull gray overall, with thin, nearly imperceptible eye ring; pale wingbars; and long, black-tipped bill that is pale underneath. Breast is a subtle dusky gray. BEHAVIOR A gift among empids, this species gives away its identity through its telltale—or "telltail"—tail dipping, pumping down and then up, sparing us the torture of painstaking scrutiny over wing length and bill size. Tends to perch low in the canopy when foraging for flying insects. VOCALIZATIONS Sweet, insectlike *chee-beet* and soft *whit* calls. ABUNDANCE Common throughout northern portion of breeding range and southern portion of winter range, with decrease in occurrence as ranges reach Central NM and AZ. Primarily observed during migration through the beltline of the range. LOOKALIKES Hammond's, Dusky flycatchers. Gray distinguished by bill color and diagnostic tail-dipping versus the upward-flicking of Dusky or the kinglet-like tail and wing flicking of Hammond's. Gray's primaries are shorter than Hammond's, making its tail appear longer. CONSERVATION STATUS Least Concern.

Dusky Flycatcher

Empidonax oberholseri

Breeds in high-elevation brushy and disturbed areas, including chaparral, riparian woodlands, pine and pine-oak forests; winters along riparian corridors
LENGTH 5–6 in. WINGSPAN 8.5 in.

Tiny flycatcher with a small, mostly all-dark bill; slight crest; and short wingtip projections that make the usually unnotched tail appear long. Greenish gray above, gray with a splash of faded yellow and a dingy vest below, and with a gray head, two whitish wingbars, and a white eye ring that is pointed behind eye. BEHAVIOR Patrols from a perch before catching insects on the wing or gleaning them from vegetation. Occasionally flicks tail and much less regularly flicks wings, rarely simultaneously. Favors brushy and disturbed areas versus the old-growth forest used by Hammond's Flycatcher. VOCALIZATIONS Song a whistled, three-part *chitip-churet-chiter*; notes sometimes disjointed or out of order. Notes can be buzzy or clear and are often indistinctly two-syllabled when compared to Hammond's. Other calls include a short, dry *whit* and, during breeding, a whistled *pee-wip*. ABUNDANCE A common migrant across both states, an uncommon breeding resident in Northeast AZ and Northwest NM, and an even less common winter visitor to extreme Southeast AZ and Southwest NM. LOOKALIKES Other *Empidonax* flycatchers. Infuriating in its similarity, Dusky best distinguished by short primaries, long and usually unnotched tail, wing and tail flicking behavior, habitat, and voice. CONSERVATION STATUS Least Concern.

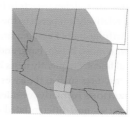

Western Flycatcher

Empidonax difficilis

Coniferous and mixed-coniferous montane forests, often near creeks or within canyons
LENGTH 5.5–7 in. WINGSPAN 7.5–9 in.

Compact flycatcher with a pointy, flattened, and bicolored bill; a slightly crested head; long wingtips; and a long tail. Both sexes are greenish gray-brown on the head and back with sunny yellow underparts, a grayish olive-green breast, deep olive-gray wings and tail, bold wingbars, and bold, white eye rings that are widest and pointed toward the rear. In heavily worn plumage, yellow tones are diminished and wingbars are more subtle. BEHAVIOR Catches insects in flight, usually from a relatively low perch in the middle canopy. Artfully crafted nest is a small cup detailed with moss, lichen, bark, and other fine natural fibers. Frequently nests in human-made structures such as the eaves of cabins. VOCALIZATIONS Formerly considered two separate species, now differentiated into two regional populations of a single species that exhibit similar but distinct vocalizations. Bird formerly known as Pacific Slope Flycatcher sings a clear, ascending *ts-SEET-tsick-tseet* and gives a *pee-DEET* call. Bird formerly known as Cordilleran Flycatcher sings a high, sweet, and three-part *persip-pseep-PSEEP*, with the second note lowest and the final note highest in pitch, and delivers calls including high chips and a high, whistled *surWEEP*, as if the bird were hailing a cab. ABUNDANCE A common breeding resident at high elevations across much of the region, avoiding the lowest elevations of Central to Western AZ and extreme Southern and Eastern NM. More widespread but less common in migration, when it is absent only from Eastern NM. LOOKALIKES Other *Empidonax* flycatchers. Western distinguished by prominent eye ring and overall yellowish coloration. CONSERVATION STATUS Least Concern.

Buff-breasted Flycatcher

Empidonax fulvifrons

Montane pine-oak canyons with sparse understory
LENGTH 5–6 in. WINGSPAN 7–8 in.

Sweet little peach of a flycatcher. Warm, buffy brown coloration with an apricot-colored wash on the breast. Bill is notably short, with pale eye ring. BEHAVIOR In flycatcher style, hunts from conspicuous perches, but will also forage along the ground and in trees. VOCALIZATIONS Quick, chirpy *chee-bit chee-boo*. Males greet the morning in song from open perches. ABUNDANCE Uncommon/rare breeding-season resident in limited range of extreme Southeast AZ and extreme Southwest NM. KEY SITES Populations limited to canyons in the Huachuca, Chiricahua, and Animas ranges of the Madrean Sky Islands. CONSERVATION STATUS Least Concern (Decreasing).

Black Phoebe

Sayornis nigricans

Varied, relatively open habitats near or along water, including natural and artificial waters such as lakes, ponds, rivers, creeks, wetlands, stock tanks, irrigation ditches
LENGTH 6–7 in. WINGSPAN 10–11 in.

Small, rotund flycatcher with a large, peaked head; a short, pointed bill; rounded wings; and a long tail. Adult's entirely black head, breast, back, wings, and tail contrast sharply with its white belly and undertail coverts. Juvenile shows slight rufous tones in wings. BEHAVIOR Flits out over water to catch passing insects in flight, tending to favor a single or small number of perches to which it repeatedly returns. Habitually pumps tail when perched. Builds a cup-shaped mud nest. VOCALIZATIONS Song is a high, clear, and repeated *pi-KEE pi-KOO*. Call is a thin, high, repeated chip. ABUNDANCE Common but local year-round resident from Northwest to Southeast AZ and along the Colorado River. Less common year round in Southern NM, ranging farther north along the Rio Grande and Pecos rivers. Expands northward in spring and summer but remains mostly absent from Northwest AZ, Northern NM, and along eastern border of NM. LOOKALIKE Eastern Phoebe. Black Phoebe is significantly darker and has a mostly distinct range, overlapping regularly only in Southeast NM. CONSERVATION STATUS Least Concern.

Eastern Phoebe

Sayornis phoebe

Woodlands and scrub, parks, farms, edge habitat, usually near water
LENGTH 6–7 in. WINGSPAN 10–11.5 in.

Stout-headed and rotund flycatcher with ashy gray-brown plumage above; pale gray, smudgy breast; and pale, dingy white belly, showing pale yellow when fresh-feathered. Lacks distinct wingbars, but shows clean white edges on wing feathers. Juvenile has subtle rufous wingbars and slightly paler head than the dark head of adults. Dark gray/black bill is petite, narrow, and sharply pointed. BEHAVIOR Perches low in tree branches and shrubs and along fence lines. Pumps tail up and down while perched. Constructs distinct nests of mud and grass, sited under eaves, bridges, and other human-made structures. VOCALIZATIONS Sings a buzzy tune for which it's named: FEE-*bee*, fee-BEE. Also high-pitched, clear chips. ABUNDANCE Uncommon/rare breeding resident in extreme Northeast NM; wintering range extends to extreme Southeast NM. LOOKALIKE Willow Flycatcher. Eastern Phoebe distinguished by vocalization, dark bill, and lack of wingbars. CONSERVATION STATUS Least Concern.

 Adult

 Fledgling

Say's Phoebe

Sayornis saya

Arid open country, deserts, grasslands, chaparral, sagebrush, agricultural areas, open suburban settings including parks, playing fields; often close to nesting sites, including rock outcrops, cliffs, canyon walls, buildings, bridges
LENGTH 7.5 in. WINGSPAN 13 in.

Medium-sized flycatcher, somewhat long-bodied for a phoebe with a sharp bill, long tail, rounded wings, and a proportionately large, round head that can appear peaked or flattened. Adult is gray above, darkest in the head and with two pale wingbars. Shows a pale gray throat, darker gray breast, bright ochre belly and undertail coverts, and black tail. Juvenile similar, with ochre wingbars. BEHAVIOR Swoops out and returns to low, open perch as it preys on flying insects. In winter or when prey is scarce, will frequently snatch insects from spiderwebs. Habitually pumps tail when perched. VOCALIZATIONS Song is a repeated *pideew-pideew-pideew-piderEEP*, starting plaintive and whistling and ending emphatic and trilling. Call is a repeated *pideew*. ABUNDANCE A common year-round resident in Northwest AZ and from the central to southern portions of both states. A common summer resident throughout the rest of the region. LOOKALIKE Female Vermilion Flycatcher. Say's Phoebe distinguished by larger size, heavier bill, more elongated shape, lack of streaking on grayer breast, and more extensive but paler wash of color below. ALTERNATIVE NAMES Aridland Phoebe, Peach-bellied Phoebe. CONSERVATION STATUS Least Concern.

Vermilion Flycatcher

Pyrocephalus rubinus

Open desert and scrub, parks, riparian woodlands
LENGTH 5–5.5 in. WINGSPAN 9–9.5 in.

Petite flycatcher with a stocky chest, slim tail, and broad, sturdy bill. Male shows brilliant vermilion coloration overall, accented by brown/black wings, back, and tail, and a black eye mask. Female is subtle but no less striking, with glowing apricot-colored belly, pale chest with taupe streaking, and gray-brown back, tail, crown, and eye mask. Juvenile female shows pale peachy coloration on belly; male shows patchy red coloration, otherwise appearing similar to adult females. BEHAVIOR Members of this species seem to know how handsome they are and enjoy showing off by perching out in the open, utilizing both low and high perches and flying out to snatch up insects, often with showy aerial acrobatics. VOCALIZATIONS Male exclusively sings a cheery, trilly *d-d-d-dik-dzee!* Both sexes give clear, squeaky *peep* calls.

ABUNDANCE Varies greatly throughout range; common resident throughout much of Southern AZ, expanding northward and eastward during breeding into North Central AZ and Central to Southern NM. CONSERVATION STATUS Least Concern.

SHRIKES AND VIREOS

Shrikes (family *Laniidae*) are songbirds, not raptors, but that means little to the varied animals that find themselves in the grasps of their heavy and sharply hooked bills. Their smaller and enthusiastically vocal relatives, the vireos (family *Vireonidae*), are similar in form and function but forgo shrikes' conspicuous perches and diverse diet in favor of dense vegetation and large insects. After watching a shrike add a horned lizard to its pantry of the impaled or observing a vireo beat a beetle to a pulp against a branch, you'll be certain of one thing: you are very fortunate to be larger than a songbird.

ABOVE Bell's Vireo

Loggerhead Shrike

Lanius ludovicianus

Open landscapes with scattered perches; deserts, grasslands, sagebrush, agricultural fields, sparse forests, suburban areas
LENGTH 8–9 in. WINGSPAN 11–12.5 in.

Chunk of a songbird with a large, squat head; a thick, hooked bill; and a long, narrow tail. Both sexes are gray above and grayish white below with a thick, black mask; white throat; black bill; black wings sporting white patches visible from above and below; and a black tail with white corners. Juvenile shows subtle brownish barring below. BEHAVIOR Hunts from a low, conspicuous perch, scanning for insects, rodents, reptiles, and songbirds. Known as butcherbirds to many, shrikes have a signature move—they impale a collection of prey on spiny shrubs, barbed wire fences, or other available spiky surfaces to mark territory or, especially when nesting, as a food cache. VOCALIZATIONS Song is a series of raspy buzzes, trills, and blackbirdlike gurgles. Call is a repeated, harsh *rehnk-rehnk-rehnk*. ABUNDANCE Uncommon and declining. Found year round across nearly the entire region, but during summer only in the Southern Rockies north of Santa Fe. LOOKALIKE Northern Shrike. Loggerhead smaller in size, with a thicker black mask without a white border above and unscaled underparts. Juvenile Loggerhead shows white wing patch, is barred rather than scaled below, has an all dark bill, and is grayish overall. CONSERVATION STATUS Near Threatened (Decreasing).

Northern Shrike

Lanius borealis

Open boreal forest/edge habitat, grasslands/meadows
LENGTH 9–9.5 in. WINGSPAN 11.5–14 in.

Relatively large, long-tailed, round-headed songbird, with stout neck and severely hooked, short bill. Silver-gray plumage on head and back, with a bold black mask, wings, and tail. White patches on wings and outer edges of tail flash in flight. Chest and belly show subtle silvery barring on pale white background. Immature shows buffy versus gray coloration and lacks white wing patches. BEHAVIOR Seemingly sweet at first glance, this little masked marauder is a slayer of small mammals, insects, and even birds. Favors impaling prey on fence wires, snags, or thorns as a means of storing its cache. Perches in the open on telephone wires, fence lines, rooftops, and treetops. VOCALIZATIONS Vocalizes sparingly in winter. Calls are mainly a harsh and scolding *rack* or *reek*. Soft warbling song given during breeding season. ABUNDANCE Uncommon to rare; most frequently observed during winter in Northern AZ and NM. LOOKALIKES Loggerhead Shrike, Northern Mockingbird. Northern Shrike distinguished by barring on chest; thinner mask versus the Loggerhead, whose mask extends above the eye to the supercilium and base of bill. Distinguished from the mockingbird by stout bill and black mask and wing feathers. CONSERVATION STATUS Least Concern.

Bell's Vireo

Vireo bellii

Most often brushy habitat, particularly within dense willow and mesquite woodlands along rivers, creeks, and washes; other dense habitat such as scrubby oak woodlands, dense tamarisk, thickets of trees within otherwise open habitat, especially during migration
LENGTH 4.5–5 in. WINGSPAN 7–7.5 in.

Pocket-sized and slender-bodied songbird with a long tail and small but somewhat heavy, hooked bill. Both sexes notably unadorned, gray above and lighter grayish white below, with a very pale eyebrow; diffuse, broken eye ring; and faint dark line through the eye. Brighter individuals may show slightly sharper facial pattern, subtle pale wingbar, and faint yellowish wash. BEHAVIOR More often heard than seen, vocalizes frequently from dense vegetation while foraging for insects and occasionally fruit. VOCALIZATIONS Extremely vocal. Song is a thin, fast, raspy, repeated, and varied *zeedazeedazee-zeedazeedazoo*. Also gives harsh, wrenlike scolding calls. ABUNDANCE A common breeding resident from Central to Southeast AZ and in Southern NM within the Pecos and Rio Grande watersheds. Less commonly encountered in extreme Western AZ and across the southern portions of both states. LOOKALIKES Other vireos, Gray Vireo. Bell's is smaller, plainer, and has a more delicate bill than other vireos. It is most similar to Gray Vireo, but distinguished by broken rather than complete eye ring. ALTERNATIVE NAME Scrub Vireo. CONSERVATION STATUS Least Concern.

Gray Vireo

Vireo vicinior

Desert mesquite and oak scrub, chaparral, pinyon-juniper woodlands
LENGTH 5–6 in. WINGSPAN 7.5–8.5 in.

Its name says it all. Pale gray overall but darker on wings and tail, with white belly and a relatively faint white eye ring, particularly when compared to other vireo species. Very subtle, single pale gray wingbar. BEHAVIOR Deftly navigates through thick, thorny vegetation foraging for insects and fruit. Is known to forage on the ground, performing towhee-like scratch-kicks to unearth tasty insects. Look for habitual tail flicking. VOCALIZATIONS Singsong, four- to six-part stanzas, delivered more quickly than Plumbeous Vireo: *che-reet...che-root...che-reet...che-row*. ABUNDANCE Uncommon but regularly observed short-distance migrant, breeding at mid-elevations across the northeastern half of AZ and much of the northwestern half of NM; also extreme Southeast NM, and wintering in Southwest AZ. LOOKALIKES Plumbeous, Bell's vireos. Gray distinguished by faint eye ring versus bold spectacles of the Plumbeous; complete versus incomplete eye ring of the Bell's. CONSERVATION STATUS Least Concern.

Hutton's Vireo

Vireo huttoni

Mostly coniferous and mixed forests, especially with a heavy oak component; occasionally observed in riparian forests, chaparral **LENGTH** 4–5 in. **WINGSPAN** 7.5–8 in.

Tiny, kinglet-like vireo with a heavy, hooked, and grayish bill; a large, rounded head; no neck; a portly little body; short wings; and medium-length tail. Both sexes are olive-gray overall, darkest in the wings and palest below, with two pale wingbars, bluish gray legs and feet, pale lores, and an an incomplete eye ring that is most prominent behind the eye. **BEHAVIOR** Gleans insects from within dense vegetation, often alongside mixed groups of other small insect-eating songbirds. **VOCALIZATIONS** Song is a harsh, rapidly repeated, and redundant *zeeoo-zeeoo-zeeoo*, *zooee-zooee-zooee*, or *zooeer, zooeer, zoooer*. Also gives raspy, wrenlike chips and scolds. **ABUNDANCE** Common in all seasons in Southwest NM and from Central to Southeast AZ. **LOOKALIKE** Ruby-crowned Kinglet. Hutton's Vireo distinguished by less wing flicking, heavier bill, paler lores, dark feet, and no black bar bordering the lower edge of lower wingbar. **ALTERNATIVE NAME** Oak Vireo. **CONSERVATION STATUS** Least Concern.

Cassin's Vireo

Vireo cassinii

Woodlands; breeds in mixed conifer forest, but is indiscriminate during migration and overwinter, in all types of canopy, from seaside to mountaintop
LENGTH 4.5–5.5 in. WINGSPAN 8.5–9.5 in.

Plump, round-headed, lovely little vireo with olive-yellow coloration overall and thick-rimmed, white spectacles. Bold yellow-white wingbars and pale belly with yellow flanks. Bill is heavy and dark. BEHAVIOR Slowly creeps through branches, foraging in low to mid-canopy. Shows preference toward true bugs—stink bugs and leafhoppers—as well as caterpillars, but will also engage in flycatching behavior from time to time. Crafts intricate nests of various plant fibers, moss, and lichen. VOCALIZATIONS Coarse, singsong stanzas with long pauses in between. Song is nearly indistinguishable from that of Plumbeous Vireo, though first notes of Plumbeous's song are clean and clear. In AZ and NM, you're likely to hear their hoarse, rapid, chattery call versus their song, which resembles those of other vireo species. ABUNDANCE Uncommon in migration throughout the region; even less commonly observed in limited AZ wintering range. LOOKALIKES Plumbeous, Bell's, Hutton's vireos. Cassin's distinguished by olive-yellow coloration versus Plumbeous's gray, and bold spectacles, whereas Hutton's lacks eye ring altogether and Bell's is incomplete. ALTERNATIVE NAMES Olive Vireo, Spectacled Olive Vireo. CONSERVATION STATUS Least Concern.

Plumbeous Vireo

Vireo plumbeus

Breeds in dry, open forests, most frequently with ponderosa pine and deciduous understory; more varied during winter and migration, in open forests, riparian corridors, suburban areas
LENGTH 5.5 in. **WINGSPAN** 10 in.

Small but hefty vireo with a large, rounded head; a heavy, hooked bill; and a medium-length tail. Lead-gray above and pale grayish white below, with two thin wingbars and a bold, white eye ring and lore stripe that together look like tiny glasses. Bright individuals may show a very slight yellowish wash in the flanks. **BEHAVIOR** Gleans insects from foliage as it moves slowly through shrubs and the middle canopy, frequently alongside other insectivorous birds. **VOCALIZATIONS** Song is a string of whistled, sweet, but somewhat nasal phrases, *chiriup-cherio-cheow*, similar to but slower than Gray Vireo song, with up to two seconds between phrases. Also harsh, wrenlike scolds. **ABUNDANCE** A common breeding resident at high elevations across the majority of the region, except Western AZ and Southern and Eastern NM, where it is a common migrant, and from Tucson west into South Central AZ, where it is an uncommon winter visitor. **LOOKALIKES** Cassin's, Blue-headed vireos. Plumbeous is similar but much duller, distinguished by little yellow in flanks, no yellow on wings or upper parts, no contrast between back and head. Very little range overlap. **CONSERVATION STATUS** Least Concern (Decreasing).

Warbling Vireo

Vireo gilvus

Deciduous and mixed-coniferous forests with tall trees; also aspen groves, sycamore-dominated riparian areas
LENGTH 5–5.5 in. WINGSPAN 8.5 in.

Small, drab vireo with a rounded head; a thick, hooked bill; and a medium-length tail. Both sexes are dingy yellowish gray above, grayest in the crown, and white below, sometimes with faintly yellow-washed flanks. Dark eyes, a white throat, pale lores, and pale eyebrows. BEHAVIOR Forages slowly and intentionally, visually inspecting each leaf for insect prey as it hops through the middle to high canopy. Inconspicuous and most often first noticed by vocalizations. VOCALIZATIONS Song consists of two short musical phrases, slow, sweet, lilting, and warblerlike, separated by a one-second pause and with the second phrase usually ending on a high note. Also gives harsh, scolding calls. ABUNDANCE A common breeding resident in middle to high elevations from Northern to Southeast AZ and from Northern to Southwest NM. A common migrant outside of this range. LOOKALIKES Red-eyed Vireo, other warblers. Warbling distinguished by smaller size, duller overall color, and lack of sharp contrast between pale eyebrow and crown. Differentiated from similar warblers by much heavier bill. CONSERVATION STATUS Least Concern.

Red-eyed Vireo

Vireo olivaceus

Deciduous woodlands
LENGTH 4.5–5 in. WINGSPAN 9–9.5 in.

Understated but exceedingly handsome, sleek-headed, and large-bodied vireo with distinct, bold, off-white supercilium; charcoal-gray eyeline; and sooty gray crown. Back is a lovely olive-green and belly is clean off-white with yellow-green wash on lower flanks. Red eye is visible only when the light catches it just right (immature shows dark eyes). Lacks wingbars, but wing feathers are finely bordered in gray. BEHAVIOR Inconspicuously forages in dense deciduous canopy, moving slowly and intentionally in search of caterpillars. VOCALIZATIONS Call is a screechy, raspy whine, much like a Gray Catbird. Song (not likely to be heard in NM) is a slurred and cheery singsong of repeated questions and responses, with ascending and descending stanzas. ABUNDANCE Uncommon. Migration corridor follows eastern boundary of NM. CONSERVATION STATUS Least Concern.

CORVIDS

Although the term "bird-brained" is usually hurled as an insult, the astoundingly intelligent jays, magpies, ravens, crows, and nutcrackers—the Southwest's representatives of the corvids (family *Corvidae*)—take it as a compliment. Gregarious, vocal, and rarely bashful, these large, charismatic, and conspicuous passerines are often the gateway bird for soon-to-be birders, offering opportunities to observe endearing antics and intricate behaviors that are more difficult to detect or absent in other families. Their exhibitionist and nonmigratory nature tend to make corvids key to a place's character, from the Mexican Jays and Pinyon Jays that define the Southwestern woodlands they inhabit, to the Blue Jays of New Mexico that can transport a homesick East Coast birder back to their old haunts in an instant.

ABOVE Common Raven

Canada Jay

Perisoreus canadensis

Boreal and subalpine conifer and mixed forests
LENGTH 10–13 in. WINGSPAN 16–18 in.

Round-headed, long-bodied, and short-billed corvid. Individuals in the Rocky Mountain region show pale gray face with charcoal-gray nape, sooty gray back, and pale gray belly. Juvenile appears darker overall, with pale mustache, much like a milk mustache. BEHAVIOR Gregarious and bold, known to eat from the hands of generous hikers. Don't let their sweet appearance fool you; these opportunistic hunter-gatherers will do just about anything to nab their next meal, including dining on baby birds and bats, attacking injured animals, and joining scavengers for a carrion feast. Flight is soft and almost silent but for a gentle whoosh. VOCALIZATIONS Departing from the raucous squawks of many of its jay relatives, gives soft but clear whistling, warbly *WEE-ah* call. Skilled mimic, giving the calls of various predators (owls, hawks, crows, Blue Jays) as a warning call. ABUNDANCE Uncommon. Isolated resident populations in subalpine zones in AZ and NM. KEY SITES AZ: White Mountains; NM: montane forests near Santa Fe. LOOKALIKE Clark's Nutcracker. Canada Jay distinguished by round head, short bill, and lack of distinct black wing feathers. CONSERVATION STATUS Least Concern (Decreasing).

Pinyon Jay

Gymnorhinus cyanocephalus

Emblematic of western pinyon-juniper woodlands; also ponderosa pine, chaparral, sagebrush, other open, scrubby habitats
LENGTH 10.5–11.5 in. WINGSPAN 18–19 in.

Medium-sized and round-headed with a relatively short tail and a long, straight, pointed, and nutcracker-like bill. Adults are pastel sky-blue with pale-gray undertones, grayest in the belly and bluest in the head, showing a pale gray-white, blue-streaked throat patch. Juvenile more gray overall. BEHAVIOR Travels in large, boisterous flocks in search of favored pinyon pine nuts. Collected nuts are stored in the ground in large granaries, helping them survive the seasons between crops and seeding the next generation of pinyon pine. VOCALIZATIONS Varied, crowlike caws, a harsh *rehnk-rehnk-rehnk*. Stiff, audible wingbeats add to the laughlike cacophony produced by groups in flight. ABUNDANCE Uncommon year round from North Central to Northern and Northwest AZ and from East Central to Northern and Northeast NM. Highly nomadic; roams across the region in winter, avoiding far Southeast NM and far Southwest AZ. Becoming less common as pinyon-juniper forests are lost to or degraded by climate change, drought, and poor land management. LOOKALIKES Mexican and other jays. Pinyon shows less gray on underparts than Mexican and has a mostly distinct range. Pinyon is stockier, shorter tailed, longer billed, and less patterned than other jays. CONSERVATION STATUS Vulnerable (Decreasing).

Steller's Jay

Cyanocitta stelleri

Montane conifer forest
LENGTH 12–13.5 in. WINGSPAN 17–17.5 in.

Large, lean corvid is jewel-toned from head to toe with an onyx head and crest, fading to shades of sapphire and deep turquoise. Interior form (those most often observed in this range, one of a dizzying sixteen subspecies) shows distinct white streaks on the front of the crest and above the eyes. Juveniles lack white markings and are slightly washed out; otherwise similar to adults. BEHAVIOR Charismatic as any corvid, bold and brash, curious and calamitous. These gregarious jays tend to linger high in conifer canopy but make their presence known with their incessant vocalizations. Not shy or skittish, they will fly low to investigate areas of human activity—picnic areas and the like—and will scour the ground for snacks. VOCALIZATIONS During courtship and breeding, males and females give a variety of rattles, clear whistles, piping calls, and pops. Call is a high-pitched, gravelly, screechy *sheck sheck sheck sheck*. ABUNDANCE Common resident in high-elevation areas of range. ALTERNATIVE NAMES Black-crested Jay, Conifer Jay, Sapphire Jay. CONSERVATION STATUS Least Concern.

Blue Jay

Cyanocitta cristata

Near edges of relatively open deciduous and mixed woodlands, often with oaks, and avoiding areas without a broadleaf component; pine-oak woodlands, forest edges, riparian areas, suburban landscapes, parks
LENGTH 10–11 in. **WINGSPAN** 14–17 in.

Large, crested, long-tailed, and painted in gray, white, black, and an impossibly broad palette of blues, this unmistakable jay is a backyard favorite east of the Rocky Mountains. Both sexes are blue above and grayish white below, with a pointed but relatively short black bill, a black necklace that borders a white face and connects to black eyelines and lores, a white or pale blue throat, white flashes in the folded wings, iridescent-blue panels in the wings and tail, and white tail corners and secondary flight feather tips, visible in flight. Juvenile is a slightly drab version of adult.
BEHAVIOR Boisterous, aggressive, territorial, and charismatic; forages in noisy flocks and readily chases other birds, even raptors, out of its territory. Omnivorous and has an extremely broad diet, but specializes in large nuts such as acorns or, when visiting feeders, peanuts. **VOCALIZATIONS** Extremely vocal, producing a dizzying variety of cries, squawks, rattles, gurgles, whistles, barks, squeaks, and shrieks. A mimic, it can produce an incredibly convincing Red-tailed Hawk cry, among other impressions. **ABUNDANCE** An uncommon but increasing year-round resident along Eastern edge of NM, observed as far as Albuquerque in winter. Increasingly rare farther west. **CONSERVATION STATUS** Least Concern.

Woodhouse's Scrub-Jay

Aphelocoma woodhouseii

Chaparral, pinyon-juniper forests
LENGTH 11–12 in. WINGSPAN 14.5–15.5 in.

Large, lean, and long jay with deep sky-blue plumage above, gray upper back/scapulars, pale gray belly, inconspicuous blue breast band/necklace, and white throat. Juvenile shows more gray on head. BEHAVIOR Curious, observant, and generally gregarious, but not as consistently as other corvids; often observed alone. An opportunistic omnivore, it forages along the ground and through canopy for acorns, insects, spiders, snails, and berries, and will sometimes snag a rodent, baby bird, or reptile. Nests low in shrubs and trees and in pairs, unlike the cooperative nesting behavior of the Mexican Jay. Flight pattern is a relatively sluggish flap and glide. VOCALIZATIONS Extremely vocal, with a diverse lexicon of songs and calls, from sweet courtship melodies to squawky alarm calls. Soft courtship song is given by both sexes, lasting several minutes. Call in flight is often a high-pitched, ascending *wreep? wreep?*, in addition to a variety of alarm and scolding calls, *sheks*, and bill claps. ABUNDANCE Common resident, with populations concentrated in arid, mid-elevation, open, pinyon-juniper woodlands. LOOKALIKES Mexican, Pinyon jays. Woodhouse's Scrub-Jay distinguished by vocalization, white necklace, long tail, and gray versus blue-gray back. ALTERNATIVE NAME Inland Scrub-Jay. CONSERVATION STATUS Least Concern.

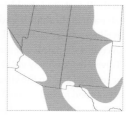

Mexican Jay

Aphelocoma wollweberi

Mixed-deciduous woodlands, almost always with an oak component, from oak-dominated grasslands to high-elevation pine-oak forests LENGTH 11–12 in. WINGSPAN 19–20 in.

Large, crestless, and long-tailed with a long, pointed, and heavy bill. Once known as the Gray-breasted Jay, is light gray below, lightest in the throat, and sky-blue above, with a grayish wash in the mantle and smudged black eyelines and lores. Juvenile is slightly grayer overall, with pale lower mandible. BEHAVIOR Travels in extremely vocal flocks, foraging in vegetation and on the ground for a wide variety of food including seeds, nuts, insects, and occasionally other small prey. Groups actively defend and are loyal to a broad territory. VOCALIZATIONS Most commonly, a loud, nasal, raspy, repetitive, and far-carrying *WAIT? WAIT? WAIT?* given by traveling flocks. Less commonly a short *chuk* or harsh scold. Flocks often produce a distinctive whoosh of wingbeats upon takeoff. ABUNDANCE A common year-round resident within a limited range in extreme Southwest NM and from Southeast to North Central AZ. LOOKALIKES Other jays. Mexican Jay is plainer, longer tailed, less stocky, shorter billed, and has grayish underparts and a mostly distinct range. CONSERVATION STATUS Least Concern (Decreasing).

Clark's Nutcracker

Nucifraga columbiana

Montane conifer forests
LENGTH 11–13 in. WINGSPAN 18–21 in.

Subtly handsome, jaylike corvid with dusky gray plumage overall and jet-black wings that contrast with white tips on the secondaries and outer tail feathers. Bill is black, sharply pointed, and slightly curved. Legs are black. BEHAVIOR Like many corvids, nutcrackers are gregarious and locally nomadic, drifting through the pine forest foraging primarily for pine cones filled with tasty seeds. A true specialist, relies almost solely on a pine seed diet to survive, in turn serving as a primary seed dispersal agent for several western pine species. Equipped with a small pouch beneath the tongue for storing extra seeds on the go. VOCALIZATIONS A wide variety of raspy, rattly, and metallic squawks, croaks, and calls. ABUNDANCE Uncommon but regular resident in preferred habitat throughout range. LOOKALIKE Canada Jay. Clark's Nutcracker distinguished by longer bill, shorter tail, and distinct bright white undertail coverts. ALTERNATIVE NAMES Montane Nutcracker, Pine Nutcracker. CONSERVATION STATUS Least Concern.

Black-billed Magpie

Pica hudsonia

Sparsely wooded settings such as grasslands, juniper woodlands, feedlots, suburban areas, parks, agricultural areas, riparian areas, roadsides
LENGTH 18–23 in. **WINGSPAN** 22–25 in.

Large, boldly patterned, and jaylike bird with rounded head, extremely long tail, and heavy, black bill. Both sexes have black head and breast, white belly, white patches near the shoulders of the folded wings, and near-black back, tail, and lower wings that iridesce in brilliant shades of blue and green in the right light. In flight, note bold white patches on the outer half of the wings and extremely long streamer of a tail. Juvenile sports a slightly shorter but still long tail. **BEHAVIOR** Forages in groups, mostly on the ground, for a wide variety of food including seeds, fruit, invertebrates and other small prey, and carrion. Often seen standing on fence posts or other low perches. Globular nests, often built in loose colonies, can reach up to 3 feet across and become obvious in winter when trees drop their leaves. **VOCALIZATIONS** A harsh, nasal *rehnk-rehnk-rehnk* and a longer, clearer, and ascending *reeehhnk* that is also repeated but more slowly. Also gives other varied barks and rattles. **ABUNDANCE** Common year round in Northern NM as far south as Santa Fe. Less commonly observed in extreme Northeast AZ. **CONSERVATION STATUS** Least Concern.

American Crow

Corvus brachyrhynchos

Open woodlands, urban and suburban areas, parks, agricultural fields, trash dumps
LENGTH 16–21 in. WINGSPAN 34–40 in.

Large and stocky but sleek bird with straight, heavy bill and lustrous black plumage from head to toe, with a bill and legs to match. In flight, wings are slightly rounded with long, fingerlike primaries. BEHAVIOR Some of the smartest birds on the planet, observed using tools and engaging in highly sophisticated social behaviors (such as organized funerals). Their smarts often lead to mischief. Extremely social, with flocks numbering in the thousands—quite a sight to see when they gather to roost. Opportunistic omnivores, they will eat just about anything, from French fries to frogs. VOCALIZATIONS Nasally, clear *caw*, as well as throaty, raspy *caw*, plus a broad repertoire of rattles, croaks, and clicks. ABUNDANCE Common resident in northern portion of range; uncommon in southern wintering range. LOOKALIKE Common Raven. American Crow distinguished by square tail, significantly smaller size, smooth, sleek throat feathers, and more slender bill. CONSERVATION STATUS Least Concern.

Showing white throat feathers

Chihuahuan Raven

Corvus cryptoleucus

Sparse habitats including grasslands, deserts, juniper forests, agricultural areas, feedlots, landfills, edges of riparian woodlands
LENGTH 18–21 in. **WINGSPAN** 41–44 in.

All black, shaggy, and small raptor-sized corvid. Both juveniles and adults have a wedge-shaped tail with a rounded tip and a somewhat short, heavy bill with feathers that extend more than halfway down the bird's length. Neck feathers have white bases that can be made visible by a stiff breeze or an agitated mood. **BEHAVIOR** Maintains flocks year round, soaring with playful agility and foraging for a wide array of food including fruit, grain, small prey, carrion, and scraps left behind by humans or other animals. **VOCALIZATIONS** Gives a variety of calls, but most frequently a repeated, relatively high, gravelly, ducklike, and honking *rohhk-rohhk-rohhk*. **ABUNDANCE** A common year-round resident in Eastern and Southern NM and in extreme Southeast AZ. **LOOKALIKES** American Crow, Common Raven. Chihuahuan Raven distinguished from American Crow by larger size, shaggier plumage, larger and heavily feathered bill, longer and more wedge-shaped tail, more soaring flight style, and distinct voice. Distinguished from Common Raven by smaller size (difficult to tell unless the two are side by side), more extensive bill feathering, and white-based neck feathers. Chihuahuan and Common ravens have distinct voices, with Chihuahuan usually higher in pitch, but they will mimic each other where ranges overlap. **CONSERVATION STATUS** Least Concern.

Common Raven

Corvus corax

A habitat generalist; grasslands, open deserts, forests, chaparral, agricultural fields, parks, garbage dumps, and more; fares particularly well in suburban and rural areas
LENGTH 22–28 in. WINGSPAN 44–48 in.

Remarkably large, stocky, heavy billed corvid with a stout neck. Black from bill to claw, with the exception of juveniles, who show a brownish hue on their chests. Throat feathers, which have gray bases, form a shaggy beard of sorts. In flight, look for wedge- or diamond-shaped tail silhouette and broad, long wings held flat and tipped by long and narrow fingerlike primaries. BEHAVIOR Incredibly clever and sly, known to use tools, play, and solve puzzles. Also known to follow humans and other predators in hopes of snagging a scrap of food. Omnivores, eating primarily animal matter but also snacking on fruits and seeds. They also happily dine on roadkill or carrion and dig through trash for snacks. Look for them cruising the sky in pairs, solo, or, less frequently, in small flocks, often joining other birds of prey to ride the thermals in kettles, flapping only intermittently between soaring glides. VOCALIZATIONS Distinct, bullfrog-like croaking; harsh, nasally squawks; and rapid rattles that sound like a piece of wood being dragged along a fence. ABUNDANCE Common resident throughout range. LOOKALIKES Chihuahuan Raven, American Crow. Common distinguished from Chihuahuan primarily by voice, slightly larger size, less extensive bill feathering extending halfway down the bill or less, gray-based throat feathers, and higher elevation habitat, though not definitive. Common Raven distinguished from crow by wedge-shaped tail, heavy bill, and shaggy throat feathers. CONSERVATION STATUS Least Concern.

LARKS AND SWALLOWS

At first glance, the larks (family *Alaudidae*) and their close relatives, the swallows (family *Hirundinidae*), have little in common. Swallows, with their slender bodies, long wings, and wide gapes at the base of their bills, specialize in capturing airborne insects, while larks, with their long legs, upright posture, and pointed bills, are ground dwellers focused mostly on seeds. However, spend some time with members of these two families and you'll discover what they have in common, beyond phylogenetic relatedness: neither is terribly concerned with offering birders easy views. Try not to strain your neck while identifying swallows zipping by overhead, and be sure not to squint too hard when trying to get a good look at a Horned Lark (North America's only native lark) as it disappears, heavily camouflaged into a field of grass.

ABOVE Cliff Swallows

LARKS AND SWALLOWS | 345

Adult male

Adult female

Horned Lark

Eremophila alpestris

Barren or sparsely vegetated areas within nearly any open habitat type, including deserts, shrublands, short grasslands, open forests, agricultural fields, lawns, parks, airports
LENGTH 6.5–8 in. WINGSPAN 12–13 in.

Cryptic, small, and slender with long wings, medium-length tail, short and thin bill, often forward-leaning posture, and round head adorned with two tiny feather tufts that can be raised to create the illusion of delightful horns. Adult male is a variable mix of gray, tan, and rusty brown above and white to buffy below, with a black, downward-curving mask; white eyebrow; pale yellow throat; two thin black stripes atop the head (the horns); and a black bib. Female is similar, but with more diffuse patterning and grayish brown instead of black. Juvenile shows white edging to gray-and-brown back feathers, white underparts, dusty gray breast, white throat, gray face, and white stippling in the head and mantle. BEHAVIOR Forages in groups for seeds and insects on open ground, frequently vanishing from view despite there being nowhere to hide. VOCALIZATIONS Song is a short, high, and cheerful jumble of accelerating, tinkling whistles. Also gives varied, high-pitched chips. ABUNDANCE Common in all seasons across both states. CONSERVATION STATUS Least Concern (Decreasing).

Bank Swallow

Riparia riparia

Open, low-elevation habitats near water
LENGTH 4.5–5.5 in. WINGSPAN 10–13 in.

Round-headed, tiny-billed swallow of petite stature, with brown plumage overall, contrasted by white throat and belly, separated by a distinct brown breast band that is clearly visible in flight. BEHAVIOR Look and listen for nesting colonies on banks of rivers and lakes, with burrows excavated into bluffs, high banks, and cliffsides. Their aerial acrobatics are a sight to behold as they switch directions on a dime to swallow up insects on the wing. VOCALIZATIONS Breeding/courtship song is a gravelly but jubilant series of twittery chittering. Call is similar in sound quality, but abbreviated, sometimes given in two notes. ABUNDANCE Common migrant; breeding populations limited to riparian corridors along the Rio Grande and Upper Pecos rivers, NM. LOOKALIKE Northern Rough-winged Swallow. Bank distinguished by clean, white underparts contrasted with dark belly band, versus dingy coloration of the Northern Rough-winged. CONSERVATION STATUS Least Concern (Decreasing in North America).

Adult male

Adult female

Tree Swallow

Tachycineta bicolor

Open habitats near water, including agricultural fields, wetlands, lakes, meadows, parks; during breeding, stays close to areas with plentiful cavity-rich trees or artificial nest boxes
LENGTH 5–6 in. WINGSPAN 12–14 in.

Medium-sized and crisply patterned with long and pointed wings, slightly notched tail, and tiny bill. Adult male is iridescent blue-green above and white below with clean white throat restricted below eyes. Adult female more grayish brown above but still slightly iridescent, and may show diffuse breast band (first year female mostly brown with sparse blue feathers). Juvenile even duller with breast band, grayish collar, and sharp contrast between dark face and white throat. BEHAVIOR Often observed peering out from cavities or foraging for insects during agile, soaring flight over water. In winter, huge flocks roost within cattail marshes along the lower Colorado River.

VOCALIZATIONS Song is a short, gurgling burst of chips and whistles with a liquid quality, reminiscent of a cowbird. Also gives short, sweet whistles and rapid strings of slurred chips. ABUNDANCE Breeds commonly in the Southern Rockies north of Santa Fe and less commonly from Central to Northwest NM and in Northern and Northeast AZ. Winters in Southwest AZ and northward along the Colorado River. A common migrant across both states. LOOKALIKE Violet-green Swallow. Tree Swallow distinguished by larger size, white throat restricted below the eye, and lack of white patches on the sides of rump in flight. CONSERVATION STATUS Least Concern.

Adult female

Adult male

Adult male

Violet-green Swallow

Tachycineta thalassina

Open, mixed woodlands, preferring areas adjacent to large bodies of water
LENGTH 4.5–5 in. **WINGSPAN** 10–11 in.

Petite, slim-bodied swallow with elongated wings that reach well beyond the tail. Clean, crisp, white face and belly contrast with jewel-toned back and head of bronze-kissed emerald-green, sapphire-blue, and deep amethyst, though colors dim to a dark gray-black when void of sunlight. Adult female shows dusky head and cheeks, and juvenile is grayish overall with clean white underparts. In flight, look for white patches on either side of a dark rump. **BEHAVIOR** This aerial acrobat takes advantage of great heights and surface-skimming lows as it forages for insects on the wing. Reaches remarkable speeds that may exceed 25 miles per hour in flight. Nests in cavities, where they build feather-lined cups using roots, twigs, and grass. Prefers to forage over or near bodies of water. Perches on exposed branches and snags. **VOCALIZATIONS** Song is a variety of sweet, clear, and raspy chirps and tweets, and flight call is a buzzy *chee*. **ABUNDANCE** Common; breeds throughout all but far-Eastern NM. **CONSERVATION STATUS** Least Concern.

Northern Rough-winged Swallow

Stelgidopteryx serripennis

Over nearly any open habitat type, especially over water and near sandbanks or mud cliffs with ample crevices and holes for nesting
LENGTH 4.75–5.5 in. WINGSPAN 11–12 in.

Medium-sized swallow, unique in its drabness, with long and pointed wings, a square-tipped tail, and a dinky bill. Adults are grayish brown and unpatterned above, darkest in the flight feathers, with a dingy, buffy gray breast and throat; dirty white belly; and white undertail coverts seen easily in flight. Juvenile is similar with subtle rust-colored wingbars. BEHAVIOR Forages for flying insects low over water or open ground with slower wingbeats than other swallows. Often perches conspicuously. VOCALIZATIONS Song is a series of short, quiet, and liquid gurgles and chips, similar to but more hoarse and less musical than a Tree Swallow. Call is a gravelly and repetitive *blurt-blurt-blurt*. ABUNDANCE A common breeding resident and migrant across the region, less common in Southeast NM. Year-round populations exist along the Rio Grande, Gila, and Colorado rivers, and a few birds linger through winter in the areas surrounding both Phoenix and Tucson. LOOKALIKE Bank Swallow. Northern Rough-winged differentiated by lack of dingy breast band. CONSERVATION STATUS Least Concern (Decreasing).

Female/immature (left) with adult male

Adult female

Purple Martin

Progne subis

Open woodlands, saguaro forests, especially along riparian corridors, where they breed in natural cavities
LENGTH 7–8 in. **WINGSPAN** 15–16 in.

Hefty swallow with wings meeting the tip of its shallowly forked tail. Male shows dark plumage overall, with dark black-brown wings and tail and deep blue-purple iridescence down the back and belly in sunlight. Female and immature male show subtler coloration and dingy white underparts with flecks of gray. Females in Sonoran Desert show significantly paler than eastern counterparts. **BEHAVIOR** Breeding populations in Desert Southwest are unique subspecies, *P. subis hesperia*. Unlike their peers to the east, desert Purple Martins don't live in lakeside/beach bird condos, but roost and nest primarily in columnar cactus cavities excavated and previously occupied by woodpeckers. Will nest in isolated pairs and small, loose colonies, another trait differentiating it from the large colonies of eastern populations. Feeds almost exclusively on the wing, with a diverse diet of insects and sometimes spiders. **VOCALIZATIONS** Various warbling and chittering, sounding both liquid and electronic in quality. **ABUNDANCE** Uncommon but regularly breeds throughout Southeast AZ and Southwest NM, with populations concentrated in far-Southeast AZ. Regularly observed during migration throughout NM. **CONSERVATION STATUS** Least Concern (Declining).

Barn Swallow

Hirundo rustica

Open areas including grasslands, meadows, wetlands, parks, ponds, agricultural fields; typically near nesting sites on human-made structures or cliffs
LENGTH 6–7.5 in. WINGSPAN 11.5–12.5 in.

Medium-sized and elegant with a boxy head, miniscule bill, long wings, and long tail streamers extending beyond the wingtips when perched and as a deep fork in flight. Adult male is oily blue-black above and pumpkin-orange below, with dark rump, orange forehead, and orange throat sometimes bordered by messy blue-black bib. Female less vibrant, sometimes nearly white below. Juvenile resembles female but shorter tailed without tail streamers. In flight, note dark flight feathers and row of white tail spots. BEHAVIOR Forages in direct flight for airborne insects, both low and high; rarely seen soaring. Glues cup-shaped mud nests under bridges or eaves, or on natural cliffs. VOCALIZATIONS Song is a reedy, rapid series of chirps, squeaks, and rattles. Also gives high chirps, sweet whistles, and raspy barks. ABUNDANCE A common breeding resident across NM and from Central to Southeast AZ. A common migrant throughout both states. LOOKALIKES Cave, Cliff swallows. Barn distinguished by dark blue nape and rump, deeply forked tail, and bicolored underwings. CONSERVATION STATUS Least Concern (Decreasing).

Cliff Swallow

Petrochelidon pyrrhonota

Cliffsides and steep embankments, highways, bridges, human-made structures in open countryside
LENGTH 5–5.5 in. WINGSPAN 11–12 in.

Small swallow with small, round head; wings roughly equal in length to its squared tail when perched. Dark blue throat abruptly contrasts with white breast and belly. Head and back show a deep, metallic navy blue and face is a deep rusty red with bright white (northern populations) or rusty (Mexican populations, extending into Southern AZ/NM) patch on forehead. Apricot-toned rump is visible while perched and in flight. Some juveniles show pale throats. BEHAVIOR Social colonial nesters, building globular mud nests on cliffsides and steep banks. Often takes advantage of vertical walls on human-made structures such as overpasses, bridges, barns, and warehouses, clustering their nests under overhangs. Aerial foragers, preferring to feed over or near bodies of water, often joining other swallow and swift species. VOCALIZATIONS Song is a squeaky, grating, rubbing sound, and call is a soft, rattly *cherr* and laserlike *chew*. ABUNDANCE Common breeding-season resident. LOOKALIKE Cave Swallow. Cliff distinguished by paler rump and darker face and throat. CONSERVATION STATUS Least Concern.

Cave Swallow

Petrochelidon fulva

Open habitats including grasslands, agricultural fields; often near water
LENGTH 5.5 in. WINGSPAN 13 in.

Small swallow with long and broad wings, square-tipped tail, and tiny bill. Adult has blue-black back, pale underside with orange and gray-washed flanks, blackish gray wings, pale nape, rufous-orange rump, spotted undertail coverts, and gray underwings. Blackish gray cap contrasts strongly with burnt-orange forehead, face, and throat. Juvenile is paler overall. BEHAVIOR Catches insects in flight over open habitats and water, often alongside other swallow species. Carries mouthfuls of mud to the walls of a sheltered site such as sinkhole, cave, bridge, or culvert, and uses it to build a shallow, bowl-shaped nest. VOCALIZATIONS Song is a fairly long string of dry and reedy whistles, squeaks, rattles, and creaks; similar to but lacking the radio static–like sounds of Cliff Swallow. Calls include a very high, nasal, and repeated *JEER*. ABUNDANCE An uncommon but increasing resident from South Central to Southeast NM. KEY SITES NM: Carlsbad Caverns National Park, lower Rio Grande River. LOOKALIKE Cliff Swallow. Cave shows darker rump, paler face that contrasts more strongly with the dark cap, and always orange (never white) forehead. CONSERVATION STATUS Least Concern.

CHICKADEES, TITMICE, AND OTHERS

Most of these birds, which include the chickadees and titmice (family *Paridae*), long-tailed tits (family *Aegithalidae*), nuthatches (family *Sittidae*), and creepers (family *Certhiidae*), are year-round residents in middle- to high-elevation woodlands. Here, they can often be spotted together in loosely associating mixed groups that benefit from a community approach to foraging and watching for predators. Although birds within these families are common across the country, the Southwest offers an additional family to this assemblage, the penduline tits (family *Remizidae*), represented by only one North American species—the Verdin, bushtit of the desert.

ABOVE Adult Verdin

Black-capped Chickadee

Poecile atricapillus

Deciduous or mixed forests, urban and suburban parks, backyards
LENGTH 4.5–6 in. WINGSPAN 6.5–8.5 in.

Little floof of a bird, with a tiny body, round head, short and pointed bill, and narrow tail. Solid black cap and chin are separated by a wedge of white on the cheek, gray back with fine white edges on gray wings, and warm, buffy flanks on a white belly. Legs and bill are black. BEHAVIOR Brimming with personality and curiosity, this tiny bird engages in complex social behaviors that include hierarchies within flocks. Extremely inquisitive and bold, they will even feed from the hands of humans. Flight is buoyant, and foraging in trees is somewhat frantic, bopping along branches in search of berries, seeds, and insects. Cavity nesters, they will make use of existing natural cavities as well as human-made birdhouses. VOCALIZATIONS A remarkably rich lexicon, from the descending, sweetly whistled *fee-bee* song to the namesake buzzy *chick-a-dee-dee-dee* and a variety of gargling, hissing, and *tsee* alarm calls, many of which are used to maintain and enforce the social hierarchy within a flock. ABUNDANCE Uncommon resident in Northern NM. LOOKALIKES Mexican, Mountain chickadees. Black-capped distinguished by short bib that ends cleanly at the throat versus the extended bib of the Mexican Chickadee, and full black cap versus the white supercilium of the Mountain. CONSERVATION STATUS Least Concern.

Mountain Chickadee

Poecile gambeli

Arid, high-elevation, relatively open pine, fir, spruce, aspen forests; sometimes lower elevation pine, juniper, oak woodlands; in winter, especially in response to winter storms, may venture into lowlands, seeking landscaped conifers
LENGTH 5–6 in. WINGSPAN 7.5 in.

Tiny and round with big, crestless head; rounded wings; long, thin tail; and small, pointed bill. Adult is dark gray above and pale gray below with gray to buffy gray flanks; a black throat, eyeline, and cap; and a white cheek and sometimes inconspicuous eyebrow. Juvenile dingier overall. BEHAVIOR Forages acrobatically and vocally, often in mixed flocks and usually high in conifers. Hops along trunks and dangles from branches in search of insects, fruit, and seeds. A frequent visitor to backyard feeders. VOCALIZATIONS Song is high, clear, whistled, and descending *fee-foh-foh, fee-fee-foh-foh, fita-fee-foh-foh,* similar to but longer and more varied than Black-capped Chickadee. *Chicka-dee-dee-dee* and other calls similar to Black-capped. ABUNDANCE Common year round at high elevations across both states, excluding AZ's Chiricahua Mountains and NM's Animas Mountains, where Mexican Chickadees are found. LOOKALIKES Black-capped, Mexican chickadees. Mountain lacks white-edged wing feathers and has grayer flanks than Black-capped, which prefers riparian habitat. Compared to Mexican Chickadee, Mountain is lighter below with a rounder head and a less extensive black bib. Mountain is the only chickadee with a white eyebrow and the sole chickadee species across much of the region. CONSERVATION STATUS Least Concern.

Mexican Chickadee

Poecile sclateri

Montane conifer forests of ponderosa pine, Douglas fir, spruce
LENGTH 4.5–5 in. WINGSPAN 7–7.5 in.

Gray overall, with fine white edging on wings, a solid black cap, and a black bib that extends to touch the breast; contrasting white cheeks and nape. Black bill is short and finely pointed. BEHAVIOR Forages in the canopy for insects and caterpillars; supplements diet with seeds. Cavity nester, using both natural cavities and nest boxes. Look for them creeping along branches to glean, often hanging upside down to pick from under branches. This species has not been observed storing food, setting it apart from other chickadees. VOCALIZATIONS Namesake *chick-a-dee-dee-dee*, slower and gravelly in comparison to other chickadee species, and a variety of buzzy chitters and *chi-dits*. ABUNDANCE Common resident in limited range of extreme Southeast AZ and extreme Southwest NM.

KEY SITES AZ: Chiricahua Mountains; NM: Animas Mountains. CONSERVATION STATUS Least Concern (Decreasing).

Bridled Titmouse

Baeolophus wollweberi

Sycamore-dominated riparian areas; mid-elevation pine-oak, oak-juniper woodlands LENGTH 5.25 in. WINGSPAN 8 in.

Resembles chickadee with short, broad, rounded wings; roundish body; small, pointed bill; and long tail, but topped with a triumphant little crest. Adults and juveniles are gray overall and lightest below—sometimes with a faint yellow wash. Affectionately referred to as zebra-faces by some, both sexes sport a bold facial pattern composed of a black throat, eyeline, crest stripe, and rear cheek border on top of a white cheek, lore, eyebrow, and crest stripe that curves down along the back edge of the head. BEHAVIOR Similar to chickadees. Travels in small flocks, often mixed with other songbirds as they investigate small branches and leaves for insects and other invertebrates. An acrobatic forager, often observed hanging upside down in search of prey. A common visitor to feeders and will make use of nest boxes. VOCALIZATIONS Song is a short, high *pita-pita-pita-pita-pita*. Calls, often heard from foraging flocks, are similar to chickadees and combine high chips and harsh, rattling scolds: *tsi-tsi-tsi-chidda-chidda-chidda*. ABUNDANCE A common year-round resident at middle elevations in extreme Southwest NM and from Southeast to North Central AZ. CONSERVATION STATUS Least Concern.

Juniper Titmouse

Baeolophus ridgwayi

Pinyon-juniper woodlands
LENGTH 5.5–6 in. WINGSPAN 8–9 in.

Formerly known as the Plain Titmouse, which was a fitting description for this little crested gray job, but it didn't capture its cuteness. Gray overall with a deep black eye and pronounced crest. Belly slightly paler gray than its charcoal-gray back. Black bill is stout and rounded, and tail is long. BEHAVIOR A foliage gleaner, this spunky little bird moves deftly about the branches of juniper and pinyon pine, sometimes hanging upside down to pick off insects and likely vocalizing along the way. Nonmigratory, it stores seeds in the crevices of tree bark to help it survive harsh, high-elevation winters. VOCALIZATIONS Song is an explosive, rapid series of *two-wee* tweets, like a shot from a tommy gun. Calls are a variety of buzzy chitters, similar to chickadee in quality, often used to scold intruders. ABUNDANCE Common resident in central portion of range, increasingly rare along the southern and eastern reaches of range. LOOKALIKE Oak Titmouse. Juniper and Oak were once considered a single species. Juniper and Oak distinguished by habitat and range. CONSERVATION STATUS Least Concern.

Adult

Juvenile

Verdin

Auriparus flaviceps

Desert scrub and chaparral with scattered, thorny shrubs; common resident in suburban areas, where it will visit hummingbird feeders and nectar-providing plants
LENGTH 4.5 in. WINGSPAN 6.5 in.

The honorary bushtit of the desert. Tiny, round-bodied, and round-headed with broad wings, a pointed bill, and a fairly long tail. Adults are pale gray overall, lightest below, with small, crimson-red shoulder patches; black lores; and a sunflower-yellow head and throat. Juvenile is unmarked gray overall. BEHAVIOR Avoids both sparse and very densely vegetated areas. Actively forages for invertebrates, bouncing around the canopies of scrubby desert trees like a chickadee. Builds globular nests both for breeding and nightly roosting, most often in palo verde. Often one of the few birds active during the heat of the day. VOCALIZATIONS Song is a high, plaintive *tsee-tso-tsoo*. Also gives extremely high and rapidly repeated chips and a low, buzzy scold. ABUNDANCE A common year-round resident across the deserts of Western AZ and central to southern portions of both states. LOOKALIKE Bushtit. Adult Verdin is distinct, but all-gray juvenile can be confused with Bushtit. Juvenile Verdin differentiated by habitat, range, behavior, association with adults, complete lack of patterning, bill shape, and obvious yellow gape (the fleshy area at the base of the bill where upper and lower mandibles meet). CONSERVATION STATUS Least Concern (Decreasing).

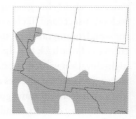

Bushtit

Psaltriparus minimus

Oak and conifer woodlands, scrub, chaparral, riparian-adjacent cottonwood-willow galleries, wooded suburban areas
LENGTH 3.5–4.5 in. WINGSPAN 5.5–6.5 in.

This is a round, pale gray, bouncing ball of a bird. It is roundbodied and round-headed, with a long and narrow tail. Plumage varies with geography, with Interior variant present in Arizona and New Mexico. Interior male shows pale gray overall, with fine white edges on wings, gray-capped head, and buffy, warm tan cheeks. Black bill is just a tiny, pointed nub. Females show pale yellow eyes, while males and juveniles are black-eyed. BEHAVIOR Foliage gleaners, almost always in flocks, in perpetual motion through trees and shrubs, picking insects off the undersides of leaves and branches. Known for their intricate, sock-like woven nests of moss and lichen, held together with spiderwebs and constructed over the span of a full month. Flight is fluttery and bobbing. VOCALIZATIONS Though they don't have a distinct song, they give a variety of soft, tinkling twitters and chips to communicate within their flocks. ABUNDANCE Common resident throughout range. LOOKALIKE Juvenile Verdin. Bushtit distinguished by habitat, range, very short bill, and longer tail. CONSERVATION STATUS Least Concern (Decreasing).

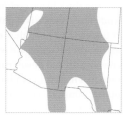

Red-breasted Nuthatch

Sitta canadensis

High elevation, mountainous, spruce-fir forests; winters in lower elevation coniferous and mixed forests; occasionally visits suburban settings with planted conifers
LENGTH 4.5 in. WINGSPAN 7.5–8.5 in.

Tiny, plump, neckless, and short-tailed with short, broad, and rounded wings; relatively long, pointed bill; crisply patterned face; and often forward-leaning posture. Adult male is steel-blue above and ochre below with a black cap and thick black eyeline on an otherwise white face. Female is slightly duller with grayer underparts. Juvenile is entirely gray below. BEHAVIOR Forages like a woodpecker, actively creeping up and down trunks and branches in search of insects to be plucked from crevices or pulled from beneath bark. Shows little regard for gravity and is often spotted upside down, sideways, or anywhere in between. VOCALIZATIONS Low, nasal, reedy, and toy trumpet–like *henk* notes, sometimes drawn out and given in repetition and sometimes combined with high chips and sped up to produce a rattling chatter. ABUNDANCE Common year round in the mountains of Northern to Eastern AZ and North Central to Western NM. In winter, and especially during population eruption years, is observable across the entirety of both states, though less commonly at lower elevations and rarely in Southwest AZ. CONSERVATION STATUS Least Concern.

Adult male | Adult female

White-breasted Nuthatch

Sitta carolinensis

Mixed mature woodlands
LENGTH 5–5.5 in. WINGSPAN 8–10.5 in.

Distinctly shaped, small bird. Plump and oval-bodied with a stout neck; large, tapered head; and a stub of a tail. Plumage is slate-blue–gray, with black edges along the wingtips, clean white belly, rufous near vent, white cheeks, sooty black cap that extends down the neck, and thin black eyeline extending from eye back toward nape. Cap on female is charcoal-gray versus black on male. BEHAVIOR Creeps stealthily along trunks and large branches, gleaning insects from grooves in the bark. Omnivorous, and will frequent backyard feeders seeking oily, meaty seeds and even suet. Look for them in high-elevation habitats, where their squeaky, nasally calls will give away their location. VOCALIZATIONS Very nasally, monotone *wenk-wenk-wenk* and two-note *wah-wah* song and calls, crowlike in quality. ABUNDANCE Common resident, observed frequently within suitable habitat. CONSERVATION STATUS Least Concern.

Pygmy Nuthatch

Sitta pygmaea

Almost exclusively in association with ponderosa pine, both in monotypic stands and in nearby mixed conifer, pine-oak, juniper-oak woodlands
LENGTH 4–4.25 in. WINGSPAN 3.75 in.

The tiniest of our diminutive nuthatches. Rotund, neckless, large headed, and extremely short tailed with a long, chisel-shaped bill; broad, rounded wings; and forward-leaning posture when perched. Adults of both sexes are bluish gray above and creamy white to buffy below, with brownish gray caps, grayish flanks, a whitish chin, and a broad dark eyeline. Juvenile is paler overall. BEHAVIOR Spritely and gregarious, this nuthatch forages in flocks, is known to share communal roost cavities, and breeds cooperatively. Highly vocal flocks remain in constant motion as they explore foliage and flaking bark for insects and seeds. Like other nuthatches, displays a highly active foraging style similar to a woodpecker and is frequently observed mixing with other songbirds including kinglets, titmice, and chickadees. VOCALIZATIONS Very high and somewhat quiet *pip* and *pipa* notes, sometimes given repetitively and sometimes combined with short squeaks and accelerated into an energetic clatter. ABUNDANCE Common year round at middle to high elevations across Northern, Central, and Eastern AZ and in all but Eastern and South Central NM. LOOKALIKE Red-breasted Nuthatch. Pygmy distinguished by smaller size, behavior, paler overall coloration, and lack of a white eyebrow. CONSERVATION STATUS Least Concern.

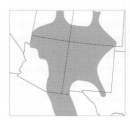

Brown Creeper

Certhia americana

Woodland generalist in mixed, deciduous, conifer, swamp forests; suburban groves; parklands.
LENGTH 4.5–5.5 in. WINGSPAN 7–9 in.

Tiny, egg-shaped, with long, narrow tail and long, slightly curved bill. Plumage is mottled brown, tan, gray-and-white overall, resembling tree bark, with a creamy white belly and buffy white supercilium. BEHAVIOR As its name would suggest, this tiny bird inches its way up the trunks of large tree specimens in a spiral pattern from the base up, using its thin, curved bill to pull insects from crevices. Will also frequent feeders for seeds and suet. In AZ, nests are uniquely constructed with two openings serving as entrance and exit. Though primarily tree nesters, they will make use of a variety of human-made structures, from cinder blocks to roof eaves. VOCALIZATIONS Distinct, high-pitched, tinkly warble, akin to the Yellow Warbler in intonation but descending and slightly different in cadence. (In areas outside AZ and NM, song sounds like "trees, beautiful trees"; however the *C. americana albescens/alticola* subspecies present in this range sings its own tune.) Call is a high, metallic *tsee*. ABUNDANCE Common resident throughout year-round range; relatively uncommon throughout wintering range. CONSERVATION STATUS Least Concern.

WRENS

The Southwest's wrens (family *Troglodytidae*) are a diverse bunch, from the uncommon, miniature, and secretive Pacific Wren, to Arizona's state bird, the abundant, comparatively giant, and extroverted Cactus Wren. Diverse as they are, they are recognizable with their stiff, banded tails; portly bodies; long and downward-curving bills; subtle shades of brown, gray, black, white, and rufous; and, in many species, particularly well-painted eyebrows. Their modest colors could help them go unnoticed, but the wrens—prolific, talented singers quick to scold those who dare invade their space—have a tendency to give themselves away.

ABOVE Adult Cactus Wren and nest

Rock Wren

Salpinctes obsoletus

From the lowest deserts to the highest peaks, in arid, rocky habitats such as canyon walls, boulders, scree slopes, crags, drainage edges; also riprap and other rocklike human-made infrastructure
LENGTH 5–6 in. WINGSPAN 9 in.

Medium-sized wren with short, rounded wings; a long tail; and a long, downward-curving bill. Adults are dusty grayish brown above with subtle speckling, white below with buffy brown flanks, and have black-and-white–banded undertail coverts. Tail is rufous-tinged with dark bands and a pale tip. Flight feathers thinly barred, breast sometimes finely streaked. Shows a faint, pale eyebrow. BEHAVIOR Hops along rocky surfaces, poking its bill into crevices for insects and other invertebrates. Performs habitual knee bends when calling. VOCALIZATIONS Song is extremely varied but always high, sweet, and somewhat buzzy, and is always composed of distinctly separated strings of repeated notes: *peera-peera, churee-churee, peeurr-peeuurr, tsoo-tsoo, deeta-deeta,* and others. Common call is a high, thin, and repeated *tickee-tickee*. ABUNDANCE A common year-round resident throughout AZ and all of NM, except for in the north-central part of the state, where it is a common summer resident, and along the extreme southeast edge of the state, where it is a common winter resident and passing migrant. CONSERVATION STATUS Least Concern (Decreasing).

Canyon Wren

Catherpes mexicanus

Rocky, arid cliffsides, canyons
LENGTH 5–6 in. WINGSPAN 7–8 in.

Finely detailed, ornate little egg-shaped bird, with a long, slender, curved bill and long, square tail. Chestnut-brown on the back and cinnamon colored underneath, with delicate spotting and barring, and distinct black barring on cinnamon-colored tail. Belly color is bright cinnamon, with a creamy white throat and bib that extend halfway up the cheek. Short, rounded wings in flight. BEHAVIOR Similar to that of a creeper, but uses cliff walls instead of tree trunks, inching along the rock in search of insects that it plucks from even the tightest crevices with its needlelike bill. Males will perch conspicuously to sing. VOCALIZATIONS Fluid, descending series of clear, whistling *peew*s, as if it were mocking something falling down, down, down into the canyon. Call is a buzzy *tsee*. ABUNDANCE Common resident throughout range. LOOKALIKE Rock Wren. Canyon distinguished by warm cinnamon coloration versus pale buffy tone. CONSERVATION STATUS Least Concern.

House Wren

Troglodytes aedon

Brushy, humid areas within middle- to high-elevation coniferous and mixed forests, riparian areas, forest edges, parks, backyards, hedges; similar habitats at lower elevations in winter LENGTH 4.5–5 in. WINGSPAN 6 in.

Small wren with broad, rounded wings; flattened head; medium-length tail; and medium-length, downward-curving bill. Both sexes are variable in color and relatively plain, ranging from rufous-brown to grayish brown, darkest above and lightest below, with dark barring on the wings, tail, undertail coverts, and sometimes flanks and with an extremely faint, pale eyebrow. BEHAVIOR Hops with great urgency through low, dense vegetation in search of insects and other invertebrates. Vocalizes frequently, often cocking its tail to the side in tandem. VOCALIZATIONS Song, given frequently, is an exuberant, warbling, sweet but somewhat raspy jumble of whistles, buzzes, rattles, and chips. Also a variety of harsh scolds, reedy chatter, and dry chips. ABUNDANCE AZ: common year-round resident in the mountains of southeastern part of the state and northwestward along the Colorado Plateau transition zone, common breeding resident to the north of this range, and common winter resident south. NM: common breeding resident throughout the state, except for the extreme south and east, where it is a common migrant. LOOKALIKES Pacific, Winter wrens. House Wren larger and longer tailed. CONSERVATION STATUS Least Concern.

Pacific Wren

Troglodytes pacificus

Breeds in mature/old-growth deciduous and conifer forests; overwinters in a variety of woodland and park habitats
LENGTH 3–4.5 in. WINGSPAN 5–6.5 in.

Small, round-bodied, and short-necked, with a short, narrow tail that it often holds upright. Warm cocoa-brown overall, with fine dark brown/black barring on back and belly, contrasted only by a subtle, pale, buffy supercilium. Its short wings flutter hastily to carry it between the safe cover of understory and brush. BEHAVIOR Inconspicuous in appearance but hardly secretive, this bird spends much of its time moving about through thick brush and along fallen logs and ground cover, making such a clatter along the way that its position is easily given up. Its movements often follow the pace of its frantic vocalizations, seemingly unable to sit still and constantly bobbing through vegetation in search of insects. VOCALIZATIONS Bubbly and chattering, singsong, tinkling trills, as if being played at double speed. Call is a one-note, metallic trill, chip, or series of eruptive chips. ABUNDANCE Small breeding population in limited range in Northeast AZ; small population overwintering along the lower Colorado River corridor. LOOKALIKES Winter, House wrens. Pacific Wren distinguished from both by warmer coloration. Distinguished from Winter Wren by range (though there is some overlap in sightings) and from House Wren by buffy supercilium. CONSERVATION STATUS Least Concern (Decreasing).

Marsh Wren

Cistothorus palustris

Dense cattail- and bulrush-dominated marshes; other brushy habitat types during winter and migration
LENGTH 4–5.5 in. WINGSPAN 6 in.

Tiny, round wren with broad wings, medium-length tail, flattened head, and long, downward-curving bill. Adults are ruddy brown overall, palest below, with a black-banded tail, dark and messily mottled flight feathers, a pale throat, a prominent white eyebrow with a dark border above, and a black-and-white–striped mantle (upper back). Juvenile is grayer overall and lacks patterned mantle. BEHAVIOR Springs around with ease in dense marsh vegetation, foraging for invertebrate prey and clinging with great flexibility to adjacent reeds, often with tail cocked high. Sings frequently and will do so, frustratingly, from around your feet, never to be seen. VOCALIZATIONS Song is mechanical composition of reedy squeaks, harsh rattles, and thin chips given at a fast pace, repeatedly, and frequently from an inconspicuous position within dense cover. Also gives thin chips and a sometimes monotonous and sometimes excited *chet-chet-chet-chet*, reminiscent of a chattering rail. ABUNDANCE Locally common; a year-round resident along the Colorado and lower Gila rivers, a breeding resident in the Southern Rockies north of Santa Fe, and a winter resident throughout the rest of the region. CONSERVATION STATUS Least Concern.

Bewick's Wren

Thryomanes bewickii

Brushy, humid thickets within otherwise arid landscapes; mesquite woodlands, riparian corridors, chaparral, oak woodlands, conifer forests, suburban areas, parks
LENGTH 5.25 in. **WINGSPAN** 7 in.

Medium-sized wren with a very long tail, portly figure, flattened head, and long, downward-curving bill. Both sexes are unpatterned brownish gray to reddish brown above and pale gray below with a white throat, black-banded tail and undertail coverts, white spotting near the tips of the tail feathers, and a bold white eyebrow. Juvenile may show subtle speckling below. **BEHAVIOR** Remains hidden in low, dense vegetation as it hops around in search of invertebrates, often cocking its tail upward, holding it sharply to the side, or fanning it to reveal white spotting. Tends to vocalize from a conspicuous perch. Also gives short, dry barks and harsh, raspy scolds. **VOCALIZATIONS** Song is sweet, whistled, trilling, sparrow-like, and varied, usually beginning with a rising whistle or short, clear notes and finishing on a clear, rapid trill: *suWEE-deedeedeedeedee*, *chut-chut-weeooo-deedeedeedeedee*, and other variations. **ABUNDANCE** A common year-round resident throughout both states, except for Southwest AZ and along the state's western edge, where it is a common winter visitor. **LOOKALIKE** House Wren. Bewick's distinguished by larger size, longer tail, and more prominent eyebrow stripe. **ALTERNATIVE NAME** Long-tailed Thicket Wren. **CONSERVATION STATUS** Least Concern (Decreasing).

Cactus Wren

Campylorhynchus brunneicapillus

Desert cactus forests, desert foothills, arid scrublands
LENGTH 7–8 in. WINGSPAN 10–12 in.

Large, thick-bodied, heavily patterned wren with long tail and long, thick, bill with slightly curved upper mandible. Brown back has fine white speckling and streaking; white breast has heavy, bold black streaking; and cinnamon-colored flanks and low belly are speckled with black. Tail is boldly barred in black and white underneath. Head is brown overall with thick, creamy white supercilium and dark brown-and-white streaking on cheeks. Juvenile shows slightly paler coloration and brown eye versus red of adult. BEHAVIOR Bold and brash, they perch high atop columnar cacti, chollas, and shrubs announcing their presence. Will often fan their tails while perched. Gregarious, often foraging along the ground and low shrub/cactus canopy in pairs or familial flocks. They build football-shaped nests among the prickly safeguards of cholla and other cactus branches, used both for raising young and for roosting outside of the breeding season. Like Canyon Wrens, they can go without drinking water; their moisture intake is derived from their food, including insect bodies and desert fruits such as hackberries and wolfberries. VOCALIZATIONS An endearing, familiar song to desert dwellers, raspy and gravelly, like a car engine fighting to start but never turning over. Calls vary from rattly *chek*s to squeals and growls. ABUNDANCE Common and abundant throughout the range, though populations are in steep decline in extended range (TX and CA) and showing decline in AZ and NM. CONSERVATION STATUS Least Concern (Decreasing).

GNATCATCHERS, DIPPERS, AND KINGLETS

The gnatcatchers (family *Polioptilidae*), dippers (family *Cinclidae*), and kinglets (family *Regulidae*) may seem like unassuming floofs, with their round bodies and pointy bills, but to small invertebrates they are predators not to be underestimated. Agile, speedy, and highly energetic, each group takes its own unique approach to the hunt. Watch for gnatcatchers as they frantically hop through low shrubs, scan a bit higher for kinglets flitting restlessly through the branches, and pay special attention to clear and swiftly flowing rivers and streams for the American Dipper, the only representative of its family in North America, as it scampers and swims among the riffles in search of aquatic prey.

ABOVE Ruby-crowned Kinglet

Female/nonbreeding male

Breeding male

Blue-gray Gnatcatcher

Polioptila caerulea

Breeds in scrubby, mid-elevation forests including pinyon-juniper, oak, pine-oak woodlands; winters in low-elevation deserts, mesquite woodlands, riparian areas, parks, other brushy habitats
LENGTH 4–4.5 in. WINGSPAN 6 in.

Tiny and nondescript with a round body, short wings, a long tail, and a thin, straight, sharply pointed, grayish bill. Adult male is bluish gray overall with a black uppertail, mostly white undertail, white-edged flight feathers, a distinct white eye ring, and, in breeding plumage, a broad black unibrow. Female and juvenile are duller and lack black unibrow. BEHAVIOR Hunts for invertebrates, usually alone, with bouncy vigor along the outer edges of trees and shrubs, rarely sitting still and frequently cocking its tail from side to side. VOCALIZATIONS Call is a short, high, thin, and nasal *feeee*. Song combines this note with high chips and whistles into short, highly variable, and marginally musical phrases. ABUNDANCE Common. AZ: year round in the southeast, a winter visitor in the southwest and along the state's western edge; a breeding resident elsewhere. NM: year round in the extreme southwest, a migrant in the lower Rio Grande Valley and along the state's eastern edge, and a breeding resident elsewhere. LOOKALIKES Other gnatcatchers. Blue-gray Gnatcatcher identification clues include undertail and facial pattern, overall color, habitat, and voice. CONSERVATION STATUS Least Concern.

Breeding male

Nonbreeding adult male

Female

Black-tailed Gnatcatcher

Polioptila melanura

Arid desert scrub
LENGTH 4–5 in. WINGSPAN 5.5–6.5 in.

Tiny, long-tailed, slender songbird with straight bill. Plumage is gray overall, darker on the back and paler on the belly, fading to a pale grayish white on the chin. Distinct, neat white eye ring and black tail with white undertail spots that flash in flight. Breeding male shows black cap. BEHAVIOR Moves about in dense, low canopy and scrub, showing discrete preference for native vegetation, so much so that it is rarely observed in developed areas. Territorial and loyal, pairs often remain together throughout the year (some for life), raising young and brazenly defending their turf against intruders, including humans. Listen for their scolding vocalizations as you approach. VOCALIZATIONS Various harsh scolds, an admonishing *chee-chee-chee*, and a variety of calls including raspy chips and *chee*s. ABUNDANCE Common resident, with frequency decreasing along the northern edge of range. LOOKALIKES Blue-gray and Black-capped gnatcatchers, Gray Vireo. Black-tailed distinguished primarily by voice, though if not vocalizing, by tail feathers appearing mostly black underneath, versus mostly white undertail feathers of Blue-gray and Black-capped. Smaller, longer tailed, and more slender billed than Gray Vireo. CONSERVATION STATUS Least Concern (Decreasing).

Breeding male

Nonbreeding adult male

Female

Black-capped Gnatcatcher

Polioptila nigriceps

Mixed riparian woodlands containing mesquite and hackberry within foothill canyons or along slow-moving rivers, creeks
LENGTH 4.25 in. WINGSPAN 6 in.

Tiny, round-bodied, short-winged, and long-tailed with a relatively long, straight, pointed, and blackish bill. Adult male is bluish gray above and pale gray below with dark wings; whitish face, throat, and breast; black uppertail; mostly white undertail; and, in breeding plumage, black cap and lores. Nonbreeding male shows weak black eyebrow and slight buffy wash below. Female and juvenile resemble Blue-gray Gnatcatcher. Tail feathers are of different lengths and are layered stepwise toward the tip. BEHAVIOR Forages actively, gleaning invertebrates from low shrubs and trees, frequently cocking and fanning its tail like other gnatcatchers. VOCALIZATIONS Call a whining *feeee* or *feeeew*, similar to Blue-gray Gnatcatcher but rougher, less thin, and rising and falling distinctively in pitch. Also gives harsh scolds, rattles, and chips that combine with its more common call to produce a complex song. ABUNDANCE A gnatcatcher of western Mexico, rare in the United States; irregular and highly sought-after from Southeast AZ to extreme Southwest NM. Nests sporadically, though permanent breeding range in AZ and NM is yet to be confirmed. KEY SITES AZ: Sonoita Creek, California Gulch, and west-facing Santa Rita Mountains; NM: Guadalupe Canyon. LOOKALIKE Blue-gray Gnatcatcher. Female/nonbreeding male Black-capped similar but distinguished by longer bill, more subtle eye ring, voice, and longer and more deeply graduated tail. CONSERVATION STATUS Least Concern.

American Dipper

Cinclus mexicanus

Woodland riparian habitat; prefers clear, rushing, rocky streams
LENGTH 6–7 in. WINGSPAN 8–9 in.

Plump, medium-sized songbird with small, round head; squat neck; hefty, straight bill; and stubby tail. Plumage is dusky gray overall, with subtle change in color on head to dusky tan, and sooty black tail. Pink legs and black bill. Juvenile shows subtle barring on underside and yellowish tinge on bill. Look for bold white plumage on eyelids when bird blinks. BEHAVIOR Little water sprites, in many ways can be likened to a duck over a songbird: aquatic, dons a thick coat of feathers, and has a low metabolic rate, enabling it to withstand long periods in frigid waters. Look for its signature whole-body bobbing behavior from its perches on boulders and logs in fast-moving streams. Forages for a variety of aquatic insects, snails, fish, and fish eggs by wading and diving through water. Nests are impressive ledge or cliffside domes of moss, twigs, roots, and grass, and they are even found in such magical places as behind waterfalls. VOCALIZATIONS Bubbly, liquidlike medleys of thrushlike, resonant notes that rise above the din of rushing water. Call is a sharp, rapid *zeek*, given in bursts of two to five repetitions. ABUNDANCE Uncommon resident throughout range. CONSERVATION STATUS Least Concern (Decreasing).

Ruby-crowned Kinglet

Corthylio calendula

Breeds in high-elevation spruce-fir forests; in other seasons, expands to lower elevations into brushy habitats including suburban areas, parks
LENGTH 3.5–4.25 in. WINGSPAN 6.5–7.5 in.

Tiny bird with a large head, no neck, medium-length tail, and a thin, needlelike bill. Adults are brownish yellow-green above and paler grayish yellow below with a black bill, pale feet on dark legs, a broken eye ring, and blackish wings with two white wingbars, the lower highlighted by a black bar just below. Adult male keeps namesake ruby crown hidden unless perturbed or looking to impress. BEHAVIOR Forages energetically in vegetation for small invertebrates, frequently taking them from the tips of branches while hovering. Seemingly nervous, constantly flicking wings and tail in unison. VOCALIZATIONS Song is loud, jubilant, and clear, starting with thin, descending whistles, continuing with accelerating chatter, and closing with neatly phrased, lilting, and singsong whistles. Call is a harsh *jidit*, sometimes given repetitively. ABUNDANCE In spring and summer, common but difficult to observe in its high-elevation breeding habitat. Outside of breeding, abandons the uppermost forests and spreads across nearly the entire region. Observed year round in the transition between these seasonal ranges and as a passing migrant in Northeast NM. LOOKALIKE Hutton's Vireo. Ruby-crowned Kinglet is more fidgety, with a thinner and blacker bill, darker lores, light feet, and a black bar highlighting the lower white wingbar. CONSERVATION STATUS Least Concern.

Golden-crowned Kinglet

Regulus satrapa

Montane conifer forests, deciduous and mixed woodlands; urban/suburban areas in winter
LENGTH 3–4.5 in. WINGSPAN 5.5–7 in.

Tiny, round-bodied songbird with bulky, round head; short, slender bill; and long, narrow tail. Compensates for petite stature with understated but striking plumage—olive-gray on the back with white wingbars and sunny yellow edges on black wing feathers, a pale gray belly, and bold black-and-white striping on face, topped with a golden crown, sometimes showing a bright, deep marigold color down the center. Juvenile lacks both facial stripes and golden crown. BEHAVIOR Flits about in high conifer canopy, gleaning insects from needles. Most often heard before seen. Will often join mixed flocks of insect-foragers such as warblers and nuthatches during migration and overwinter. Nests are hanging cups of moss and other materials, usually high (more than 50 feet) in conifer canopy, especially spruce and fir. VOCALIZATIONS Sweet, high-pitched series of ascending *tea-tea-tea-tea*s, accelerating in progression, and punctuated by bubbly, descending chatter. Call is a buzzy, high-pitched *tsee*. ABUNDANCE Uncommon throughout both year-round and winter ranges. LOOKALIKE Ruby-crowned Kinglet. Golden-crowned distinguished by lack of eye ring, presence of black-and-white facial stripes, and golden versus ruby crown. CONSERVATION STATUS Least Concern.

THRUSHES

The thrushes (family *Turdidae*) include the bluebirds, solitaires, nightingale-thrushes, and robins. Whether you bird by eye or by ear, this family provides many of the avian world's most recognizable species. Take the American Robin, for example, a common and conspicuous backyard classic, or the Hermit Thrush, a secretive bird whose haunting voice permeates western forests. These chubby, medium-sized woodland birds subsist on a diet of mostly insects and other invertebrates, but come winter, many species switch things up, offering birders a show when large flocks descend upon berry-rich trees and shrubs.

ABOVE Western Bluebird exhibiting foraging behavior

Adult male

Female/immature

Juvenile

Eastern Bluebird

Sialia sialis

Open areas with adjacent trees, such as oak grasslands, prairie edges, suburban areas, parks
LENGTH 6.5–8 in. WINGSPAN 10–13 in.

Cheerfully chubby with a rounded head, thick neck, long wings, short tail, short legs, and short, sturdy bill. Adult male is cobalt-blue above with white belly and undertail coverts and rusty orange throat, breast, sides of neck, and flanks. Adult female is grayish blue above with the most prominent blue in wings and tail, shows paler orange coloration, and has a white belly, pale throat, and partial white eye ring. Juvenile shows heavy brown mottling above and below. BEHAVIOR Takes small invertebrates from the ground in quick dives from a usually low and conspicuous perch. VOCALIZATIONS Song is a series of short and crisply phrased whistles and dry chattering; more musical than Western Bluebird. Calls include nasal chatter and a whistled *too-wee* or *tula-wee*. ABUNDANCE Uncommon. Exists as two subspecies in the Southwest: the Eastern Bluebird, *S. sialis sialis*, a winter visitor to the eastern half of NM; and the Azure Bluebird, *S. sialis fulva*, a year-round resident in Southeast AZ and extreme Southwest NM. LOOKALIKE Western Bluebird. Eastern distinguished by orange throat and sides of neck in male and by white belly and undertail coverts in both sexes. CONSERVATION STATUS Least Concern.

Female/immature
Adult male

Juvenile

Western Bluebird

Sialia mexicana

Mixed open woodland, including parks, backyards; open burned areas, agricultural fields

LENGTH 6.5–7.5 in. WINGSPAN 11.5–13.5 in.

Small, thick-bellied thrush with short tail and straight bill. Adult male plumage is brilliant cobalt-blue on the head, back, and wings; with blue throat; deep apricot-colored breast that extends up the shoulders to the scapulars; and pale gray-blue belly. Female plumage varies, with some more subdued than others, but typically a duller wash of blue that shows brightest on the wings; a washed-out, apricot-colored belly that extends to scapulars; and a pale blue to buffy gray throat. Juvenile shows heavy brown mottling above and below. BEHAVIOR Perches low in the canopy, scoping out the ground for insects, swooping down for a meal. Often gathers in flocks outside of breeding season. Forages for insects and berries, with an affinity for mistletoe, juniper berries, elderberries, and hackberries. Cavity nesters, will make use of well-placed nest boxes. VOCALIZATIONS Song involves whistling *pew* notes interspersed with rapid chatter. Call is a similar soft *pew*, as well as harsh chip notes. ABUNDANCE Common throughout resident range; common in NM wintering range, while less common in Western AZ wintering range. LOOKALIKE Eastern Bluebird. Western distinguished by blue throat, extension of rusty breast back to the scapulars, and gray-blue lower belly, versus white belly and rusty throat of Eastern. CONSERVATION STATUS Least Concern.

Adult male

Female/immature

Juvenile

Mountain Bluebird

Sialia currucoides

Open habitats with scattered trees at middle and upper elevations, including grasslands, meadows, open woodlands, sagebrush, pastures, agricultural fields, parks, suburban areas; similarly structured habitats at lower elevations in winter
LENGTH 6.5–8 in. WINGSPAN 11–14 in.

Endearingly plump with a round head, sturdy neck, relatively long wings and tail, and a thin, pointed bill. Adult male is bright sky-blue above, pale sky-blue below, and has dark wingtips and white undertail coverts. Adult female shows a partial white eye ring and is grayish blue overall, with blue most concentrated in the wings and tail. Juvenile is more brownish overall with coarse brown streaking below. BEHAVIOR Hovers low over open areas, surveying for invertebrate prey, which it takes with quick, downward plunges. Forms larger flocks than other bluebirds. VOCALIZATIONS Song is lower and more robinlike than Eastern Bluebird, a warbling series of rough whistles and dry chips sung in messily separated phrases that are sometimes rushed together. Common call is a short *choo* or *choo-choo*.
ABUNDANCE Common; year-round resident from Central to Northeast AZ and Central to Northwest NM, a breeding resident north of Santa Fe, and a winter visitor in the rest of the region, most commonly at middle and upper elevations but also straying into the lowlands.
CONSERVATION STATUS Least Concern.

Adult

Juvenile

Townsend's Solitaire

Myadestes townsendi

High-elevation pine, fir, spruce forests throughout breeding season; overwinters in pinyon-juniper forests at lower elevations **LENGTH** 8–9 in. **WINGSPAN** 13–14.5 in.

Medium-sized, short-billed, sleek and slender thrush with large black eyes contrasted by bold white eye rings, buffy gray plumage overall, with warm, buffy peach wing patches and clean white outer tail feathers that are most prominent in flight. Bill and legs are black. Juvenile is deep charcoal-gray with fine, heavy, warm buff-and-cream spotting. **BEHAVIOR** Look for them perched atop snags and exposed branches high in the canopy, flying out to take insects on the wing; also forages for insects on the ground. Overwinter populations move to lower elevation pinyon-juniper woodlands, feasting primarily on berries. **VOCALIZATIONS** Frantic and melodic song consists of various warbles and chirps, reminiscent of a House Finch or Gray Catbird, ranging widely in pitch. Calls include a high-pitched, harsh screech used to defend territory; a similar harsh call given during breeding; a low, one-note call given during feeding; and most commonly, a clear, resonant one-note call, sounding much like a metal mechanical component that needs oiling (such as a squeaky wheel). **ABUNDANCE** Common throughout resident range and interior of wintering range. Less frequent in outer reaches of wintering range. **LOOKALIKE** Northern Mockingbird. Townsend's Solitaire distinguished by bold white eye ring, darker belly, and lack of white wingbars. **ALTERNATIVE NAMES** Juniper Solitaire, Warbling Solitaire. **CONSERVATION STATUS** Least Concern.

Swainson's Thrush

Catharus ustulatus

Breeds in dense, mature, relatively humid coniferous and mixed forests with brushy understories; during migration, nearly any dense, brushy, forested setting
LENGTH 6.5–7.5 in. **WINGSPAN** 11.5–12 in.

Medium-sized and barrel-chested with a round head, stout and pointed bill, long tail, long and often drooping wings, long legs, and frequently forward-leaning posture. Olive-backed form is olive-gray above and pale gray below with olive-gray–washed flanks, heavily spotted and buffy breast, buffy throat bordered by two dark stripes, curved and buffy border below the cheek, tail the same color as the back or only slightly more rufous, and buffy eye ring and lore that give a be-spectacled look. Russet-backed form is more rufous above, with less distinct breast spotting and facial markings. **BEHAVIOR** Inconspicuous; spends most of its time foraging for invertebrates on or near the ground under the cover of dense understory vegetation. **VOCALIZATIONS** Song is bubbly, flutelike, and always ascending in pitch. Call types are many and include a flat *whip*, a rising whistle, and reedy chatter. **ABUNDANCE** Breeds uncommonly in North Central NM and locally at high elevations in Northern and Eastern AZ. A common migrant across both states. **LOOKALIKE** Hermit Thrush. Swainson's has a buffier eye ring, more buff in the breast, and a tail color that matches the back color. **ALTERNATIVE NAME** Buff-spectacled Thrush. **CONSERVATION STATUS** Least Concern.

Adult

Juvenile

Hermit Thrush

Catharus guttatus

Woodland understory
LENGTH 5–7 in. **WINGSPAN** 10–11.5 in.

Medium-sized, rotund thrush with straight bill and long tail. Plumage is cool cocoa-brown (as opposed to warmer tone of Eastern and Pacific populations) on upper parts/back, leading to a warm, diagnostic rufous-brown tail and subtle rufous wings. Thin, white eye ring and spotted throat, with heavily and boldly spotted breast. Deep chocolate-brown smudgy spots fade along a pale, creamy belly. Juvenile is deeper brown and spotted on the head, breast, and back, but rufous along wings and tail. **BEHAVIOR** Look for the Hermit Thrush hopping along the forest floor, doing its best towhee impression as it hop-scratches through leaf litter and organic matter in search of insects, worms, and even small amphibians. Winter diet consists primarily of berries such as elderberries, mistletoe, grape, and others. Sometimes displays subtle tail-bobbing behavior. **VOCALIZATIONS** Hauntingly lovely song, woodwind-like in quality, starting with one plaintive introductory note followed by a series of melodic, echoing notes: *ohhh, eerily eerily*. Call is a quick *chup*. **ABUNDANCE** Common throughout resident and breeding range; uncommon throughout wintering range, with the exception of lower Colorado River corridor. **LOOKALIKE** Swainson's Thrush. Hermit is less buffy overall, with a whiter eye ring and a tail more rufous than its back. **CONSERVATION STATUS** Least Concern.

Rufous-backed Robin

Turdus rufopalliatus

Dense woodlands
LENGTH 8–9.5 in. WINGSPAN 15–16 in.

Medium-sized thrush with pale lower belly and washed-out blue-gray plumage on head, wings, and tail; contrasted with warm rufous back, flanks, and breast. Throat is creamy white with deep brown streaks. Black-tipped, yellow bill. Eyes are deep rufous, thinly outlined in gold. BEHAVIOR Elusive, but seasonally gregarious, gathering in winter flocks. Can be spotted in mixed flocks with American Robins, typically moving about in dense canopy or brush. VOCALIZATIONS Song consists of bubbly, melodic, warbly phrases, similar to American Robin. Alarm call a descending *cheeuu* or *tseeuu*. ABUNDANCE Vagrant; regular rarity, with a small handful of individuals spotted in AZ and NM annually, typically during winter. Historical range along western coast of Mexico. LOOKALIKE American Robin. Rufous-backed distinguished by namesake rufous back, slender silhouette, and blue-gray coloration versus American's brown. CONSERVATION STATUS Least Concern.

American Robin

Turdus migratorius

Mixed woodlands, parks, urban/suburban landscapes, grasslands with some shrub/tree canopy
LENGTH 8–11 in. WINGSPAN 13–16 in.

Large and somewhat potbellied with a round head, a relatively long and heavy bill, long legs, a long tail, and long, sometimes drooping wings. Plumage is variable, but in general adult male is deep coffee-brown above and bright rufous below with a blackish head and nape, limited white eye arcs above and below the eyes, a small white lore spot, a yellow bill, sometimes white tail corners, and sometimes faint white streaking in the throat. Female is paler overall with a lighter head, more white in the throat, dark malar stripes, a faint white eyebrow, and a dingier bill. Juvenile resembles female with heavy spotting above and below. BEHAVIOR Flocks forage for invertebrates and fruit on the ground in open areas or crowding into low trees and shrubs. Individuals may be seen running in short bursts, standing tall with head tilted, or squabbling over captured prey. VOCALIZATIONS Song is varied, flute-like, whistled, and delivered in short phrases: *cheery-chireoo-cheery-chirup*. Also gives nasal laughter, abrupt yelps, and a *nuk-nuk-nuk* (like Curly of The Three Stooges comedy team). ABUNDANCE Common year round at middle and upper elevations across Northern and Eastern AZ and throughout all but extreme Southeast NM. Expands into lower elevations in winter and observable throughout the region. CONSERVATION STATUS Least Concern.

THRASHERS AND OTHER MIMIDS

In the Southwest, the mimids (family *Mimidae*) are represented by members of the thrashers, mockingbirds, and New World catbirds. Although all are similarly built with long and often decurved bills, long and powerful legs, earth-toned plumage, and in many species, brightly colored eyes, none of these traits earned them their family name. *Mimid* means "mimic" in Latin, and these birds sing complex, musical, and often distinctly phrased songs, with some species having the impressive ability to imitate the songs of other birds, the vocalizations of nearby mammals, and the clamors of urban spaces.

ABOVE Curve-billed Thrasher

Adult

Adult in typical hunched posture

Gray Catbird

Dumetella carolinensis

Dense thickets, shrubs; dense, low canopy
LENGTH 8–9.5 in. WINGSPAN 9–11.5 in.

Medium-sized, slender, long-tailed songbird. Simply and beautifully adorned in solid steel-gray plumage with a smart little black cap, sooty black tail, and bright cinnamon-colored undertail coverts, with straight black bill and black eyes. In flight, look for long, rounded tail and rounded wings. BEHAVIOR Often heard before seen. Catbirds occupy dense vegetation as they forage for insects and berries, often making a ruckus with their movements and erratic vocalizations as they navigate through thickets. Males will perch atop shrubs and on open branches during breeding season to sing. VOCALIZATIONS Unmistakable lexicon includes a song of chattering, mechanical, and musical notes—as if a robotic bird were malfunctioning in the thickets—and the distinct, harsh *mew* call that sounds like a scolding cat meow or the harsh cry of an infant. ABUNDANCE Uncommon throughout breeding and migratory range. LOOKALIKES Northern Mockingbird, Townsend's Solitaire. Gray Catbird distinguished by vocalizations, black cap, solid gray coloration, and rufous undertail coverts. CONSERVATION STATUS Least Concern.

Adult (Chihuahuan)

Adult (Sonoran)

Curve-billed Thrasher

Toxostoma curvirostre

Brushy, arid habitats including deserts, scrublands, mesquite and pine-oak woodlands, suburban areas, parks
LENGTH 11 in. WINGSPAN 13.5 in.

Medium-sized, short-winged, and gangly, with a long tail, long legs, and a long and deeply curved bill. Sonoran form is dusty gray-brown above and pale gray to off-white below, with orange-gold eyes, a pale throat, a faint malar stripe, buffy undertail coverts, grayish flanks, and faded, round spots arranged in messy streaks on the breast and belly. Chihuahuan form is more ornate with white throat, wingbars, and tail corners and paler underparts with heavier spotting. Sonoran and Chihuahuan forms blend where the two deserts collide. Juvenile has a shorter, straighter bill. BEHAVIOR Forages on the ground, using its bill to rake soil and flip cover in search of invertebrates. Also takes berries, cactus fruit, seeds, and nectar from large flowers. Frequently found among plentiful cholla cactus. VOCALIZATIONS Song is clearer and more distinctly phrased than Bendire's Thrasher, with a musical series of harsh whistles. Classic call is a loud *WHIT-WHEET*. ABUNDANCE Common year round across Eastern and Central to Southern NM and across most of Central to Southern AZ. LOOKALIKE Bendire's Thrasher resembles Sonoran form. Curve-billed prefers more heavily vegetated habitat and has round spotting below; a longer, more deeply curved, and all dark bill; and a distinct voice. CONSERVATION STATUS Least Concern (Decreasing).

Brown Thrasher

Toxostoma rufum

Dense woodland scrub, forest edges with dense underbrush, riparian areas
LENGTH 9.5–12 in. WINGSPAN 11.5–13 in.

Sturdy bird with long bill, long tail, and long body. Chock-full of bold features, but keeps a low profile. Deep, ruddy coloration overall, with pale tan face; striking golden eyes; long, slightly downcurved black bill with pale lower mandible; and two delicate white wingbars. Creamy white underparts are boldly streaked in warm dark brown. BEHAVIOR Forages in low, dense canopy and on the ground in dense vegetation or just along the edges of hedgerows and shrubs, scraping and picking through organic matter for arthropods, earthworms, and small amphibians. Also forages for berries. During breeding season, both males and females engage in the endearing behavior of passing along small twigs and other gifts. VOCALIZATIONS Song is a complex string of singsong phrases, each typically repeated twice and varying from individual to individual. Mimics, they often incorporate their interpretations of other bird species' songs into their repertoire. Calls are a variety of harsh, gravelly *churp*s and *chup*s and smacking, kisslike chips. ABUNDANCE Uncommon but regularly observed within wintering range. Typically an East Coast species, with one isolated wintering population in Central NM. CONSERVATION STATUS Least Concern (Decreasing).

Bendire's Thrasher

Toxostoma bendirei

Arid, brushy, sparse habitats including dry grasslands, open deserts, scrublands, pastures, abandoned agricultural lands
LENGTH 10 in. WINGSPAN 13 in.

Medium-sized, short-winged, and spindly with a long tail, long legs, and a relatively long and subtly curved bill showing a usually light base. Both sexes are a soft gray-brown above and pale grayish, buffy-brown, or off-white below with straw-yellow eyes, a pale throat, a faint malar stripe, buffy brown flanks and undertail coverts, pale tail corners, and faded, triangular spots arranged in tangled rows on the breast and belly. BEHAVIOR Forages on the ground, using its bill to poke into soil and flip debris in search of invertebrates. Will also forage in low vegetation for seeds and fruit. VOCALIZATIONS Song, a melodious series of buzzy whistles, is harsher than that of Curve-billed Thrasher, lacks distinct pauses between phrases, and repeats phrases two or four times each. Call is a rough and sometimes rattling *chek*. ABUNDANCE Common year round in a limited range from West Central to South Central AZ. Expands northward and eastward in summer and can be found, though less commonly, in Northern and Eastern AZ and from North Central to Southwest NM. LOOKALIKE Sonoran form of Curve-billed Thrasher. Bendire's uses starker habitats and has triangular spotting below; a shorter, straighter, and pale-based bill; and a unique voice. ALTERNATIVE NAMES Scrubland Thrasher, Scrub Thrasher. CONSERVATION STATUS Vulnerable (Decreasing).

LeConte's Thrasher

Toxostoma lecontei

Low and extremely arid deserts, preferably with saltbush scrub, cholla, mesquite
LENGTH 9.5–11.5 in. **WINGSPAN** 12 in.

Long and lean, with long, narrow tail and thick, decurved, black bill. Plumage is pale sandy color overall, with grayish tone on back and warmer tone underneath. Pale belly and creamy throat accented by thin, subtle malar stripe. Subtle, peachy undertail coverts. Eyes are dark. **BEHAVIOR** Ground forager, uses its long, curved bill to unearth prey, including arthropods and other small animals such as reptiles and amphibians; sometimes hunts in pairs. Also feeds on berries and seeds. Often runs along the desert floor in the manner of a roadrunner, with tail held high. Like some other desert thrasher species, prefers to build bulky nests in dense cholla cactus. A true desert creature, it rarely partakes of water but gleans most of its moisture from insect bodies. **VOCALIZATIONS** Song varies remarkably by individual, with a variety of musical phrases sung in a high, squeaky pitch. Call is an ascending, whistling *tooWEEP*. **ABUNDANCE** Uncommon resident throughout range. **KEY SITE** AZ: west of city of Buckeye, Mohawk Valley. **LOOKALIKE** Crissal Thrasher. LeConte's distinguished by dark eyes and paler color overall, including pale peach undertail coverts versus cinnamon-colored undertail of Crissal. **ALTERNATIVE NAMES** Sickle-billed Sand Thrasher, Saltbush Thrasher, Pallid Thrasher. **CONSERVATION STATUS** Least Concern.

Crissal Thrasher

Toxostoma crissale

Washes, canyon bottoms, foothill drainages within arid, brushy, mid-elevation habitats such as high deserts, saltbush communities, chaparral, shrubby mesquite, pine-oak and pine-juniper woodlands
LENGTH 11.5 in. **WINGSPAN** 12.5 in.

Medium-sized, short-winged, and lean, with a long tail, legs, and exaggeratedly curved bill. Both sexes are a rich gray above and below with a bright white throat, black-and-white malar stripes, pale yellow eyes, soft black tail, and a burnt-rufous crissum, or undertail coverts. **BEHAVIOR** Usually under dense cover; uses its bill to disturb soil and toss debris in search of invertebrates, occasionally hopping into low vegetation to forage for fruit. Often takes a conspicuous midlevel perch to scan the area or vocalize, providing clear but brief viewing opportunities. **VOCALIZATIONS** Song is a musical series of whistling, trilling, and rattling phrases, delivered relatively slowly with distinct pauses in between and repeated two or three times each. Common call is a high, clear *churee-churee* or *churee-churee-churee*. **ABUNDANCE** Uncommon and inconspicuous year round in Western and from Central to Southern AZ; in NM along the Rio Grande as far north as Albuquerque and across the southernmost portion of the state as far east as the Pecos River. **LOOKALIKE** Le Conte's Thrasher. Crissal distinguished by pale eyes and darker plumage. **CONSERVATION STATUS** Least Concern.

Sage Thrasher

Oreoscoptes montanus

Sagebrush, desert scrub, gentle slopes, mesas
LENGTH 8–9 in. WINGSPAN 11.5–12.5 in.

Unique among thrashers, with diminutive, thrushlike stature (straight posture, with wings held low) and relatively short, straight bill. Plumage is sandy gray overall, with heavy deep brown spotting on pale underparts, becoming more streaked along lower belly and peach-colored flanks. Outer tail feathers have subtle white tips, and wings show two faint white bars. Eyes are pale straw color. BEHAVIOR Forages through dense brush, eating a varied diet of insects and berries. Will concentrate in groups during winter months to dine on a variety of fruits—mistletoe, elderberries, hackberries, and sometimes cultivated crops—and nests primarily in sagebrush, saltbrush, or rabbitbrush. Males vocalize incessantly during breeding season, perching atop shrubs. VOCALIZATIONS Song is a lovely, rambling, clear, and sweet series of warbling phrases, occasionally including mimicked songs of other birds. Call is a low, gravelly cluck or *chuck*. ABUNDANCE Uncommon throughout range with exception of northernmost breeding range in NM and AZ. LOOKALIKES Bendire's, Curve-billed thrashers. Sage distinguished by short, straight bill; heavily patterned breast and wings; and silhouette that resembles that of a mockingbird. CONSERVATION STATUS Least Concern (Decreasing).

Northern Mockingbird

Mimus polyglottos

Suburban settings including parks and backyards; low- to middle-elevation and relatively open habitat with plentiful low, dense shrubs LENGTH 8.5–10 in. WINGSPAN 12.5–14 in.

Medium-sized and slender with a small, rounded head and elongated, slightly curved bill; long legs; long tail; and short, sometimes drooping wings. Adults are brownish gray above and grayish white to buff below and have dark lores, black flight feathers with white edging, two white wingbars, and a tail that is soft black above and mostly white below. In flight, note bold white wing patches and tail edges. Juvenile is indistinctly spotted below. BEHAVIOR Conspicuous in all ways, with occasional flashes of the bold white wing patches, singing from obvious perches long into the night, foraging on open ground, and aggressively defending territories. Consumes mostly insects in spring and summer, switching mostly to fruit in fall and winter. VOCALIZATIONS A talented mimic, delivers imitations of birds and other sounds, each repeated in always even-numbered sets of two to six before moving on to the next set. Call is a harsh *CHEK*, similar in quality to two rocks being hit together. ABUNDANCE Common; a breeding resident in Northeast AZ and Northern NM and a permanent resident throughout the remainder of the region. CONSERVATION STATUS Least Concern.

STARLINGS, WAXWINGS, SILKY-FLYCATCHERS, AND OLIVE WARBLER

A seemingly motley grouping, there is a common theme among these disparate families: each is represented in the Southwest by only a single species. The waxwings (family *Bombycillidae*) and silky-flycatchers (family *Ptiliogonatidae*) come from extremely small families of just two and three species, respectively. Olive Warblers (family *Peucedramidae*) are a family of one; and the starlings (family *Sturnidae*), while a very diverse family, are represented in North America by only a single species brought to the continent by misguided colonialists. As the sole representatives of their families, these birds are as unique as they are difficult to organize.

ABOVE Male Phainopepla

Breeding adult

Nonbreeding adult/immature

European Starling

Sturnus vulgaris

Areas near human developments, including parks, agricultural fields, urban and suburban areas, sports fields, lawns
LENGTH 8–9 in. WINGSPAN 12.5–16 in.

Small but stocky songbird with long, straight bill and short tail. Dull black at first glance; a closer look at breeding/summer plumage reveals a glossy, iridescent palette of teal, purple, and sapphire, with a waxy yellow bill and orange legs. Nonbreeding adults and juveniles show fine white speckles from head to tail, including underparts. Look for star-shaped bird in flight, formed by sharply pointed wings, tail, and head. BEHAVIOR Oft maligned for their introduced status, these colorful, intelligent, and social characters engage in mimicry, murmuration, and other irresistible antics that earn most birders' eventual admiration. Cacophonous flocks gather in sometimes spectacular numbers, foraging along the ground and perching on wires, and never being quiet about it. Will often join mixed flocks of blackbirds and/or grackles. Opportunistic omnivores (insects, fruit, seeds) and cavity nesters, using human-made or natural cavities, including large columnar cactus species such as the saguaro. VOCALIZATIONS The starling's illustrious lexicon is difficult to narrow down, as it is a talented mimic with great vocal ability, from dripping water to the harsh cries of Red-tailed Hawk. Calls vary from a whirring, wheezy *weee* to rattling chatters and low purrs. ABUNDANCE Common resident throughout range, with the exception of decreased abundance in the southwesternmost portion of AZ. Most prevalent near human development. CONSERVATION STATUS Least Concern (Introduced).

Adult

Juvenile

Cedar Waxwing

Bombycilla cedrorum

Woodlands, parks, urban and suburban areas with fruiting trees, orchards
LENGTH 5.5–7 in. WINGSPAN 8–11 in.

Medium-sized, compact bird with short neck, crested head with a flat top and anvil-like silhouette, pointed bill, and short, squared tail. Sleek, distinguished plumage, with jet-black mask bordered in snow-white, and warm tan head fading to soft cocoa-brown down the scapulars to cool gray on the lower back and tail, tipped in bright yellow. Warm tan extends from the head down the belly, fading below to a pale wash of yellow. *Waxwing* is a literal reference to the lipstick-red, waxy tips that form on the secondaries of the wing. BEHAVIOR Vocal, fruit-seeking flocks course through woodlands on the hunt for juicy serviceberries or juniper berries, though just about any fruit or berry will do. Aside from its primary food source of fruit, it dines on insects when the opportunity arises. Highly social, and a delight to welcome to backyard fruit trees. VOCALIZATIONS High-pitched, whispery trills and soft, tinkling *tsee*ps and *tsee*s. ABUNDANCE Common throughout winter range. CONSERVATION STATUS Least Concern.

Adult male | Adult female

Phainopepla

Phainopepla nitens

Mesquite, palo verde, ironwood, other shrubby tree–dominated deserts, often near washes or streams and in areas with plentiful desert mistletoe; in chaparral in late summer **LENGTH** 7–8 in. **WINGSPAN** 10.5–11.5 in.

Medium-sized and elegant with slender body, long tail, scraggly crest, red eyes, and short, pointed bill. Adult male is entirely silky black with bold white wing patches visible in flight. Female is dark gray to grayish brown, lacks white wing patches, and has a subtle dark mask, a dark tail, and black flight feathers edged in white. **BEHAVIOR** A mistletoe specialist, forages heavily on the berries of and nests within clumps of this parasitic plant. Also forages on other berries and on insects, which it either gleans from vegetation or takes in flight. Both sexes take obvious high perches and males perform circling, fluttering flight displays during courtship and nesting. **VOCALIZATIONS** Song, reminiscent of a very slow thrasher, is a series of short and distantly spaced whistles, scolds, and blackbirdlike phrases. Common calls include a short, rising, breathy, and single-note whistle and a harsh, buzzing scold. **ABUNDANCE** Common year round in Southwest NM; also in Western and from Central to Southern AZ. Pushes slightly farther north during the heat of summer. **CONSERVATION STATUS** Least Concern.

Adult male

Adult female

Olive Warbler

Peucedramus taeniatus

Open conifer forests, preferring stands of tall pine
LENGTH 5–5.5 in. WINGSPAN 9–9.25 in.

Recently removed from the warbler family, yet similar to a warbler in appearance, with slender, elongated bill and notched tail. Male shows marigold-orange head with black ear patch that extends forward over the eye like a partial mask, charcoal-gray back and pale gray belly, with two bold white wingbars on sooty gray wings. Female shows paler yellow head with gray ear patch/mask. Immature shows faint wash of yellow on face, breast, and back, with greenish tinge on outer wing feathers. BEHAVIOR Unless you spot this bird from a ridge or high peak, be prepared to identify it from its underparts, as it spends nearly all of its time in pine canopy, gleaning insects as it hops about and occasionally flying out to nab one midair. Diet believed to consist nearly exclusively of insects. Nests high in pine canopy, with females typically constructing nests at least 30 feet up. VOCALIZATIONS Song is a repeated series of two-note phrases, reminiscent of a titmouse or ovenbird: *teacher-teacher-teacher* or *peter-peter-peter*. Call is a sweet, high-pitched *keew*. ABUNDANCE Uncommon throughout both summer and resident ranges. KEY SITES High-elevation pine forests throughout the Madrean Sky Island Region of Southeast AZ and Southwest NM. Near Phoenix, breeds in highest elevation pine forest of the Mazatzal Mountains and atop the Mogollon Rim. LOOKALIKES Hermit, Townsend's warblers. Olive distinguished by vocalization, full hood, bold ear patch, and clean chest. CONSERVATION STATUS Least Concern (Decreasing).

OLD WORLD SPARROWS, PIPITS, FINCHES, AND LONGSPURS

Each of these plump-bodied, small to medium-sized, and frequently flocking birds has unique adaptations that help it thrive in a particular niche. The Old World sparrows (family *Passeridae*) are represented in North America by only the House Sparrow, an imported species with an unsurpassed ability to flourish in urban spaces. The pipits (family *Motacillidae*), which spend their days pursuing insects on the ground in open habitat, are similar in appearance to shorebirds, with cryptic plumage, long legs, and pointy bills. The finches (family *Fringillidae*), compact birds with heavy bills, are mostly seed specialists, with the shape of the beak often indicative of their seeds of choice. Lastly are the longspurs (family *Calcariidae*), buntinglike and ground-dwelling birds that, when they visit the Southwest in their drab winter plumage, vanish effortlessly into open grasslands.

ABOVE Nonbreeding Adult American Pipit

Nonbreeding male

Breeding male

Adult female

House Sparrow

Passer domesticus

Human-modified landscapes, from low deserts to high forests; densely urban, suburban, or sparsely inhabited landscapes; largely absent from unmodified natural habitats
LENGTH 6–6.5 in. WINGSPAN 7.5–9.5 in.

A beefy sparrow with a large and rounded head, sturdy body, short tail, and heavy bill. Breeding male is fancily patterned with a black bill, gray crown, black lores and throat/bib, white cheeks and sides of neck, rich brown behind the eye and into the neck and nape, gray underparts, and a brown back marked with black-and-buff streaking. Nonbreeding male is similar, but less richly colored and with a pale bill. Adult female is paler overall, with a light brown crown; buffy stripe above and behind the eye; dark eyeline; grayish brown cheeks and underparts; a heavily streaked brown, black, and buff back; and a pale or bicolored bill. Juvenile resembles female but more drab. BEHAVIOR An impressive generalist, may be seen foraging for seeds, scratching for insects, or pecking at French fries in your local fast-food parking lot. Usually found in raucous flocks. VOCALIZATIONS Calls and songs consist of harsh and somewhat nasal *choop* and *chir-up* notes. Less varied and musical compared to House Finch. ABUNDANCE Common year round throughout the entire region. CONSERVATION STATUS Least Concern (Introduced).

Breeding adult

Nonbreeding adult

American Pipit

Anthus rubescens

Mudflats, agricultural fields, muddy shorelines, plains, meadows
LENGTH 5.5–6.5 in. WINGSPAN 10–11 in.

Dainty, lark-like songbird with slender bill and medium-length tail. Plumage is a sandy tan overall, with solid taupe back and pale belly that shows a peachy buff wash during breeding and heavy brown streaks during winter months. Bill is darker during breeding season. Cheek is deep taupe or tan with pale supercilium and bold white eye ring. BEHAVIOR Look for pipits foraging through fields and on mudflats, cocking their heads back and forth like chickens, and bobbing their tails upward as they take a pause. VOCALIZATIONS Song is a sweet, repetitive, gradually quickening *twee*. Call is a namesake *pip-it, pip-it*, as well as soft *pee-pee-pee*. ABUNDANCE Common throughout breeding and wintering range, though populations limited to migration seasons in northernmost areas of AZ and NM. LOOKALIKE Sprague's Pipit. American distinguished by diagnostic tail-bobbing, bold eye ring, plain or faintly patterned back, and darker face. ALTERNATIVE NAME Buff-bellied Pipit (by which it is recognized throughout much of its global range). CONSERVATION STATUS Least Concern (Decreasing).

Sprague's Pipit

Anthus spragueii

Open, grassy habitats including short grasslands, pastures, agricultural fields
LENGTH 6.5 in. WINGSPAN 10 in.

Small, slender, and lark-like, with a relatively short, thin bill; a small, rounded head; a medium-length tail; and fairly long legs. Adult is dark and heavily streaked or scalloped in brown, black, and buffy white above and pale whitish buff below with a dark-streaked crown; a necklace of thin, dark streaks confined to the breast; pale pinkish legs; white outer tail feathers; a very faintly patterned face showing a light malar stripe and dark ear; and a very faint eye ring. Juvenile is similar with obvious white wingbars. BEHAVIOR In winter, spends its days alone or in small flocks, foraging inconspicuously for insects and seeds under low, grassy cover, rarely mixing with other pipits. When flushed, has a tendency to shoot straight up, circle, and dive sharply back into grass. VOCALIZATIONS Song, almost always given in flight, is a high, raspy, and rising then downwardly cascading *psew-psew-psew-psew*. Call is a high squeak or chip similar to that of woodpecker. ABUNDANCE Local and uncommon to rare in winter from Central to Southeast AZ and across Southern NM. Rare farther north during migration from late summer to fall, most commonly in Eastern NM. LOOKALIKE American Pipit. Sprague's distinguished by less heavily streaked underparts, pinkish legs, streaked and scalloped back and crown, plainer face, and fainter eye ring. ALTERNATIVE NAME Great Plains Pipit. CONSERVATION STATUS Vulnerable (Decreasing).

Adult male

Adult female

Evening Grosbeak

Coccothraustes vespertinus

High-elevation woodlands, primarily pinyon-juniper, pine-oak, spruce-fir
LENGTH 6–7 in. WINGSPAN 12–14 in.

Large, beautiful, heavy billed, and thick-bodied finch. Female shows pale taupe plumage overall, with a golden wash around the nape and flanks, black-and-white wing feathers, and pale, waxy, thick conical bill. Male shows striking deep golden plumage overall, with jet-black wings and tail and a bold, pure-white wing patch. Head is golden brown with black cap and bright golden stripe over the eyes. BEHAVIOR A nomadic, irruptive, and gregarious species, gathering in flocks in winter and descending on backyard seed feeders and fruit-bearing trees. Feeds on insects in summer months, though primarily granivorous year round, with its conical bill designed for easy seed retrieval. Male gives a delightful courtship display during breeding season, complete with dance moves. Both sexes also engage in bowing behavior during courtship. VOCALIZATIONS Lacking a true song, its coarse, brief chirps are met with long pauses between notes. Calls are typically clear, high-pitched, one-note whistles. ABUNDANCE Uncommon throughout both resident and winter ranges. CONSERVATION STATUS Vulnerable (Decreasing).

Adult male

Adult female

Pine Grosbeak

Pinicola enucleator

Spring through fall in humid valleys within open montane spruce-fir forests at elevations approaching tree line; in winter, may move to lower elevation coniferous and mixed forests of spruce, fir, maple, ash
LENGTH 9 in. **WINGSPAN** 14.5 in.

A heavy bodied, robin-sized finch with a short but substantial seed-crunching bill and a long, subtly notched tail. Both sexes have a base coat of slate-gray with dark wings and white wingbars. Adult male has deep red to pinkish red wash over head and, in varying amounts, breast, belly, flanks, and rump. Adult female and immature birds show yellow to orangish wash on head and rump. **BEHAVIOR** In small, slowly moving flocks, forages for seeds, fruit, and fresh green growth in trees and shrubs. Productive breeding seasons can lead to population irruptions that send birds wandering outside their usual range in winter. **VOCALIZATIONS** Songs and calls reminiscent of other finches but relatively slower, richer, and clearer. **ABUNDANCE** A year-round but uncommon and irregular resident found most reliably from NM's Sandia Crest northward into the Southern Rockies. Even less commonly encountered in AZ, with scattered records in the northern and eastern portions of the state at sites including the Mount Baldy Wilderness, the San Francisco Peaks, and Grand Canyon National Park. **CONSERVATION STATUS** Least Concern (Decreasing).

Breeding adult

Nonbreeding adult

Gray-crowned Rosy-Finch

Leucosticte tephrocotis

High-elevation montane forests, including alpine tundra, snow fields, meadows; in winter, will occupy open suburban habitat, regularly frequenting feeders
LENGTH 6–8 in. WINGSPAN 12–13 in.

Small, sturdy, short-tailed finch. Interior form (one of six) inhabits this range, showing gray crown with distinct black forehead, brown cheek, and warm brown coloration overall, accented by blushing pink wash on the flanks, belly, and wings (pink coloration more prominent in males). Tail feathers are deep sooty gray. Conical bill shows yellow in winter (black in breeding). Juvenile brown overall, with gray flanks and dark edges on primaries, lacking rosy coloration and gray crown entirely. BEHAVIOR Ground forager, with a primarily granivorous diet in winter (shifts to insects in breeding/summer range), may be seen foraging in the harshest of settings and conditions, from steep slopes to frigid tundra. VOCALIZATIONS Calls are rising, buzzy chirps and *cheer* notes. Song is quite similar but with chirps varying in pitch and cadence. ABUNDANCE Uncommon, but regularly observed overwintering in high-altitude habitats throughout northernmost NM. KEY SITES NM: Sandia Crest, Cibola National Forest. LOOKALIKES Black, Brown-capped rosy-finches. Gray-crowned distinguished by pale gray cap and warm brown coloration on body, versus sooty black coloration and brown cap, respectively. CONSERVATION STATUS Least Concern.

Breeding adult

Nonbreeding adult

Black Rosy-Finch

Leucosticte atrata

Breeds above the tree line in areas with plentiful cliffs, talus slopes, boulder piles; in harsh winters, moves lower into open habitat such as montane meadows, high deserts, short grasslands, parks
LENGTH 5.5–6.25 in. WINGSPAN 13 in.

Medium-sized bird with a long, notched tail; pointed, conical bill; and head that can appear flattened or somewhat peaked. Breeding adult is mostly black with a black bill, a strongly contrasting gray crown on the back of the head, and bright salmon-pink in the wings and belly. Nonbreeding adult is grayish black and sports a yellow-orange bill. Juvenile is grayish brown with buffy wingbars and a yellow-orange bill. BEHAVIOR In winter, forages on open ground for insects and seeds within large mixed flocks of other rosy-finch species. VOCALIZATIONS High, raspy chips and a buzzy slur reminiscent of House Sparrow. ABUNDANCE Uncommon to rare in winter from NM's Sandia Crest northward into the high peaks of the Southern Rockies. Rare in Northern AZ, with scattered winter records mostly on the Navajo Nation between Page and Bitter Springs. LOOKALIKES Other rosy-finches. Black Rosy-Finch distinguished by contrasting gray crown, black cheeks, and overall black color in breeding adult; grayer overall color and limited pink in nonbreeding adult; and buffy wingbars in juvenile. CONSERVATION STATUS Endangered (Decreasing).

Breeding adult

Nonbreeding adult

Brown-capped Rosy-Finch

Leucosticte australis

Rocky cliffs, alpine meadows; will descend to forests/woodlands during winter storms
LENGTH 5.5–6.5 in. **WINGSPAN** 12–13 in.

Gorgeous finch, short-tailed and stocky. Breeding adult shows rich brown above and deep pink below, like a cocoa-dusted raspberry, with subtle charcoal-brown cap, dark conical bill, and dark legs. Nonbreeding adult similar in appearance, with muted coloration and yellow bill outside of the breeding season; female coloration far more subtle overall. Juvenile shows cocoa-brown coloration overall, with heavily patterned, rosy edged wings. **BEHAVIOR** Typically ground foragers, plucking seeds, insects, and arachnids from snow fields, particularly along melting edges. Driven to lower elevations in winter, with large flocks gathering along roadsides, fields, and other developed areas to forage; also known to frequent feeders. Crevice nesters, they often build nests in remarkably narrow cliffside cracks, mine shafts, or even beneath rocks on rocky cliff faces. **VOCALIZATIONS** Male song is a series of sweet, high-pitched *chews* and raspy chirps. Calls are low, sweet chirps. **ABUNDANCE** Endemic to Rocky Mountain high peaks, with a range that just barely extends into North Central NM; breeding populations have declined drastically in NM portion of range. Migration is altitudinal: high elevation in breeding season, lower elevation in winter. **KEY SITES** NM: Sandia Crest, Cibola National Forest. **LOOKALIKES** Gray-crowned, Black rosy-finches. Brown-capped distinguished by lack of distinct gray crown and overall cocoa-brown coloration. Juveniles distinguished by lack of wingbars. **CONSERVATION STATUS** Endangered (Decreasing).

Female/immature

Male

House Finch

Haemorhous mexicanus

Open areas with widely spaced trees and shrubs, including deserts, wooded grasslands, chaparral, sparse woodlands, forest edges, agricultural fields, yards, parks, other urban and suburban settings
LENGTH 5–6 in. WINGSPAN 8–10 in.

Relatively small with a flattened head, a medium-length and barely notched tail, and a short, heavy bill with a deeply curved culmen (upper edge of upper mandible). Adult male is grayish brown and streaked above and white with coarse, dusty brown streaks below. Shows two pale wingbars and red, or more rarely yellow or orange, in the head, face, breast, and rump. Female and juvenile lack red coloration and have indistinctly patterned faces. BEHAVIOR Almost always in boisterous flocks. Frequently observed at backyard seed feeders, perching in clusters on high branches, or foraging together on the ground, in shrubs, or in trees. VOCALIZATIONS Song is a meandering, cheerful, and musical jumble of short, clear chips and whistles, often ending in a short buzz. Call is a high, clear, and slightly nasal *CH-WEEP*, often given in repetition and as a chorus from flocks. ABUNDANCE A common, permanent resident across both states. LOOKALIKES Cassin's, Purple finches. House Finch distinguished from Cassin's by lack of deeply saturated, contrasting crown, streakier underparts, and round versus peaked head, diffusely patterned faces, and curved culmens in both sexes; from Purple by less extensive red in male, especially on back and wings, and by lack of obvious white eyebrow and malar stripes in female. CONSERVATION STATUS Least Concern.

Adult male | Female/immature

Cassin's Finch

Haemorhous cassinii

Montane aspen/conifer forests; occasionally lower elevations throughout winter
LENGTH 6–6.5 in. **WINGSPAN** 10–11 in.

Compact, stocky finch with thick, conical bill; short, notched tail; and subtly peaked crown. Notably long wings. Adult males show rosy mauve and brown overall, with a bright berry-red crown, pink breast, and pale gray belly with pink wash. Females and immature males show brown overall, heavily and cleanly streaked with brown on creamy white belly, and subtle tonal streaking on head and back. **BEHAVIOR** Primarily an arboreal forager, will take advantage of food along the ground and in low vegetation, feeding mainly on seeds, buds, and berries (the carotenoid source that results in the male's bright red crown), Drawn to black oil sunflower seeds at backyard feeders. Often seen in mixed flocks with other montane finch species; usually forages in flocks. **VOCALIZATIONS** Male gives hurried series of warbles; capable of mimicking other birds' songs, often inserting them into its own. Both sexes give fluid, two- or three-syllable calls: *two-lip* or *too-A-loo*. **ABUNDANCE** Common throughout range, diminishing in abundance toward southern extent of wintering range. **LOOKALIKES** House, Purple finches. Male Cassin's distinguished by peaked, strawberry-red crown and clean, pink-washed underparts versus streaked bellies of House and Purple. Immature male and female Cassin's distinguished by clear, bold streaking versus blurry streaking, with Purple showing brown cheek and contrasting white supercilium that is more subtle and mottled on Cassin's. **ALTERNATIVE NAMES** Mountain Berry-crowned Finch, Western Red-crowned Finch. **CONSERVATION STATUS** Least Concern (Decreasing).

Adult Female

Adult Male

Immature

Red Crossbill

Loxia curvirostra

Mature coniferous forests of spruce, fir, pine
LENGTH 5.5–7.5 in. **WINGSPAN** 10.5–11.5 in.

Medium-sized and chunky with a short, notched tail; long, pointed wings; and a relatively long, heavy bill with thin, crossing tips. Occurs in at least eleven different varieties, each with slight differences in bill shape and overall size. Adult male is ferrous-red overall with dark brownish red wings and tail; a subtle brownish eyeline and ear patch; and occasionally faint wingbars. Female is dusky yellow overall with a more obvious ear patch. Juvenile is brownish gray above and white with coarse brown streaking below. **BEHAVIOR** Forages in flocks high in conifers, using specialized bill to extract seeds from cones. **VOCALIZATIONS** Song is a series of phrases with abrupt, buzzy blurts, sandwiched between strings of high chips and whistles. Call, often given in flight, is a high-pitched *kip-kip-kip*. Vocalizations vary subtly between populations. **ABUNDANCE** Uncommon, nomadic, and irregular year round at high elevations from Central to Western NM and in Northern and Eastern AZ. In winter, sometimes wanders to lower elevations and can be found anywhere planted conifers and cones are present. **LOOKALIKE** White-winged Crossbill. Red Crossbill lacks sharply contrasting white wingbars in all plumages, shows grayish wings and tail in adult male, and lacks heavy streaks below in adult female. **CONSERVATION STATUS** Least Concern.

Pine Siskin

Spinus pinus

Spring to fall, in relatively open coniferous or mixed forests; winter flocks in habitats including open woodlands, chaparral, wooded grasslands, scrubby fields and roadsides, suburban settings such as parks, grassy fields, backyards
LENGTH 4.5–5.5 in. WINGSPAN 7–9 in.

Small with long, pointed wings; a short, notched tail; and a slender, pointed bill. Adult male is grayish brown above, whitish below, and heavily streaked throughout with bold yellow wingbars and varying amounts of yellow in the wings, undertail, tail edges, flanks, breast, back, and head. Yellow undertail, tail edges, and wing stripe become clearly visible in flight. Female and juvenile show significantly less yellow. BEHAVIOR Forages for small seeds, especially those of coniferous trees, in talkative, wandering flocks. Frequently observed dangling from branch tips like a chickadee or clamoring around backyard thistle feeders. VOCALIZATIONS Song is a thrasher-like series of short phrases of reedy buzzes, whistles, and chirps. Calls include a buzzy, ascending *zreeeeeee* and a short *chit-a-chit* given in flight. ABUNDANCE Common but nomadic and irregular year round at high elevations from Central to Western NM and in Northern and Eastern AZ. Found regionwide in winter, when flocks move into atypical habitat in search of food. Rare in Southwest AZ. CONSERVATION STATUS Least Concern.

Lesser Goldfinch male (right) with females

Lesser Goldfinch

Spinus psaltria

Open, shrubby areas in suburbs, parks, open fields rich in seed heads, urban canopy, montane canyons, lush deserts
LENGTH 3.5–4.5 in. WINGSPAN 6–8 in.

Small, slight finch with squat bill and short, notched tail. Male shows yellow underneath with slick, dark cap; plumage on back ranging from olive-green to solid black (individuals from eastern portion of range tend to show darker back); and bold white corners on tail. Female/immature male shows faded olive-green and yellow coloration overall, with faded gray-black wings and white, rectangular wingbars/patches. BEHAVIOR Extremely social granivores, regularly frequenting backyard feeders in small to impressively large, chattering flocks, sometimes intermingling with other species. VOCALIZATIONS Song is a rambling series of sweet twitters and *chits*, with a stuttering, whiny quality. Call is a clear, descending *tee-ler* or a quick *chit*. ABUNDANCE Common throughout their year-round and breeding ranges, but somewhat less so in Southwest AZ. LOOKALIKES American, Lawrence's goldfinches. Lesser distinguished from American by smaller size, full black cap in breeding males, white rectangular wing patch in female/immature birds, and yellow undertail coverts; distinguished from Lawrence's by yellow versus gray overall plumage. CONSERVATION STATUS Least Concern (Decreasing).

Adult male (left) and female

Lawrence's Goldfinch

Spinus lawrencei

Unpredictably in a variety of relatively open, shrubby habitats including deserts, open woodlands, mesquite floodplains, agricultural fields, brushy riparian areas, scrubby fields; also suburban settings with plentiful seeds LENGTH 4–4.75 in. WINGSPAN 8.25 in.

Tiny with a small, pointed bill and a medium-length and slightly notched tail. Adult male is light gray, more brownish outside of breeding, with a black mask, yellow central breast, black wings with extensive yellow, and pale, unmarked underwings. Female and juvenile lack mask and are browner and more subtly yellow overall. BEHAVIOR Forages in low weeds and shrubs, often alongside Lesser Goldfinch or other small seed-eating songbirds. Roaming winter flocks appear and vanish with little warning and are always a delight to birders lucky enough to cross their paths. VOCALIZATIONS Song is higher and thinner than other goldfinches—a cheerful burst of rapidly delivered whistling, chirping, and trilling phrases. Common call is a breathy, descending, and whistled *too-do*. ABUNDANCE Uncommon, irregular, and nomadic in winter when they expand from their mostly Californian range into Central and Southern AZ and as far east as Southwest NM. ALTERNATIVE NAME Wandering Goldfinch. CONSERVATION STATUS Least Concern (Decreasing).

Flock of nonbreeding

Immature

Breeding male (top) and female

American Goldfinch

Spinus tristis

Fields, plains, meadows, roadside thistle patches, suburban parks, urban/suburban backyards
LENGTH 4–5 in. WINGSPAN 7.5–8.5 in.

Slight but sturdy little songbird with short, rounded head; short, conical bill; and short, notched tail contrasted by long, pointed wings. In flight, look for dramatically undulating, bobbing flight pattern. Breeding male shows sunshine-yellow overall, with neat black forehead cap, white undertail coverts, and clean, jet-black wings with bold white wingbars. Adult female shows duller straw-gold/olive-green on back and lacks bold forehead cap. Both sexes show orange bill when breeding, silvery gray during winter. Immature individuals are golden tan overall with dark wings and warm, buffy wingbars. Nonbreeding adults and some immature individuals show golden straw-brown or pale yellow coloration and buffy tan underparts. Molting males show a combination of breeding and winter plumage in spring and fall. BEHAVIOR These seed lovers will gather in flocks large and small to feed on the seed heads of thistles, echinacea, asters, and sunflowers, much to the amusement of gardeners and backyard birdwatchers, and they can hardly resist a feeder filled with nyjer or black oil sunflower seeds. VOCALIZATIONS Flight call is a sweet "potato chip!" or an ascending, clear whistling *tee-wee?* Male courtship song is a series of jumbled warbles and tweets. ABUNDANCE Common throughout much of overwintering range, decreasing in frequency throughout southernmost AZ and NM. CONSERVATION STATUS Least Concern.

Breeding male

Nonbreeding adult

Lapland Longspur

Calcarius lapponicus

Winters in open, treeless habitats such as plowed agricultural fields, sparse grasslands, barren fields
LENGTH 6–6.25 in. WINGSPAN 9–11.5 in.

Small and sparrow-like with a small, pointed bill; short legs; elongated wings; and a rounded head often held high. Breeding male is striking, with a black-tipped, yellow bill and black crown, mask, and breast, contrasted by a white-and-yellow facial border and a chestnut-colored nape. Breeding female similar but with less black. Wintering birds of both sexes are similar, richly colored overall with a pinkish orange and dark-tipped bill; heavy buff, black, and brown streaking above; and bold, pale eyebrow on buffy face with a dark ear border and crown. Rufous patch on lower wing, and dark tail with thin white edges. Buffy breast and flanks show coarse streaking on white belly, and blackish diffuse breast band is darker in males. BEHAVIOR Often found singly or in small groups mixed with Horned Larks and others, but occasionally forms large flocks. Forages on sparse to barren ground mostly for seed. VOCALIZATIONS Song, not given in the Southwest, is reminiscent of Western Meadowlark. Calls are varied and include a high-pitched, whistled *see-you*; rapid chatter; and high-pitched chips. ABUNDANCE Rare in winter throughout the region, with greatest abundance in Northeast NM. LOOKA-LIKE Thick-billed Longspur. Nonbreeding Lapland has a more contrasting facial pattern, a smaller bill, streaked flanks, and does not show Thick-billed's dark T shape on white tail. CONSERVATION STATUS Least Concern.

Breeding male

Nonbreeding adult

Chestnut-collared Longspur

Calcarius ornatus

Shortgrass prairies; open, grazed areas
LENGTH 5–6.5 in. WINGSPAN 10–11 in.

Small but stout, short-tailed songbird. Breeding male shows solid black underside with straw-yellow throat and cheek, chestnut collar, bold white supercilium and malar stripe on black head, and golden brown back. Nonbreeding male and female show subtle and relatively dull, pale brown coloration overall with smudgy underparts. In all plumages, look for small gray bill and mostly white tail with a dark triangle at its center. BEHAVIOR During breeding season, males perch atop tall grasses and fence wires to sing. In winter, look for small flocks foraging along the ground, often endearingly bobbing their heads as they go. VOCALIZATIONS Song is a sputtering, descending warble with a mechanical sound quality, reminiscent of a meadowlark. Calls vary, including a rapid *chee-de-de-deet* and a sweet, finchlike whistle. ABUNDANCE Uncommon but regularly observed throughout migratory range. More commonly observed in southeast corners of wintering range in both AZ and NM. KEY SITES AZ: San Rafael Grasslands and Las Cienegas National Conservation Area. LOOKALIKES Other Southwestern longspurs. Nonbreeding Chestnut-collared adult distinguished by small, gray bill and duller coloration overall with a less distinct facial pattern. CONSERVATION STATUS Vulnerable (Decreasing).

Breeding male

Nonbreeding adult male

Thick-billed Longspur

Rhynchophanes mccownii

Grasslands, agricultural fields; breeds in shortgrass prairie; winters in areas dominated by short grasses, sparse shrubs
LENGTH 6 in. **WINGSPAN** 11 in.

Sparrow-like and small with large, conical bill; short legs; longish wings; round head; and often upright posture. Breeding male, not likely to occur in the Southwest, is gray with a black, gray, and white face; a black breast; a reddish wing patch; and a gray bill. Sexes are similar in winter, with streaked, brownish gray back; a buffy eyebrow; a relatively plain, brown face; plain, buffy breast; dingy white belly; and pinkish bill. In males, sometimes remaining hints of the rufous wing patch and black breast. In flight, dark tip and central stripe on white uppertail form a T shape. **BEHAVIOR** Found singly or in flocks, often with other ground-feeding songbirds of barren country. Walks on open ground seeking seeds and insects. **VOCALIZATIONS** Song, not sung in the Southwest, is warbling, liquid, and flutelike. Calls include a sharp chip and a dry *chit-ip*, often repeated rapidly. **ABUNDANCE** Uncommon in winter along the eastern and southern edge of NM and in extreme Southeast AZ. More widespread but still uncommon in migration, found mostly from Central to Eastern and Southern NM. Rare elsewhere during winter and migration. **LOOKALIKE** Lapland Longspur. Nonbreeding Thick-billed distinguished by more diffuse facial pattern, larger bill, unstreaked flanks, and distinct tail pattern. **CONSERVATION STATUS** Least Concern.

NEW WORLD SPARROWS

Though superficially similar to Old World sparrows and finches, New World sparrows (family *Passerellidae*) are actually more closely related to the Old World buntings. An incredibly diverse family, New World sparrows of the Southwest are small to medium-sized birds, are inclined to stay near the ground, and sport cryptically colored plumage, frequently longish tails, often intricately patterned faces, and conical bills. Although plumage patterns can be intricate, they can also be very similar between species, and many of these birds are reluctant to offer unobstructed views. To overcome the most difficult identification challenges, ignore the commonalities and pay special attention to behavior, habitat, key field marks, and voice.

ABOVE Adult Chipping Sparrow

Rufous-winged Sparrow

Peucaea carpalis

Thornscrub and desert grasslands with bunchgrasses, cacti, abundant spiny shrubs including mesquite, hackberry, graythorn LENGTH 5–5.75 in. WINGSPAN 7.5–8 in.

Small, slender, and long-tailed with a stout bill and a squared or slightly crested head. Adult is brown and streaky above, unmarked whitish gray below, with a gray face, rufous crown and eyeline, mostly pale bill, paired black malar stripes, and pale throat. Wingbars are dingy off-white to buffy, with a sometimes visible rufous patch near the shoulder. Juvenile is similar but buffier overall and with a more diffuse facial pattern and coarse streaking on the breast, flanks, and nape. BEHAVIOR Seen singly, in pairs, or sometimes in loose foraging groups but almost never mixed with other sparrows. Forages on the ground and in low vegetation for seeds and insects. VOCALIZATIONS Songs include a series of sweet chips, starting slow and accelerating with the cadence of a ball bouncing to a buzzing stop, and a fast trill preceded by two clear notes, reminiscent of Bewick's Wren. Calls include clear, high chips and whistles. ABUNDANCE Uncommon year round in a limited range from South Central to Southeast AZ. LOOKALIKES Rufous-crowned, Chipping sparrows. Rufous-winged distinguished from Rufous-crowned by paired malar stripes and mostly pale bill; from Chipping by bill color (all dark in breeding Chipping, pinkish in nonbreeding), malar stripes, and rufous (not black) eyeline. CONSERVATION STATUS Least Concern (Decreasing).

Botteri's Sparrow

Peucaea botterii

Open semi-arid grasslands and Gulf Coast prairie; areas with sacaton grass in AZ range
LENGTH 6–7 in. WINGSPAN 8–9 in.

Sizable for a sparrow, with a noticeably flat head, heavy bill, and long tail. Streaked plumage on adults is a warm gray-brown with a tinge of russet. Throat is warm white, lower belly is pale. Thin brown eyeline with thin white highlight on the brow, and subtle brown crown. Immature shows brown on the back with a boldly streaked, buffy belly. BEHAVIOR To spot this sparrow, don't look up. In AZ, look for them in or adjacent to stands of tall sacaton bunchgrass, where only breeding males perch conspicuously to serenade during breeding season. Otherwise, as ground foragers, they dine on a mix of seeds and insects, frequently choosing to run rather than fly after prey and generally living life low to the ground. Quite vocal during dusk hours, but easier heard than seen. VOCALIZATIONS Song like a bouncing Ping-Pong ball, starting with thin, sputtering tweets; bubbling with increased tempo; and then coming to a sudden halt. Call is a faint *tsit*. ABUNDANCE Uncommon resident; limited to southeast corner of AZ, with populations showing potential signs of northward range extension. KEY SITES AZ: Appleton-Whittell Research Ranch, San Rafael Valley. LOOK-ALIKE Cassin's Sparrow. Botteri's distinguished by streaked versus scalloped plumage pattern on back, heavier bill, distinctly flat head, and unmarked breast and flanks. ALTERNATIVE NAMES Flat-headed Sparrow, Sacaton Sparrow. CONSERVATION STATUS Least Concern

Cassin's Sparrow

Peucaea cassinii

Tall, dense, arid grasslands with cacti, shrubs, scrubby trees including oak, mesquite; other native grass and shrub-dominated habitats, especially in summer
LENGTH 5–6 in. WINGSPAN 7.75–8 in.

Hefty sparrow with a longish bill; long, rounded tail; and a flattened head sometimes showing a hint of a crest. Adults are scalloped brown, gray, and rust-colored above and grayish below, with faint streaking in the flanks and breast, white edging in wings, dark stripes bordering a pale throat, and a relatively plain face with a pale eye ring, faded eyebrow, indistinct dark ear patch, and brownish red, streaky crown. BEHAVIOR Forages in dense grass for insects and seeds. During courtship, male sings loudly during high, fluttering, and conspicuous skylarking displays. VOCALIZATIONS Song is a plaintively whistled and trilling *tsee-tseeeee-tsee-tsooee*. Also gives a more complex song reminiscent of a very high-pitched House Wren. Call is a clear, high chip. ABUNDANCE Common year round in extreme Southeast AZ and Southwest NM. Expands its range in summer, moving farther north and west into AZ and spreading across all but Northwest NM. Nomadic and opportunistic, is sometimes found outside of this range following summer rain. LOOKALIKE Botteri's Sparrow. Cassin's distinguished by subtly streaked breast and flanks, streaked versus scalloped back, and shorter and less bulky bill. ALTERNATIVE NAME Skylark Sparrow. CONSERVATION STATUS Least Concern (Decreasing).

Grasshopper Sparrow

Ammodramus savannarum

Grasslands, prairie, pastures
LENGTH 4–4.5 in. WINGSPAN 7–8 in.

Small, squat, thick-necked, and flat-headed sparrow with stubby tail. A large, deep bill breaks from collection of otherwise diminutive features. Handsome plumage, showing clean, warm, but pale buffy belly contrasted by heavily patterned chestnut, espresso-brown, and buffy back. Look for marigold-orange lore patch and distinct white eye ring. Sometimes shows gold patch along shoulder edge of the wing. BEHAVIOR Aptly named, this little bug of a bird forages through pasture for grasshoppers, its meal of choice. Spends most of its time on the ground. A weak flyer, its movement on the wing is labored and heavy, with tail drooping, typically moving over short distances from one patch of grass to another. Very difficult to detect if not vocalizing. VOCALIZATIONS Song is a high-pitched, metallic, and buzzy *tink-tinkle-dzzzzzz*, as if someone shrunk a Red-winged Blackbird into grasshopper form. Call is a brief, sweet *tseep*. ABUNDANCE Uncommon throughout wintering and resident range, limited to extreme Southern AZ, though sporadically observed in suitable habitat throughout northern areas of AZ and NM. LOOKALIKE Baird's Sparrow. Grasshopper distinguished by lack of streaking on breast and lack of malar/mustache stripe. CONSERVATION STATUS Least Concern (Decreasing rapidly).

Five-striped Sparrow

Amphispiza quinquestriata

Steep, rocky, vegetated slopes within canyons; along washes and creeks dominated by dense desert thornscrub
LENGTH 6 in. WINGSPAN 8 in.

Medium-sized with a flattened head, a fairly long and heavy bill, short wings, and a medium-length, rounded tail. Adult is dark gray overall with an unmarked brown back and uppertail, a thin white eyebrow and malar stripe, a white throat bordered by black, a black central breast spot, a contrasting white belly, and dark underwings. Juvenile is paler overall, buffy below, and lacks black markings. BEHAVIOR A skulking sparrow, spends much of its time foraging for seeds and insects under dense vegetation. Often sings from an exposed perch, allowing for good views. VOCALIZATIONS Song is a series of high, thin whistles, chips, and twitters, organized into short phrases separated by long pauses. Pairs communicate with nasal chirps and high, clear chips. ABUNDANCE Extremely local and uncommon year round. Found only in Mexico and a very limited portion of Southeast AZ, from the Atascosa Highlands west of Nogales to the Santa Rita Mountains southeast of Tucson. Sightings outside of this known range are rare but increasing. KEY SITES AZ: California Gulch (Atascosa Highlands) and Box Canyon (Santa Rita Mountains). LOOKALIKE Black-throated Sparrow. Five-striped is darker overall, with thinner facial stripes and a longer bill, and has brown back, white throat, and contrasting white belly. CONSERVATION STATUS Least Concern (Decreasing).

Adult

Juvenile

Black-throated Sparrow

Amphispiza bilineata

Desert scrub and chaparral, canyons, washes
LENGTH 4.5–5.5 in. WINGSPAN 7–8 in.

With a sweet face and striking plumage, this unmistakable medium-sized sparrow shows a jet-black throat, like a long beard extending down to meet the breast, accented by bright white supercilium and mustache stripe, black lores, and thick, gray bill. Warm, buffy underparts contrast the cool gray-brown crown and back. Juvenile lacks black throat patch and shows a warmer version of plumage overall, with faint streaking on underparts and pale tan supercilium and mustache. BEHAVIOR Look down for this sparrow as it hops from the ground to low, shrubby perches, seeking insects and seeds and making itself known along the way with its metallic-sounding calls. Nesting activity coincides with the coming of summer monsoon rains; low-slung nests are well protected by pokey desert scrub. VOCALIZATIONS Tinkling, metallic song starts a two-note introduction, followed by a tinny buzz and clear trill. Call sounds like the tinkling of a jingle bell. ABUNDANCE Common throughout resident range and breeding range, except in Northeast NM, where breeding populations are more scarce. CONSERVATION STATUS Least Concern (Decreasing).

Lark Sparrow

Chondestes grammacus

Relatively open areas with widely spaced trees and shrubs such as grasslands; openings in sparse, grassy woodlands; agricultural areas; roadsides; suburban settings including parks, fields
LENGTH 6–6.5 in. WINGSPAN 11 in.

Large with a long, rounded tail; relatively long and somewhat pointed wings; a rounded head; and an often extended neck. Adults are gray, brown, and buff above with black streaking, whitish below with dirty flanks and a black central breast spot, and have a white undertail, white outer tail edges, and a uniquely crisp facial pattern with a rufous crown and ear patch, pale eyebrow and arc below the eye, black eyeline, and black malar stripes on a white throat. BEHAVIOR Forages on relatively open ground for insects and seeds. In winter, often observed in single-species clusters on the fringes of flocks of other foraging sparrows. Displaying males sing from a relatively low but conspicuous perch. VOCALIZATIONS Song is a choppy series of high, clear chips and whistles mixed with mechanical-sounding buzzes and rattles. Call is a high, thin chip. ABUNDANCE Common breeding resident throughout NM, though less commonly observed in the mountains north of Santa Fe and in Eastern and Northern AZ. Winters in extreme South Central AZ and along the state's western edge. Common migrant in Central and Western AZ. CONSERVATION STATUS Least Concern (Decreasing).

Breeding adult male

Nonbreeding adult male

Female/immature

Lark Bunting

Calamospiza melanocorys

Grasslands/plains, sagebrush, agricultural fields
LENGTH 5.5–7 in. WINGSPAN 9.5–11 in.

A chunk of a sparrow with a bill to match. Male is a deep inky black, contrasted by bright white wing patches and edging on the primaries. Female/immature male shows gray-brown above and pale belly with bold brown streaks. Look for white patch along upper coverts, white-tipped inner tail feathers, and white mustache stripe. Nonbreeding male appears similar to female, but with more extensive deep espresso-brown on the head and back. Bill on both sexes is steely blue-gray. BEHAVIOR Displays complex mating behaviors including colonial nesting and unmated individuals contributing to brood rearing. Adapted to arid environments, they can withstand long periods of drought. VOCALIZATIONS Males are known to give two different songs throughout the breeding season: a premated song and a postmated song, with the former version harsh and the latter jubilant and buoyant. Songs are a lengthy combination of bubbling and buzzy trills and clear notes. Both sexes give a bubbly two-note whistling call. ABUNDANCE Central NM makes up much of the species' overall migration range, and breeding range is concentrated in the extreme northeast corner of the state. Commonly observed throughout winter range. CONSERVATION STATUS Least Concern (Decreasing Rapidly).

Breeding adult

Nonbreeding adult/immature

Juvenile

Chipping Sparrow

Spizella passerina

Suburban parks, backyard feeders; during breeding, in grassy, brushy, open habitat within and along edges of pine, oak, other relatively open woodlands; in winter, similarly structured habitat at lower elevations, favoring evergreen trees where available
LENGTH 4.75–6 in. WINGSPAN 8.5 in.

Tiny and delicate with a small bill, rounded head, and long, notched tail. Breeding adults are tan with dark streaks above, grayish white and unmarked below, and have white wingbars, white throat, and gray face painted with a rufous crown, white eyebrow, and black eyeline that reaches a dark gray bill. Nonbreeding adult and juvenile are similar, but are more buffy brown overall and have a pinkish bill. BEHAVIOR Forages for insects and seeds on the ground in sprightly and often fairly large flocks. VOCALIZATIONS Song is composed of high, flat chips delivered rapidly in an up to four-second trill. Calls include thin chips and cheerful twittering. ABUNDANCE Common during breeding at middle and high elevations across much of Northern NM and in Northern and Eastern AZ. In winter, migrates to lower elevations in Southern NM and Western, Central, and Southern AZ. Populations persist year round across Southern NM and into Southeast AZ. CONSERVATION STATUS Least Concern (Decreasing).

Breeding adult

Nonbreeding adult/immature

Clay-colored Sparrow

Spizella pallida

Shrublands, thickets, prairies, sparse conifer stands
LENGTH 4.5–6 in. WINGSPAN 8–8.5 in.

Pretty and petite sparrow with a lean silhouette; slender, pointed bill; and long, notched tail. Warm, pale tan, and taupe overall, with diagnostic taupe-gray and unstreaked collar; pale, crisp mustache stripe; buffy brow stripe; and crown adorned with thin brown streaks and a pale central stripe. Bill is pinkish orange. During breeding, facial marking and dark brown streaks on back are more contrasting and distinct than in nonbreeding plumage. BEHAVIOR Look for small flocks foraging on the ground along shrubby roadsides and in open fields dotted with shrubs and thickets, sometimes in mixed sparrow flocks. VOCALIZATIONS A series of from two to eight (though usually somewhere in between) extremely insectlike buzzes, with little to no intonation. Call is a light, sweet chip. ABUNDANCE Uncommon but regularly observed throughout migration. LOOKALIKES Chipping, Brewer's sparrows. Clay-colored distinguished from Chipping by browner cap, paler gray coloration, and eyeline that does not extend past the bill to the eye; from Brewer's by larger size, more contrasting facial pattern, unstreaked gray collar, and streaked crown with pale central stripe. CONSERVATION STATUS Least Concern (Decreasing).

Breeding adult male

Female/nonbreeding male

Black-chinned Sparrow

Spizella atrogularis

Chaparral and adjacent desert scrub; arid, rocky hillsides with heavy shrub cover from species such as scrub oak, acacia, manzanita **LENGTH** 5.75 in. **WINGSPAN** 7.75 in.

Small, delicate sparrow with a long tail, rounded head, and a small, pink bill. In all plumages, is dark gray overall, lightest below, with dark streaks on a buffy brown and slightly rufous back. In breeding plumage, adult male develops a black throat. **BEHAVIOR** Males will sing from conspicuous perches. Species more frequently hidden under dense cover while foraging for insects and seeds. As a chaparral specialist, is often among the first to take advantage of recently burned habitat. **VOCALIZATIONS** Song is a series of clear, whistled notes followed by a fast trill that accelerates to a buzz, like a bouncing ball; similar to but faster than Field Sparrow song. Calls include very high-pitched chips and trills. **ABUNDANCE** Local and common. During breeding, ranges from Northwest through Central and Southeast AZ into Southwest, Central, and pockets of Northern NM. Expands further into Central and Southern AZ in winter. Persists year round in Southeast AZ and across much of extreme Southern NM. **LOOKALIKE** Dark-eyed Junco. Black-chinned Sparrow distinguished from similar forms by combination of plain gray underside; streaked, brownish wings; relatively bright pink bill; and lack of white outer tail feathers. **CONSERVATION STATUS** Least Concern (Decreasing).

Field Sparrow

Spizella pusilla

Brushy scrub, heavily vegetated/bushy meadows
LENGTH 4.5–6 in. WINGSPAN 7.5–8.5 in.

Petite, slim sparrow with short, conical bill; round head; and long, notched tail. Warm rufous and gray overall, with back showing warm reddish brown streaked with espresso-brown; underside is buffy gray with patchy, pale washes of cinnamon-brown throughout, and wings show two faint white bars. Face is gray with a distinct rusty crown and ear patch (which juveniles lack), as well as a white eye ring. Bill and legs are deep coral-pink. Great Plains individuals show far less rufous coloration, instead showing a warm taupe with pale apricot-colored crown. BEHAVIOR Tends to frequent thorny, mature shrubs, with flocks gathering to forage but staying low to the ground. Relatively quiet and subdued outside of breeding season, during which males perch atop tall grasses and shrubs to announce their availability. VOCALIZATIONS Another "bouncy ball" sparrow mating song (not likely to be heard in AZ or NM), this one starting slow with clear whistles accelerating into a quick, intense trill, and then fading off. Dawn song is complex, with a trill-whistle-trill pattern. Calls range from a single note *tseep* when foraging to territorial chips and quick, high *tzee* chirps of alarm. ABUNDANCE Uncommon, with wintering range just barely peeking into the eastern edge of NM. LOOKALIKES Chipping, Swamp, White-throated, White-crowned sparrows. Field distinguished by pink bill and legs, white eye ring, gray face, and rufous ear patch. CONSERVATION STATUS Least Concern (Decreasing).

Adult

Juvenile

Brewer's Sparrow

Spizella breweri

Breeds in sagebrush, nearby sagebrush-dominated habitats; in winter, similarly structured habitats such as shrubby deserts, saltbush communities
LENGTH 5–6 in. **WINGSPAN** 7–8 in.

Small and delicate with a long, notched tail; a tiny, pointed bill; relatively short wings; and a rounded or slightly crested head. Breeding adults are exceedingly plain, showing streaked grayish brown above and plain grayish brown below, with streaked nape and back. Plain face shows a subtle gray eyebrow and cheek, accented by dark lines behind and below eye and contrasted by a thin white eye ring. Throat is pale with very subtle throat stripes; streaked brown crown usually lacks central stripe. Nonbreeding adult and juvenile are even more indistinctly marked. **BEHAVIOR** Forages in dense shrubs for insects and seeds. Forms large, vocal flocks in winter. Male delivers song from an obvious perch. **VOCALIZATIONS** Song is a canarylike and somewhat mechanical series of buzzes, trills, chips, and whistles lasting up to fifteen seconds. Calls include high chips and trills. **ABUNDANCE** Common during breeding in Northwest NM and extreme Northeast AZ, during winter in Western AZ and across the central to southern portions of both states, and during migration throughout the region. **LOOKALIKES** Chipping, Clay-colored sparrows. Brewer's is less distinctly patterned and shows a streaked nape. **ALTERNATIVE NAMES** Undecorated Sparrow, Humble Sparrow, Pallid Sparrow. **CONSERVATION STATUS** Least Concern (Decreasing).

Adult (Slate-colored/Interior West)

Adult (Slate-colored/Interior West)

Fox Sparrow

Passerella iliaca

Forests, woodland scrub, edge habitat, densely vegetated streamside thickets
LENGTH 6–7.5 in. **WINGSPAN** 10–11.5 in.

A hefty sphere of a sparrow with distinctive, thrushlike plumage. The Slate-colored form that inhabits the Interior West shows a steely gray head and back that transitions to warm rusty colors on the wings and tail. Belly is creamy white with bold, deep rufous spotting and streaking from breast to flanks. **BEHAVIOR** Look—or listen—for them making a ruckus in the leaf litter or carpet of conifer needles, kicking it out of the way to get to the good stuff (insects and seeds). Will frequent backyard feeders in winter. Variable migration patterns, with some migrating in elevation only and others traveling great distances. **VOCALIZATIONS** Cheerful song opens up with clear whistles (reminiscent of a catcall) followed by *churr*s and trills. Calls range from sweet, quiet *tsip*s to territorial calls known as "smacks," though we would just as soon call it a *chit* or scolding chip. **ABUNDANCE** Uncommon but regularly observed throughout migratory and winter ranges in AZ and NM, typically in high-elevation habitat. **CONSERVATION STATUS** Least Concern (Decreasing).

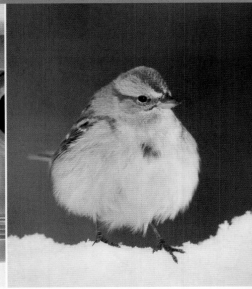

American Tree Sparrow

Spizelloides arborea

Brushy, weedy fields adjacent to forests, other wooded habitats, wetlands; readily visits backyard seed feeders
LENGTH 5.5–6.25 in. WINGSPAN 9.5 in.

Medium-sized and full-bodied with a long, thin, and slightly notched tail; a rounded or slightly crested head; a small, conical bill; and fairly short, rounded wings. Adult is streaky reddish brown above, clean gray below. Rufous crown, gray face and nape, rufous line behind the eye, and bicolored yellow-and-black bill. Two white wingbars, with sometimes rufous-washed flanks or a rufous patch on the upper breast near the shoulder, and usually a dark central breast spot. BEHAVIOR Forages in small flocks on open ground or in low vegetation for seeds and the occasional insect. Often perches conspicuously atop short, weedy plants. VOCALIZATIONS Song is a short series of whistles that opens with one or two clear notes and closes with a descending warble. Calls include high chips and a squeaky *teedoo* or *teedledoo*. ABUNDANCE Uncommon to rare winter visitor to Northeast AZ and Northern NM, most frequently encountered in northeastern corner of NM. LOOKALIKES Other sparrows. American Tree Sparrow easily distinguished by bicolored bill. Misguidedly named for its resemblance to the Eurasian Tree Sparrow. CONSERVATION STATUS Least Concern (Decreasing).

Adult (Oregon)

Adult (Slate-colored)

Dark-eyed Junco

Junco hyemalis

Breeds in coniferous, mixed-coniferous, high-elevation deciduous forests with a mix of shrubs, open ground; in winter, lower elevations in open woodlands, parks, backyards, scrubby fields
LENGTH 5.5–6.25 in. WINGSPAN 7.5–9.5 in.

Medium-sized, chunky, round-headed, and stocky necked with an elongated tail, a pinkish conical bill, and white undertail coverts, lower belly, and tail edges. Six uniquely featured regional populations occur in the Southwest. Oregon form shows contrasting black or grayish hood, rusty flanks, and reddish brown upper wings and back. Slate-colored form is slate-gray to brownish gray overall, with little contrast between the hood, breast, and back. Gray-headed form is light gray overall with a black mask and burnt-rufous back. Red-backed form is similar to Gray-headed, but with dark upper mandible. White-winged form is gray overall with two faint wingbars. Pink-sided form shows a light blue-gray head, black mask, and salmon-pink to dusty mauve on flanks and back. Juvenile is similar across populations: heavily streaked, with grayish head and rufous-brown tones throughout. BEHAVIOR Forages in groups for seeds and invertebrates either on the ground with vigorous scratching or while bouncing through low trees and shrubs. VOCALIZATIONS Song, similar to but shorter than Chipping Sparrow's and varying between populations, is a short and sometimes trilling series of whistled notes. Calls include high chips and twittering. ABUNDANCE Common year round across the northern portions of both states to Southeast AZ and Southwest NM; also in the Lincoln National Forest. Common in winter across both states. Each regional population occupies a distinct range. Oregon Junco winters across both states. Slate-colored winters across both states but is rare in Southwest AZ. Gray-headed breeds from North Central to Northwest NM and in extreme Northeast AZ; in winter, ranges across most of AZ and all but the eastern edge of NM. Red-backed

(continued)

Adult (Gray-headed)

Adult (White-winged)

Adult (Red-backed)

Adult (Pink-sided)

juvenile

resides year round in montane habitat from North Central AZ to South Central NM. White-winged, our rarest junco, is encountered in Northern NM and Northeast AZ, with scattered records in the southern portions of both states. Pink-sided winters across all of NM and from Central to Eastern AZ; during migration can be found from North Central to Northeast AZ. **LOOKALIKE** Yellow-eyed Junco similar to Gray-headed and Red-backed forms. Gray-headed distinguished by darker underparts and pinkish bill; Red-backed distinguished by silvery bill and limited or no rufous in wings; and both forms distinguished from Yellow-eyed by dark eyes and the absence of a yellow lower mandible. **CONSERVATION STATUS** Least Concern (Decreasing).

Adult

Juvenile

Yellow-eyed Junco

Junco phaeonotus

Mountainous pine-oak, pine, mixed-coniferous forests; in winter, lower altitude wooded or brushy habitats
LENGTH 5.5–6.25 in. WINGSPAN 9.5–10 in.

Medium-sized and potbellied with short legs, long tail, rounded head, stout neck, and conical bill. Adult is slate-gray overall, lightest below, with bright yellow eyes, a bicolored bill with black upper and yellow lower mandible, black lores and eyeshadow, white outer tail feathers, and rufous on the back and in the wings. Juvenile is rufous-brown with heavy dark streaks above, buffy with dark spotting below, and has a gray head, pale throat, and, by late summer, yellow eyes and a bicolored bill. BEHAVIOR Forages in small groups for invertebrates and seeds, either with towhee-like scratching or while hopping through trees and shrubs. VOCALIZATIONS Song consists of several sweet, clear notes followed by a rolling trill, similar to but clearer and slower than Spotted Towhee song. Calls include high chips and twittering. ABUNDANCE Common year round in the mountains of Southeast AZ and extreme Southwest NM. In AZ, regularly occurs as far north as the Pinal Mountains, and less commonly in the Sierra Anchas north of Phoenix. LOOKALIKES Red-backed and Gray-headed forms of Dark-eyed Junco. Yellow-eyed identifiable by rufous in wings, yellow eyes, black-and-yellow bill, and voice. CONSERVATION STATUS Least Concern.

Adult

Immature

White-crowned Sparrow

Zonotrichia leucophrys

Breeds in relatively open habitats with plentiful nearby brush, such as chaparral, alpine meadows, tundra, and along forest edges; winters in any habitat with mixed brush and open space
LENGTH 6–7 in. **WINGSPAN** 8.5–9.5 in.

Large but not quite towhee-sized, with a long tail, conical bill, and flattened or peaked head. Adult is mottled brown, black, and buff to rufous above; plain gray below; and has a gray face, throat, and nape; a streaked upper back; dusty brown flanks; two subtle white wingbars; a pinkish to yellow-orange bill; and a black-and-white–striped crown. Juvenile is more brownish overall with warm brown-and-gray crown stripes. **BEHAVIOR** Forages in groups for seeds, other plant matter, and insects. Rarely ventures far from low, dense cover, hopping along the ground, scratching in loose dirt, and gleaning from low shrubs. Regularly visits backyard feeders. **VOCALIZATIONS** Song varies between subspecies, but always begins with clear, whistled, and slightly descending notes; continues with a messy arrangement of rougher whistles; and closes on a downward buzz: *sweeeee-swiddle-a-dee-dee-doo*. Most common call is a high-pitched chip. **ABUNDANCE** Common year round in the Southern Rockies north of Santa Fe and in winter throughout both states. **LOOKALIKE** White-throated Sparrow. White-crowned lacks white throat and yellow above and in front of eyes. **CONSERVATION STATUS** Least Concern.

Adult

Immature

White-throated Sparrow

Zonotrichia albicollis

Forests, woodlands, edge habitat, burn scars; frequents suburban areas, parks
LENGTH 6–7 in. **WINGSPAN** 8–9 in.

Hefty and heavy billed sparrow, showing warm brown above and gray-tan below, with cool taupe flanks and faint streaking. Bold, bright facial markings with yellow lores, white throat, striped black-and-white head. In the tan-colored morph, buff and brown replace the typical black-and-white facial coloration. Immature shows prominent streaking and less yellow behind eyes. **BEHAVIOR** Ground foragers, traveling in small, loose flocks that flit and forage on the ground along woodland edges and shrubby roadsides. Look for their towhee-like, double-footed kick-hop as they poke through leaf litter. In spring, look for them low in shrubs, gleaning insects and flower buds. Omnivores, preferring insects throughout breeding season; shifting to a primarily granivorous diet of grass seed in winter.

VOCALIZATIONS A sweet, melancholy serenade, often one of the first to start the dawn chorus (the most lovely song to greet one's ears—this is a fact). The song sounds similar to "Oh, sweeeeeet, Canada, Canada." A treat for observers in this range, they are known to sing even outside the breeding season, a haunting song to hear on a cloudy winter morning. **ABUNDANCE** Primarily a species of Eastern North America; uncommon but regularly observed throughout both wintering and migration ranges. **LOOKALIKE** White-crowned Sparrow. White-throated distinguished by yellow lores and white throat patch, lacking on White-crowned. **CONSERVATION STATUS** Least Concern (Decreasing).

Sagebrush Sparrow

Artemisiospiza nevadensis

Open plains, sagebrush flats, scrub
LENGTH 5–6 in. WINGSPAN 8–9 in.

Midsized sparrow with a squat, round head and long tail. Gray- and sandy hued plumage overall, with pale, creamy white belly; brownish back; buffy and faintly streaked flanks; and a dark spot in the middle of the chest. Gray head is contrasted by bright white eye ring, white lores, and a white throat. Gray malar stripes roughly as dark as head. Upper back shows distinct streaks. Tail is deep gray-brown with pale outer edges. Juvenile shows heavily streaked belly and lacks gray head. BEHAVIOR Ground foragers, gleaning insects as the climate permits and shifting to seeds as winter sets in. Moves about in small, loose flocks outside breeding season. Males return to claim the same nesting territory year after year. Often seen running with tail held high. VOCALIZATIONS Song (male only) is a quick up-down-up-down series of chips and trills—short and sweet. Call given by both sexes is a light, metallic, high-pitched chip. ABUNDANCE Uncommon but regularly observed in preferred habitat throughout wintering and breeding range. Sparse resident population in Northern AZ. LOOKALIKE Bell's Sparrow. Sagebrush and Bell's sparrows were once considered a single species (Sage Sparrow). Sagebrush distinguished from Bell's (not very easily, with winter ranges overlapping) by slightly paler overall plumage, slightly paler/less pronounced malar stripe, and presence of distinct streaks on mantle (upper back). A challenge to differentiate unless they are side by side in the field. CONSERVATION STATUS Least Concern.

Bell's Sparrow

Artemisiospiza belli

Breeds in CA in brushy habitats including coastal sagebrush, chaparral; in the Southwest, winters in similarly structured habitats including grasslands, creosote flats, saltbush communities

LENGTH 5–6 in. WINGSPAN 7.75–8.25 in.

Medium-sized with a rounded head, conical bill, fairly long wings, long tail, and forward-leaning, tall-standing posture. Adult is dark brown and unstreaked or nearly so above, grayish white below with a black central breast spot, and has buffy to pale brown–washed flanks and underwings. Head is dark gray with dark bill, white eye ring and lore spot, and broad, sooty black malar stripe that is usually darker than the head and borders a white throat, forming the lower edge of a white swoosh below the cheek. BEHAVIOR Forages for insects and seeds in loose groups on the ground or under the cover of low shrubs. More inclined to run than hop or fly, darts quickly across open ground with tail cocked upward. VOCALIZATIONS Song is a short, whistled, trilling warble. Common call is a high chip similar to that of Black-throated Sparrow. ABUNDANCE Local and uncommon in winter from Western to Central AZ. KEY SITE AZ: Robbins Butte Wildlife Area. LOOKALIKE Sagebrush Sparrow. Though Sagebrush and Bell's are sometimes indistinguishable in the field, in general, Bell's has a darker head; a darker, thicker, and more contrasting malar stripe; and no or less distinct back streaking. ALTERNATIVE NAMES Coastal Sagebrush Sparrow, California Sagebrush Sparrow. CONSERVATION STATUS Least Concern.

Vesper Sparrow

Pooecetes gramineus

Open, arid, grassy areas including grasslands, meadows, sagebrush, shortgrass prairies
LENGTH 5–6.5 in. **WINGSPAN** 9–9.5 in.

Large sparrow with a long tail and small bill. Heavy brown-on-sand streaking overall, with warm cinnamon-colored tone on scapulars, white eye ring, pale malar stripe, dark ear patch with pale center, and white outer tail feathers that flash in flight. Belly is creamy white with finely streaked breast. Juvenile shows heavier streaking on lower belly. **BEHAVIOR** A ground dweller, moves about in low grass and scrub, foraging for insects and seeds. Is far more tolerant of human observation than its other not-so-exhibitionist peers. Male will prominently perch to sing over its territory during nesting. **VOCALIZATIONS** Song begins with a clear-noted *cue-cue-tee-tee* followed by frantic, buzzy trills. Call is a blunt chip. **ABUNDANCE** Commonly observed throughout breeding range across the northern portions of both states, during winter in the southern portions, and during migration in Central NM. **CONSERVATION STATUS** Least Concern (Decreasing).

Baird's Sparrow

Centronyx bairdii

Arid grasslands, prairies
LENGTH 4.5–5.5 in. WINGSPAN 8.5–9 in.

Small but stout, with a shallow-notched, short tail; flat head; and heavy, pointed bill. Plumage shows deep coffee-brown and chestnut-colored streaks on warm, sandy brown back, with a clean, creamy white belly accented by a fine necklace of brown and chestnut-colored streaks. Face has thin deep brown malar and mustache stripes (sometimes bold and clean, sometimes broken and faded in appearance). Buffy golden supercilium and ear patch. Juvenile shows similar plumage, but not as sharp and clean. BEHAVIOR One of the trickiest sparrows to spot in their wintering range in the Southwest, as they spend their time with their feet on the ground, running about in the grasses, foraging solo. Bring your patience and keen eyesight to their preferred grassland habitat and you might just get lucky. VOCALIZATIONS Song (unlikely to be heard in this range) beginning with clear tinkling notes and ending in a descending, buzzy trill. Call is a barely detectable, high-pitch *peep*, reminiscent of a bat. ABUNDANCE Uncommon; overwinters only in the far reaches of Southeast AZ and Southwest NM grasslands. KEY SITES AZ: grasslands of Sonoita and Elgin, San Rafael Grasslands; NM: Animas Valley, Otero Mesa. LOOKALIKE Grasshopper Sparrow. Baird's distinguished by necklace and mustache/malar stripe. ALTERNATIVE NAME Great Plains Ground Sparrow. CONSERVATION STATUS Least Concern.

Savannah Sparrow

Passerculus sandwichensis

Open, relatively treeless habitat such as grasslands, meadows, roadsides, agricultural fields
LENGTH 4.5–5.5 in. **WINGSPAN** 8–8.75 in.

Medium-sized with a small, pointed bill; often-raised crest; and medium-length, notched tail. Adult is grayish brown to rufous-brown with dark streaks above, white below, with fine streaks on the breast and flanks, and a dark crown with pale middle stripe and dark lines behind the eye, below the ear, and bordering the throat. Pale malar stripes, a pale eyebrow accent the head, with usually a splash of yellow in front of and above the eye. Legs are pinkish. **BEHAVIOR** Forages on open ground and takes low, conspicuous perches, which can make this sparrow easy to observe compared to other grassland denizens. **VOCALIZATIONS** Song begins with high, accelerating, clear notes; continues with a harsh buzz; and concludes on a low chip. Common call is a dry chip. **ABUNDANCE** Common in winter in Western AZ, Eastern NM, and across the central and southern portions of both states; during breeding in AZ's White Mountains and in the Southern Rockies north of Santa Fe; and during migration across both states. Less commonly breeds from Northeast AZ through Northwest NM. **LOOKALIKE** Vesper Sparrow. Savannah is smaller, has a more distinct eyebrow, frequently shows yellow lores, and lacks Vesper's white eye ring, pale center on a dark ear spot, and white outer tail feathers. **CONSERVATION STATUS** Least Concern (Decreasing).

Song Sparrow

Melospiza melodia

Open habitats, including wetland, stream, and forest edges; open scrub and meadows; suburban parks; residential areas
LENGTH 4.5–6.5 in. **WINGSPAN** 7–9.5 in.

Stout and stocky, with a short bill, a long and rounded tail, and broad wings. Bold brown streaking overall, with a dark central breast spot and an extensively streaked belly, from breast to flanks. Rusty crown stripes, eye stripes, and malar contrast pale gray eyebrow and cheek. A highly variable species, with more than twenty recognized subspecies and dozens of regional forms. Southwest individuals typically show pale taupe/gray and rufous coloration. Look for characteristic tail-pumping flight pattern. **BEHAVIOR** Ground foragers, hunting and pecking for a wide variety of insects and seeds, with the latter serving as their primary food source in winter. Look for them moving through low, dense vegetation along wetland and forest edges and in low canopy and scrub along desert washes. Males perch conspicuously on reeds to belt out their ballads. **VOCALIZATIONS** Consistent with high intraspecies variability, song is regionally specific but consistent in overall structure, commencing with two or three clear, truncated notes followed by one or more trills, often adding some embellishments in between. **ABUNDANCE** Common and abundant in suitable habitat throughout resident and winter ranges, with frequency heavily dwindling in arid regions of Southwest AZ. **LOOKALIKE** Lincoln's Sparrow. Song Sparrow distinguished by heavier streaking overall, central breast spot, and lack of eye ring and buffy tones. **CONSERVATION STATUS** Least Concern (Decreasing).

Lincoln's Sparrow

Melospiza lincolnii

Breeds in high-elevation montane settings along streams, in meadows, in other moist and brushy habitats; during winter and migration, in open, shrubby habitats including fields, forest edges, open woodlands, water edges LENGTH 5–5.75 in. WINGSPAN 7.5–8.5 in.

Medium-sized and delicate with a rounded to slightly crested head; medium-length, notchless tail; and thin, conical bill. Adult is brown, buff, gray, and streaked above with white belly. Buffy breast and flanks painted with thin, dark streaks. Crown is rufous-brown with gray central stripe. Face shows buffy eye ring; gray eyebrow; dark rufous lines behind the eye, below the cheek, and on the throat; and buffy malar. BEHAVIOR A skulking sparrow, spends much of its time searching for insects and seeds in low shrubs or scratching like a towhee under dense cover; rarely ventures into the open. VOCALIZATIONS Song is warbling and wrenlike, opening with clear notes and continuing with a series of trills that bounce up and down in pitch. Common call is a harsh *cheep*, similar to but shorter and drier than Song Sparrow call. ABUNDANCE Common in winter in Western AZ, Eastern NM, and across the central and southern portions of both states; during breeding near and north of Santa Fe and in AZ's White Mountains; and during migration widely across both states. LOOKALIKE Song Sparrow. Lincoln's distinguished by eye ring, buffy tones, finer streaking, and lack of central breast spot. ALTERNATIVE NAME Buff-breasted Sparrow. CONSERVATION STATUS Least Concern.

Breeding adult

Nonbreeding adult/immature

Swamp Sparrow

Melospiza georgiana

Densely vegetated, waterlogged habitat including wetlands, edges of riparian areas, ponds, occasionally flooded fields LENGTH 4.75–5.75 in. WINGSPAN 7–7.5 in.

Medium-sized and portly with a long tail, rounded to slightly crested head, relatively long legs, and stout, conical bill. Breeding adult often appears very dark and has a gray eyebrow and face, rufous cap with pale central stripe, dark line behind the eye, and dingy bill with a yellow base. White throat is bordered on each side by a dark stripe and pale malar. Gray, unstreaked nape. Gray underparts, ochre-colored flanks, and coarsely streaked upper back with extensive rufous-brown on the back and wings. Juvenile and nonbreeding adult have a brown cap and faint, blurry streaking below. BEHAVIOR Forages along the water's edge or within dense vegetation. Rarely observed far from water. When threatened, is more inclined to run for cover than to fly away. VOCALIZATIONS Song is a steady, clear, and evenly pitched series of notes: *chee-chee-chee-chee-chee-chee*. Calls include an abrupt buzz and sharp chips. ABUNDANCE Uncommon in winter from Central to Southeast AZ, in Southeast NM, and northward along the Pecos and Rio Grande rivers. Uncommon in migration along NM's eastern edge. Rare elsewhere. LOOKALIKES Song, Lincoln's sparrows. Swamp distinguished by ochre-colored flanks, indistinct streaking below, and extremely thick black stripes on the mantle. CONSERVATION STATUS Least Concern.

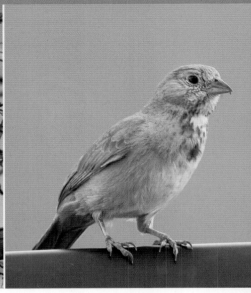

Canyon Towhee

Melozone fusca

Arid and relatively open, brushy areas including desert grasslands, chaparral, mesquite washes, rocky drainages, open woodlands; uncommon backyard bird except in rural areas
LENGTH 8.5–9.5 in. **WINGSPAN** 11.5 in.

Large for a sparrow, with a long tail, plump body, short wings, rounded head, and a stout, conical bill. Adult is grayish brown overall, lightest below, with a subtle rufous crown and stripe behind the eye, burnt-orange undertail coverts, a messy necklace of short and dark streaks, often a central breast spot, and buffy eye rings, underwings, face, breast, and lores. Juvenile is buffy brown overall with coarse streaking below. **BEHAVIOR** In typical towhee fashion, scratches at the ground in search of seeds and insects, staying low and rarely flying more than a short distance. **VOCALIZATIONS** Song starts with a squeaky chip and continues with a series of repeated whistles: *jip-pew-pew-pew-pew*. Calls include high-pitched tinkling and a short, harsh chip similar to that of Ladder-backed Woodpecker. **ABUNDANCE** Common year round across Central and Southeast AZ and in all but extreme Eastern and Northwest NM. **LOOKALIKES** Abert's, California towhees. Canyon distinguished from Abert's by lack of black mask, paler coloration, buffy tones, rusty crown, and messy necklace; from California by buffy versus rusty coloration, more distinct necklace, central breast spot, and range. **CONSERVATION STATUS** Least Concern (Decreasing).

Abert's Towhee

Melozone aberti

Sonoran Desert specialty, frequenting low-elevation cottonwood/willow galleries, lush washes, mesquite bosque; also heavily watered urban and suburban areas **LENGTH** 8–9 in. **WINGSPAN** 10.5–11.5 in.

Large, long-tailed, heavy billed sparrow with the gestalt of a cardinal. Warm taupe all over, with sooty black at the base of the bill extending to the eyes to create a small mask, and deep apricot-colored undertail coverts. Bill is pale gray. **BEHAVIOR** Endearingly referred to as the "twohee," these monogamous sparrows form a lifelong bond. Active ground foragers, intently hop-kicking up organic matter to unearth insects. Typically avoids open areas, staying under or close to the cover of dense desert thickets. **VOCALIZATIONS** Song is a sweet set of three to five high-pitched *beeps* followed by rapid, rambling chatter. Duet call, used as a contact call by mated pairs, is a single, high-pitched *peep*, reminiscent of the sound of sneakers squeaking on a basketball court. **ABUNDANCE** Common resident throughout range. **LOOKALIKE** Canyon Towhee. Abert's distinguished by habitat, black at base of bill, and lack of rusty crown, streaky necklace, and buffy coloration. **ALTERNATIVE NAMES** Masked Towhee, Gila Towhee. **CONSERVATION STATUS** Least Concern.

Rufous-crowned Sparrow

Aimophila ruficeps

Arid, rocky, relatively open hillsides with sparse grass, shrubs, small trees within deserts, grasslands, chaparral, open woodlands LENGTH 6 in. WINGSPAN 7.75 in.

Large sparrow with a long tail, rounded head, and a heavy, somewhat long and pointed bill. Adult is brownish gray with rufous-brown streaking above, unmarked gray below, and has a gray bill, a rufous crown and stripe behind the eye, a gray face and eyebrow, a pale malar and spot above the lores, thin white eye rings, and a black stripe on either side of the throat. Juvenile is similar but with faded, less crisp patterning. BEHAVIOR More often heard than seen, often stays high up on inaccessible slopes, foraging for seeds and insects within and under vegetation. More inclined to run than fly away. VOCALIZATIONS Song is a buzzy, squeaky, warbling jumble of chips and whistles, similar to but shorter than those of House Wren. Common call is an emphatic *deer-deer-deer*. ABUNDANCE Common year round in Northwest, Central, and Southeast AZ and in Northeast and Central to Southern NM. LOOKALIKES Rufous-winged, Chipping sparrows. Rufous-crowned distinguished from Rufous-winged by all-dark bill, single malar stripe instead of paired, white eye ring, and rufous instead of brown back streaking. Rufous-crowned distinguished from Chipping by dark malar stripes and rufous eyeline that does not extend in front of the eye. CONSERVATION STATUS Least Concern (Decreasing).

Adult

Juvenile

Green-tailed Towhee

Pipilo chlorurus

Dense desert scrub and mesquite bosque, lower elevation disturbed montane forest slopes, chaparral

LENGTH 6.5–7.5 in. **WINGSPAN** 9–11 in.

Distinct sparrow, taking the guesswork out of identification for many grateful birders. Large, long-bodied and tailed, spry sparrow, showing warm gray plumage overall, contrasted by bright chartreuse wash on wings and outer tail feathers, and a showy, bright rusty cap and matching eye color. White throat is bordered by dark mustache stripe. Juvenile shows streaked plumage and lacks rufous cap. **BEHAVIOR** Ground foragers, though a bit less aggressive in their organic litter thrashing than other towhees, often foraging through low brush. Dines primarily on seeds and a plethora of insects, with fruit/berries as an occasional treat. **VOCALIZATIONS** Song is a jubilant melody of whistles and slurred trills. Call is remarkably similar to an inquisitive kitten's *mew*. **ABUNDANCE** Common in preferred habitat throughout resident, winter, and migratory ranges, with decreasing abundance throughout Northern AZ. **CONSERVATION STATUS** Least Concern.

Adult male

Female/immature

juvenile

Spotted Towhee

Pipilo maculatus

During breeding, scrubby, middle to upper elevation habitats with plentiful ground cover, such as forest edges and openings, chaparral, riparian corridors, weedy fields, shrublands, suburban areas; in winter, similar habitats at lower elevations
LENGTH 7–8.5 in. **WINGSPAN** 10.5–11 in.

Large sparrow with a slightly pointed head, thick neck, plump body, long tail, and a heavy, pointed bill. Adult male wears a black hood and has red eyes, black wings with white spotting, a clean white belly, rufous flanks and undertail coverts, and a black tail with white corners. Adult female is dark brown above with paler undertail coverts. Juvenile is streaky brown overall. **BEHAVIOR** Like other towhees, forages for insects, seeds, and occasionally fruit through vigorous hop-scratching or by popping up into low vegetation. Rarely strays far from dense cover. **VOCALIZATIONS** Song begins with several clear chips and concludes with a speedy trill: *chut-chut-chut-zeeeee*. Call is a short, raspy, and jaylike scold. **ABUNDANCE** Common year round at middle to upper elevations across Northern and Eastern AZ and across all of NM, excluding its extreme eastern edge and the lower Rio Grande Valley south of Las Cruces. Can be found across both states in winter. **LOOKALIKE** Eastern Towhee (extremely rare in the Southwest). Spotted distinguished by white spots above. **CONSERVATION STATUS** Least Concern.

BLACKBIRDS, ORIOLES, AND OTHERS

Originating from the tropics, the icterids (family *Icteridae*) are a diverse grouping that in the Southwest includes the New World blackbirds, New World orioles, meadowlarks, cowbirds, and grackles. Birds in this family are often clothed in iridescent black plumage, but bright yellows and reds are also common, most notably in the orioles and meadowlarks and more subtly in many of the blackbirds. Shared across the family is a specialized bone structure that enables these birds to force their bills open as powerfully as they snap them shut, an adaptation that offers access to a wide variety of food resources. Closely related is the Yellow-breasted Chat (family *Icteriidae*), a bird with mockingbird-like habits that blurs the lines between blackbirds and warblers.

ABOVE Flock of Yellow-headed Blackbirds

Yellow-breasted Chat

Icteria virens

Brushy habitat within mature riparian woodlands, near wetlands; also drier settings such as forest edges, shrub-rich washes LENGTH 7.5 in. WINGSPAN 9.75 in.

Similar to (and once classified as) a warbler, but larger and heftier with a rounded head, long tail, short and rounded wings, and a fairly long, heavy bill. Adult is soft gray-green above and has a bright, sunny yellow throat and breast; clean white lower belly with olive-brown–washed flanks; a gray face; bold white eye rings and lores; and a gray bill. Juvenile is drab gray overall with a pale bill, a yellow splash in the chest, and facial patterning that becomes more distinct with age. BEHAVIOR Extremely vocal but incredibly secretive, rarely offers clear views as it skulks through dense vegetation in search of insects and the occasional fruit. VOCALIZATIONS Song resembles that of a mockingbird, with a string of harsh and cleanly separated phrases, each composed of repeated whistles, chucks, rattles, buzzes, and barks, often concluding with a harsh scold. Also like a mockingbird, sings deep into the night during breeding season. ABUNDANCE Uncommon breeding resident and migrant in suitable habitat across the region. CONSERVATION STATUS Least Concern (Decreasing).

Female/immature

Juvenile

Adult male

Yellow-headed Blackbird

Xanthocephalus xanthocephalus

Marshes, wetlands, agricultural fields, pastures
LENGTH 8.5–10.5 in. WINGSPAN 16–18 in.

Big, beefy blackbird with large, rounded head and pointed bill. Male shows deep black overall, with bold white wing patches, yellow head, and yellow extending down the chest. Female and immature male show a lovely ash-brown, with splotchy yellow coloration on head and chest. Female lacks white wing patch. BEHAVIOR A marvel to observe in winter, as huge flocks gather to roost, often numbering in the thousands and engaging in waves of murmurations before settling into slumber. In summer, loose breeding colonies form, where males sing aggressively to defend their patch of marsh habitat. Males engage in polygyny, nesting with several different females within one breeding season and territory; a male typically helps carry the parenting load only with the first nesting female of the season. Female-constructed nest, usually located in aquatic reeds or cattails, is a deep woven cup secured firmly to marsh vegetation. VOCALIZATIONS Male song follows general structure of several hurried musical notes followed by a harsh, metallic creak, but individual expression ranges broadly. Female song is a clattery chittering. A diverse lexicon of cacophonous calls, with large roosting flocks making a boisterous racket. ABUNDANCE Common and abundant in preferred habitat throughout wintering range, excepting Southeast NM where numbers wane. Breeding populations are concentrated in North Central NM and East Central AZ. CONSERVATION STATUS Least Concern.

Chihuahuan Meadowlark

Sturnella lilianae

Arid grasslands, prairies, agricultural fields
LENGTH 7.5–10 in. WINGSPAN 14–16 in.

Round-bodied, stout-necked, sharp-billed, and stubby-tailed tennis ball with wings. Plumage is sandy tan with black speckles above, sunny yellow below, with white flanks adorned with black spots, and with a perfectly centered black "bandana" around the neck. Cheeks are pale taupe with prominent dark eyeline, yellow lores, buffy to white malar, and white supercilium. Bill is a silvery blue-gray. Broad white outer tail edges conspicuous in flight. Coloration on breeding individuals is slightly more saturated with subtly cleaner plumage. BEHAVIOR Ground-dwelling birds primarily, though look for their white outer tail feathers flashing in flight when they flush (quite easily). Males perch on fence posts and low branches to sing. VOCALIZATIONS Your best bet at a clear identification outside of range, song is a clear and resonant whistling. A helpful mnemonic device might be "three cheers for three beers!" *Meep* call is similar to Common Nighthawk. ABUNDANCE Uncommon resident in appropriate habitat throughout AZ and NM, with populations concentrated in the south. KEY SITES AZ: Elgin/Sonoita/San Rafael Valley in southern part of the state; NM: Animas Valley. LOOKALIKES Western, Eastern meadowlarks. Chihuahuan distinguished from Western by distinct voice, more extensively white tail, and whitish malar; from Eastern by voice, range, and more extensively white flanks with fainter spotting. CONSERVATION STATUS Near Threatened (Decreasing). (Conservation status for Eastern Meadowlark [*S. magna*], as Chihuahuan Meadowlark was split from this species in 2022.)

Breeding adult

Nonbreeding adult/immature

Western Meadowlark

Sturnella neglecta

Open, treeless areas including grasslands, roadsides, agricultural fields, meadows
LENGTH 7–10 in. WINGSPAN 14.5–16 in.

Mockingbird-sized, with a heavy body, short tail, short wings, a flat-topped head, a long and pointed bill, and frequently hunched and neckless posture. Breeding adult is heavily patterned in black, gray, brown, and buff above, bright yellow below. Yellow lores fade into white eyebrows, with dark line behind the eye, pale gray face, yellow throat and malar. White flanks with rows of black spots contrast with black, V-shaped bib. Bill is silver-gray. White outer tail feathers prominent in flight. Nonbreeding adults and immature birds are less strikingly patterned, brown instead of black, with a washed-out, yellow belly and faded bib (absent in very young birds). BEHAVIOR Forages in groups on open ground, picking up seeds and probing into soil for insects. Males pop above the grass onto short perches to sing and defend territory. VOCALIZATIONS Song is similar to that of blackbird, beginning with clear whistles before falling into a complex warble. Call is a dry *chup*. ABUNDANCE Common year round across both states, except from South Central to Southeast AZ and in extreme Southwest NM, where it is a winter visitor. LOOKALIKES Chihuahuan, Eastern meadowlarks. Western distinguished by voice, thinner white tail edges, and yellow malar. CONSERVATION STATUS Least Concern (Decreasing).

Adult male

Female

Immature male

Orchard Oriole

Icterus spurius

Open woodlands, riparian woodlands, edge habitat, suburbs/parks, orchards
LENGTH 6–7 in. WINGSPAN 9–10 in.

A gorgeous species, slender but subtly curvaceous, with a long, tapered bill. Adult male is dressed in rich burnt sienna on rump, belly, and chest, extending to a wing patch at the bend of the wing, with deep black head and upper back. Fine white edging on wings. Female is straw-yellow tinged with green overall, with a pinkish bill and gray wings accented by two creamy white wingbars. Immature male shows similarly to female, but with a prominent black throat and dark bill. BEHAVIOR Look for them skulking through the canopy, picking off insects and berries and imbibing on nectar-rich flowers. Like other oriole species, they weave pouch-shaped nests, though they are far less dramatically elongated than their relatives' socklike abodes. VOCALIZATIONS Song is often remarkably similar to that of the American Robin—a series of warbles—but with a sweet, finchlike quality, and highly variable among individuals. Calls include various *chut*s, chatters, and short, inquisitive whistles. ABUNDANCE Uncommon. Breeding range just peeks into southeastern corner of NM. KEY SITE Rattlesnake Springs area of Carlsbad Caverns National Park, NM. LOOKALIKES Baltimore, Hooded orioles. Compared to Baltimore, adult male Orchard shows burnt sienna versus bright orange plumage; female Orchard is more greenish yellow and sports a pinkish bill. Orchard best distinguished from female and immature male Hooded by slightly smaller size, shorter bill, shorter tail, and bolder wingbars. CONSERVATION STATUS Least Concern.

Female

Immature male

Adult male

Hooded Oriole

Icterus cucullatus

Open and suburban woodlands, desert riparian woodlands with willows, walnut, cottonwoods, sycamores
LENGTH 7–8 in. WINGSPAN 9–11 in.

Slim and sleek, from its long and sharply pointed bill to its long and rounded tail. Adult male shows deeply saturated bright yellow to brilliant flame-orange, with jet-black throat that extends to just above the eyes. Wings and tail are black, with bright white wingbars and edging. Female is buttery yellow overall with an olive-green tinge, an olive-gray back, and delicate white wingbars. Immature male shows similar coloration to female, but with black throat. BEHAVIOR Deftly dangles from branches and twigs to glean insects and berries, slowly moving through the canopy and darting forcefully as it moves from tree to tree. Male courtship posturing includes dramatic bowing and stretching routines combined with song. Like other orioles, female weaves impressive pouch-shaped nests attached firmly to the undersides of palm fronds, yucca leaves, mistletoe clumps, or deciduous branches. VOCALIZATIONS Series of chatters and warbles similar to mockingbird, often interspersed with mimicked sounds of other bird species. Both male and female sing, but female's song is on the simpler side. Calls vary, with most common being a sweet, inquisitive, whistling *weet* or a scolding, rapid *chit-it-it*. ABUNDANCE Breeds in preferred habitat throughout Western, Central, and Southeast AZ; in Southwest NM; and less frequently in suitable habitat along the southernmost NM boundary. LOOKALIKES Female Scott's and Bullock's orioles. Hooded distinguished by unstreaked back versus streaking on Scott's, and yellow belly versus white/grayish white of Bullock's.
CONSERVATION STATUS Least Concern.

Adult male

Female/immature

Streak-backed Oriole

Icterus pustulatus

Dry, open tropical forests and savannah; riparian areas; suburban areas with plentiful flowering plants and hummingbird feeders LENGTH 8.25 in. WINGSPAN 12.5–13 in.

Slender but fairly large with a rounded head, long tail, and a relatively long, straight, and sharply pointed bill. Adult male is rich orange overall and has a gray bill with a dark tip to the lower mandible, a black mask and bib, a black tail, dark wings showing two white wingbars and extensive white edging, and dark streaks on the upper back. Adult female and immature birds are yellow-orange overall with light-colored tails and less distinct streaking. BEHAVIOR Forages in trees for insects and nectar, using its long bill to probe blooms and glean prey from foliage. VOCALIZATIONS Song is a choppy series of simple whistles and distinct pauses. Calls include a dry rattle and a clear, rising whistle. ABUNDANCE A very rare visitor from Mexico to Southeast and Central AZ, Southwest AZ along the lower Gila and Colorado rivers, and even more rarely to NM along the Rio Grande and lower Pecos rivers. Sightings occur less than annually. Has nested in AZ. LOOKALIKES Hooded Oriole, female/immature Bullock's Oriole. Streak-backed distinguished from Hooded by more white in wings; a broader, shorter, and straighter bill; and streaked back. Streak-backed distinguished from female/young Bullock's by yellow-orange versus pale belly. CONSERVATION STATUS Least Concern.

Bullock's Oriole

Icterus bullockii

Open woodlands, riparian woodland/cottonwood galleries, lush riparian-adjacent desert, orchards
LENGTH 6.5–7.5 in. Wingspan:11–12.5 in.

Slim-bodied songbird with a short to medium-length tail and a sturdy bill that comes to a fine, sharp point from its thick base. Adult female shows a deep golden head, throat, and tail, with a gray back and pale whitish belly. Adult male shows bright marigold-orange to flame-orange overall, with clean, black eyeline; black throat (more like a goatee); black back; and large, white wing patch. Immature male appears similar to female, but often shows darker throat. **BEHAVIOR** Expert gleaners, dangling from the thinnest branches to pluck berries and insects from the undersides of leaves and twigs. Will sip nectar from flowers, and will frequent backyard jelly, orange, and hummingbird feeders to satisfy their sweet tooth. Their nests are marvelous woven pouches, securely attached to branches, 10–50 feet from the ground. Highly vocal species is often heard before seen. **VOCALIZATIONS** Song is a rapid back and forth of squeaky whistles and harsh rattles, a few seconds in length. Call is a harsh series of *chut*s or a single, scolding *chut*. **ABUNDANCE** Common and breeding resident throughout most of the region, preferring riparian corridors and lush, riparian-adjacent deserts, including saguaro forest. Nonbreeding migrant in southwest corner of AZ. **LOOKALIKE** Baltimore Oriole. Bullock's distinguished by slightly paler golden orange coloration, white wing patch versus wingbars, and black eyeline and throat versus black head. Female Bullock's also shows more white on wing than Baltimore. **ALTERNATIVE NAMES** Marigold Oriole, Cottonwood Oriole, White-winged Oriole, Western Oriole. **CONSERVATION STATUS** Least Concern.

Adult male

Female

Immature male

Scott's Oriole

Icterus parisorum

Areas with high densities of yucca (mostly Soaptree and Torrey's yucca) within arid, mid-elevation habitats including high deserts, grasslands, pine-oak and pinyon-juniper woodlands, occasionally chaparral **LENGTH** 9 in. **WINGSPAN** 12.5 in.

Medium-sized and athletic with a round head, long tail, and a relatively long, straight, and extremely sharply pointed bill. Adult male is sunflower-yellow below with yellow shoulders, white wingbars and flight feather edges, and a black head, breast, back, wings, and tail tip. Adult female and immature birds are dull yellow-green overall, with grayish head and face, dark gray wings, white wingbars and flight feather edging, and dingy stripes on the upper back. **BEHAVIOR** Uses its finely tipped bill to catch insects, pluck fruit, sip nectar, and build tightly woven, sock-shaped nests from yucca fibers, grasses, and other desert plants. **VOCALIZATIONS** Song, a short melody of clear whistles, is more easily confused with a Western Meadowlark than another oriole. Common call is a harsh *CHEK*. **ABUNDANCE** Uncommon during breeding across most of AZ and from Central to Western NM, with greatest abundance at middle elevations from Northwest to Southeast AZ and from Central to Southwest NM. **LOOKALIKES** Female/immature Hooded and Bullock's orioles. Scott's distinguished from Hooded by grayish face, from Bullock's by yellow belly, and from both by streaked back, distinct wing bars, and straighter bill. **ALTERNATIVE NAME** Yucca Oriole. **CONSERVATION STATUS** Least Concern (Decreasing).

Red-winged Blackbird

Agelaius phoeniceus

Varied, brushy, often waterlogged habitats including wetlands; marshy edges of rivers, lakes, and ponds; irrigated agricultural fields; feedlots; grasslands; meadows; golf courses; parks
LENGTH 7–9 in. WINGSPAN 12–15 in.

A chunky blackbird with a medium-length tail and a sharply pointed, conical bill. Adult male is solid black except for a bright red shoulder patch bordered below in yellow. Adult female is heavily streaked in brown, buff, rufous, and white; shows a pale eyebrow, malar, and throat; tiny red shoulder patch sometimes visible when perched; and often has a buffy yellow or pinkish wash on the throat and around the bill. Juvenile resembles adult female, but young males can be difficult to identify as they transition to their darker adult plumage. BEHAVIOR A communal rooster and gregarious in all seasons, but most social in winter when they join massive, murmurating flocks, often with other species of blackbirds. Anything but inconspicuous, males sing loudly and frequently from high perches with feathers ruffled. VOCALIZATIONS Song is variable with cowbirdlike liquid notes preceding a harsh, buzzy, and mechanical-sounding *konk-ola-REEEE* or *cla-REEEE-ah*. Calls include an abrupt *chup*, high chips, reedy whistles, and scolding chatter. Groups are extremely vocal and raucous throughout the year, often drowning out the voices of other wetland birds. ABUNDANCE Common year round across both states; most abundant in winter. CONSERVATION STATUS Least Concern (Decreasing).

Adult male

Female

Bronzed Cowbird

Molothrus aeneus

Open habitat, including prairies, pastures, chaparral, mesquite bosque, parklands
LENGTH 7–8 in. **WINGSPAN** 12–13 in.

Sturdy, stout, thick-necked songbird with a heavy, sharply pointed bill and rounded tail. Male is sleek, matte-black with bright red eyes and blue-black iridescent wings. Female is ashy brown with a wash of graphite-gray—darker as range extends eastward—also with bright red eyes. Juvenile shows dark eyes and streaky gray plumage. **BEHAVIOR** Moves in small flocks, ground foraging for both insects and seeds, but primarily granivorous. Look for males performing their "helicopter" courtship displays and strutting their stuff with their neck feathers fluffed. Like the Brown-headed Cowbird, Bronzed Cowbirds are nest parasites, often seeking out and sneaking into oriole nests, where they lay their "gifts." **VOCALIZATIONS** Perplexing when first encountered, the distinct liquid, metallic sound of the male's song, like a yo-yo sound effect accented by a dripping faucet, is strikingly similar to that of the Brown-headed Cowbird but slightly deeper and more resonant. **ABUNDANCE** Uncommon, with limited breeding range throughout the southeastern quadrant of AZ and peeking into Southwest NM; another breeding population along the lower Colorado River corridor along the Western AZ boundary. Small resident population in Atascosa Highlands area, west of Nogales, AZ. **LOOKALIKE** Brown-headed Cowbird. Male Bronzed distinguished by black head with red eye, bluish iridescence on wing; female Bronzed distinguished by larger size, red eye, and hefty bill. **CONSERVATION STATUS** Least Concern.

Female

Adult male

Juvenile

Brown-headed Cowbird

Molothrus ater

Open settings with widely spaced trees and scattered brush, including grasslands, agricultural fields, edges of riparian and other open woodlands, suburban areas
LENGTH 7.5–8.5 in. WINGSPAN 12–14 in.

Small and stocky with a short tail, bulky head, and a finchlike conical bill. Adult male is black with a green, iridescent sheen and chestnut-brown head. Female is dusty brown overall, lightest below and on the face, with indistinct streaking below and a very faint pale eyebrow and malar. Juvenile similar but with scalloped pattern above and streakier below. Molting young males may be awkwardly patchy. BEHAVIOR Forages on open ground for seeds and insects, usually in groups and often mixed with other blackbirds, grackles, and starlings. Though often maligned for being an obligate brood parasite (they lay eggs in the nests of other birds, often to the hosts' detriment), human alteration of habitat has enabled this species to become a conservation challenge in many settings. VOCALIZATIONS Song is short, beginning with a distinctive, metallic, and leaky faucet–like gurgle and closing on a high, rising whistle. Calls include rapid chatter and piercing whistles. ABUNDANCE Common regionwide during breeding and migration and year round across Southern AZ and Southern to Eastern-Central NM. LOOKALIKE Bronzed Cowbird. Brown-headed distinguished by lighter bill, rounder head, and dark eyes. CONSERVATION STATUS Least Concern (Decreasing).

Adult male

Female

Brewer's Blackbird

Euphagus cyanocephalus

Open settings including grasslands, meadows, wetlands, sparse woodlands, chaparral; human-modified landscapes such as parks, agricultural fields, lawns, parking lots
LENGTH 8.5–9.5 in. WINGSPAN 15–15.5 in.

Medium-sized with a chunky body, rounded head, thick neck, longish legs, a long tail with a wide and rounded tip, and a medium-length, stout-based, and pointed bill. Adult male has yellow eyes, is black overall, and when well-lit shows a glossy green body and bluish purple head and breast. Adult female and juvenile have dark eyes and are unmarked and brownish overall. BEHAVIOR Forages for seeds and insects in groups on open ground with a jerky, chicken-like gait. VOCALIZATIONS Song is a simple, raspy, and mechanical *la-REE*, often preceded by a sometimes repeated *chuck*. Calls include a harsh *chuck*, whistled gurgles, and squawking chatter. ABUNDANCE Most widespread in winter when they range across both states. Common during breeding in the Southern Rockies north of Santa Fe and year round from East Central to Northern AZ, from West Central to Northern NM, and in the Lincoln National Forest of NM. LOOKALIKES Cowbirds, grackles, Rusty Blackbird (rare in the Southwest). Brewer's Blackbird distinguished from cowbirds by thinner bill and longer tail; from grackles by shorter bill and shorter tail; and from Rusty by glossier plumage in male, less rufous plumage and dark eyes in female, and a heavier bill in both sexes. ALTERNATIVE NAME Glossy Blackbird. CONSERVATION STATUS Least Concern (Decreasing).

Adult male

Adult male

Female

Common Grackle

Quiscalus quiscula

Open habitat including woodlands, suburban parks, agricultural fields, urban lawns, grassy marshes
LENGTH 11–13.5 in. WINGSPAN 13–17.5 in.

Elegant, elongated blackbird, from bill to tail. Adult male appears black at first glance, but upon closer inspection in good lighting, a palette of colors is apparent, with a blue head, purple/green/bronze metallic wash on the back and wings, and eyes of pale gold. Male's tail is longer and more paddle-shaped than female's. Female plumage is slightly less slick in appearance, with variable coloration including dull black, purplish black, and bronze, with pale golden eyes. Immature birds are deep brown-black with dark eyes. BEHAVIOR Gregarious and gangly, strutting around on their lanky legs. Flocks occupy lawns and open fields, backyard feeders, and marsh edges. Opportunistic omnivores, eating anything from small fish (such as shiners/minnows) to insects and seeds—even other birds. Known to be a backyard feeder bully, swiftly evicting competitors. VOCALIZATIONS Raucous series of screechy, rusty door–like squeaks and squawks, croaks, and clear, resounding whistles. Call is a quick, harsh, nasally *chit*. ABUNDANCE Broad breeding range in suitable habitat from Central to Eastern/Northeast NM, yet relatively uncommon. Rare in AZ. LOOKALIKE Great-tailed Grackle. Common distinguished by smaller overall size and lack of exaggeratedly long, keeled tail. CONSERVATION STATUS Near Threatened (Decreasing).

Adult male

Female

Great-tailed Grackle

Quiscalus mexicanus

Well-watered habitat—natural and human-made—in developed settings such as parks, irrigated suburban areas, playing fields, agricultural fields; in natural settings such as wetlands, edges of ponds and lakes, marshy riparian areas
FEMALE LENGTH 15 in. **WINGSPAN** 19 in.
MALE LENGTH 18 in. **WINGSPAN** 23 in.

A large and long-bodied blackbird with a flattened head; long legs; a long, wedge-shaped tail; and a bill that is straight, heavy, and longer than that of other blackbirds. Adult male is entirely iridescent blue-black with an exaggeratedly long tail and yellow eyes. Adult female is rich brown overall, lightest below, and has a pale throat and eyebrow. Juvenile resembles female but with light streaking below and dark eyes. **BEHAVIOR** Intermingles with other blackbirds in natural settings but is more common than its compatriots in developed areas, where it employs astoundingly creative foraging tactics that impress even the nonbirders among us. **VOCALIZATIONS** Song is varied but always extremely loud and consists of loud shrieks, rapid chatter, harsh rattling, and rusty gate–like creaking. Most common call is an abrupt *CHUT*. **ABUNDANCE** Common year round throughout both states, except from Northern NM to Northeast AZ, where it is a breeding resident only. Once a Central American species, was brought to Mexico in the late fifteenth century and has been expanding northward with agriculture and development ever since. **LOOKALIKE** Common Grackle. Great-tailed is significantly larger and longer tailed. **CONSERVATION STATUS** Least Concern.

NEW WORLD WARBLERS

Small, round, and technicolored insectivores with needlelike bills, the New World warblers (family *Parulidae*) can be difficult to spot as they bounce energetically through mostly middle and upper canopies. Just as they're reluctant to sit still in the foliage, they are reluctant to linger geographically, with many warblers present in the Southwest only during the breeding season, others migrating through during spring and fall, and a handful of species arriving and camping out throughout winter. Fortunately, these tiny migrants are also prolific songsters. By learning their songs and calls, birders can save themselves significant struggle and neck strain.

ABOVE Painted Redstart

Ovenbird

Seiurus aurocapilla

Dense, deciduous hardwood forests
LENGTH 4.5–5.5 in. WINGSPAN 7.5–10 in.

A delight of a woodland warbler, with long wings and a chunky body. Unique among warblers, it shows thrushlike plumage: olive-brown on the back and creamy white on the belly, with bold, deep espresso-colored streaks and spots; a large, dark eye with a bold white eye ring; and a clean, black-bordered orange crown that it often holds peaked. Deep coffee-colored mustache stripe on clean, creamy white throat. Coral-pink legs. Often holds tail upright in a wrenlike posture, but this forms more of a smooth, U-shaped back compared to the sharp V shape of wrens. BEHAVIOR Look for this little cutie hunting and pecking along the forest floor, searching among the leaf litter for its next snack, primarily insects and larva, worms, and spiders. It walks, not hops, along branches and across the ground, like the world's tiniest chicken. VOCALIZATIONS Breeding season song is the notorious summer song of the eastern woods—*teacher teacher teacher*. Male's dusk flight song is a jubilant series of bubbling warbles punctuated by a softer version of its trademark *teacher teacher* song. Typical call (among many) is a harsh, pointed chip; migration call is a sweet *tseep* (heard primarily at night). ABUNDANCE Rare. Migration range cuts just slightly into the extreme northeastern corner of NM. CONSERVATION STATUS Least Concern.

Northern Waterthrush

Parkesia noveboracensis

During migration, in shaded, brushy, wooded habitats near stagnant or slow-moving water such as wetlands, ponds, sluggish streams LENGTH 5–6 in. WINGSPAN 9–10 in.

Large, long-legged warbler with a fairly short tail; thin, pointed bill; and forward-leaning posture. Both sexes are dark grayish brown above; white with coarse, dark streaks and often a yellow wash below; and with a thin, whitish to yellowish eyebrow and malar. Pale throat marked with fine spotting and bordered by dark stripes. Dingy, pinkish legs. BEHAVIOR Hunts for insects, arthropods, and occasionally vertebrates including small fish and salamanders while walking near the water's edge or hopping atop partially submerged rocks and debris. Similarly to Spotted Sandpiper, rear end bounces while in motion. VOCALIZATIONS Song is a choppy and accelerating series of clear, descending, and whistling notes: *sweet-sweet-sweet-chew-chew-chew*. Common call is a clear, sharp chip. ABUNDANCE Uncommon during migration across both states, with observations concentrated mostly from Central to Eastern and Southern NM and less frequently from Central to Southeast AZ. LOOKALIKE Louisiana Waterthrush. Northern distinguished by preference for stagnant or slow-moving water, eyebrow that does not widen toward the rear, yellowish wash below, smaller bill, spotted throat, dingier legs, and flanks lacking a buffy wash. CONSERVATION STATUS Least Concern.

Adult male

Adult female

Black-and-white Warbler

Mniotilta varia

Deciduous forests, mixed conifer forests; any wooded area during migration
LENGTH 4–5 in. **WINGSPAN** 7–8.5 in.

Small, round warbler with long, slightly downcurved bill, short neck, and short tail. Long wings extend past the body. Adult male is heavily streaked in black and white overall, with bold black ear patch. Female sports similar plumage, but with a washed-out, grayish ear patch; paler overall coloration (more white); sometimes showing buffy wash on the flanks. White undertail coverts show distinct black spots, a good identification mark when craning your neck up from under a tree. **BEHAVIOR** A trunk creeper, much like a nuthatch, thanks to an extra hind claw that offers it stability as it picks insects from the bark. If you're in their nesting range (Eastern/Northeastern United States) and happen upon a small, cup-shaped nest tucked into the ground or into the depression of a stump, leave it be! These little creeping warblers build nests on or near the ground, though they are seldom seen. **VOCALIZATIONS** Song is a high-pitched, thin, rapid, and repetitive *wee-see wee-see wee-see*, like a squeaky shopping cart wheel at high speeds. Call is a sweet, gentle chip or *tsip*. **ABUNDANCE** Uncommon. Migrates mostly through Eastern NM, but regularly spotted as a rarity during migration throughout AZ and NM. **CONSERVATION STATUS** Least Concern.

Adult male

Female/immature

Orange-crowned Warbler

Leiothlypis celata

Breeds in mid- to high-elevation habitats from chaparral to coniferous forests, always with low, dense shrubs, often near riparian corridors; similarly structured habitat during winter and migration
LENGTH 4.5–5.5 in. **WINGSPAN** 7.25–7.5 in.

Petite and lean with a notably thin bill, short wings, and short tail. Adults of both sexes are tennis ball yellow-green overall, brightest in the undertail coverts, and have dingy streaks below, a pale eyebrow, a blackish, smudgy eyeline, a broken eye ring, a grayish undertail, and an often hidden orange crown patch. Highly variable, may show a grayish head and breast. Female and juvenile are similar, but duller overall. Female's crown patch may be absent or brownish. **BEHAVIOR** Forages for insects by gleaning in low, dense vegetation, sifting through leaf litter, flycatching, or hovering beneath large leaves. **VOCALIZATIONS** Song is a simple, clear, two-part, and trilling *chee-chee-chee-chew-chew-chew-chew*. Call is a clear, high chip. **ABUNDANCE** Breeds uncommonly in the mountains of Central to Western and Northern NM, Northwest AZ, and in East Central AZ along the Colorado Plateau transition zone to the White Mountains. Common in migration across both states and in winter from Central to Southern/Southwest AZ and in Southern NM, mostly along the Rio Grande. **LOOKALIKES** Tennessee, Nashville, Yellow warblers. Orange-crowned distinguished from Tennessee by weaker eyeline, streaked underparts, and yellow undertail coverts; from Nashville by less distinctly gray head and broken eye ring; and from Yellow by duller overall color and thinner bill. **CONSERVATION STATUS** Least Concern (Decreasing).

Adult male

Female/immature

Lucy's Warbler

Leiothlypis luciae

Low-desert mesquite bosque and desert riparian cottonwood/willow galleries; higher elevation riparian areas with sycamore/oak LENGTH 3.5–4.5 in. WINGSPAN 5.5–7 in.

A dainty little tink of a warbler, compact and slender with thin bill and short tail. Adult male is dusty gray overall, with deep russet crown and rump and buffy gray chest. Female/immature shows pale, warm, dusty gray overall, lacking russet coloration but sometimes showing warm yellow-brown or yellow-orange on cap and rump. BEHAVIOR Flits through the canopy foraging for insects, with rarely an idle moment. Will make use of species-specific nest boxes, but typically nests in mesquite cavities or abandoned woodpecker holes. Males are extremely vocal throughout spring/breeding season. VOCALIZATIONS Song is similar to Virginia, Nashville, and Yellow warblers—a set of two or three clear, clean trills, each varying in pitch. Call is a snappy *tsip*. ABUNDANCE Common in suitable habitat throughout AZ breeding range. ALTERNATIVE NAMES Mesquite Warbler, Bosque Warbler, Rufous-rumped Warbler. CONSERVATION STATUS Least Concern.

Adult male

Female/immature

Nashville Warbler

Leiothlypis ruficapilla

Varies across range: western populations breed in brushy, oak-dominated forests; Southwest migrants prefer brushy, shrub-rich desert flats, densely vegetated washes
LENGTH 4.5–5 in. **WINGSPAN** 7–8 in.

Small and round-bodied with a short tail and a short, sharply pointed bill. Adult is greenish yellow above, brightest on the rump, yellow below fading to white on the lower belly, with yellow undertail coverts, a bluish gray head and upper back, and a bold white eye ring. Male is brighter overall and sports a small, rusty red crown patch. **BEHAVIOR** Gleans insects at middle heights near the tips of branches, often flicking its tail as it goes. **VOCALIZATIONS** Song is a sweet, clear, two-part *tsee-tsee-tsee-too-too-too-too*, sometimes with distinctly two-part introductory notes: *tasee-tasee-tasee-too-too-too-too*. Common call is a sharp, dry chip. **ABUNDANCE** An uncommon to rare migrant across both states. Most widespread in autumn and occurs most reliably in Southwest and Southeast AZ. **LOOKALIKES** MacGillivray's, Mourning, Virginia's, Tennessee, Orange-crowned warblers. Nashville distinguished from MacGillivray's and Mourning by daintier bill, complete eye ring, and white lower belly; from Virginia's by more yellow plumage overall; from Tennessee by lack of pale eyebrow, complete eye ring, and brighter underside with yellow undertail coverts; and from Orange-crowned by unstreaked underparts and true yellow (versus yellow-green) plumage. **CONSERVATION STATUS** Least Concern.

Adult male | Female/immature

Virginia's Warbler

Leiothlypis virginiae

Arid, rugged, mountainous, open woodlands of pinyon-juniper, oak, pine-oak
LENGTH 4–5 in. WINGSPAN 6.5–7.5 in.

Compact, short-tailed warbler with slender, sharp bill. Male shows gray overall with buttery yellow wash on chest and bright yellow rump. Large, dark eyes encircled by bold, bright white eye rings. Inconspicuous cinnamon-colored crown patch is usually hidden. Female/immature birds show similar plumage, with paler or no yellow on chest. BEHAVIOR Moves quite exclusively through dense, scrubby midlevel canopy searching for insects. Frequently displays tail-bobbing behavior, a nice hint for field identification. Nests made with grasses, bark, roots, moss, and animal hair are extremely difficult to locate, cryptically hidden within intricately built hollows of leaves within dense brush and concealed by mats of grass. VOCALIZATIONS Similar in quality and form to Nashville, Lucy's, and Yellow warblers, song is a series of clear, resonant, whistled warbles accented by a fading, lower pitched, faster paced sputter. Call is tiny, sweet *tsip* or chip. ABUNDANCE Common in preferred habitat but has patchy, sparse breeding ranges and broad migratory range. LOOKALIKES Nashville and Lucy's warblers, Common Yellowthroat (female). Virginia's Warbler distinguished by bold eye ring, gray belly, and bright yellow undertail coverts. Look for tail-wagging behavior. ALTERNATIVE NAMES Tail-wagging Warbler, Pine-oak Warbler, Pinyon Warbler. CONSERVATION STATUS Least Concern (Decreasing).

Adult male

Female/immature

MacGillivray's Warbler

Geothlypis tolmiei

Breeds in dense streamside vegetation within disturbed, second-growth coniferous and mixed forests typically with maple, cottonwood, spruce, fir; migrants prefer shrubby, wooded areas, often near water and with relatively open canopies
LENGTH 4.5–5.5 in. WINGSPAN 7.5 in.

Small but heavy bodied with a fairly long tail and pointed bill that is, for a warbler, relatively thick. Adult male is deep greenish yellow above and bright yellow below, with dingy flanks, a dark lead-gray hood and breast, black lores, and bold white arcs above and below the eye. Female and immature similar but with paler gray hood and usually whitish throat. BEHAVIOR Bounces through low vegetation in search of insect prey, sometimes dropping to the ground to make a catch. VOCALIZATIONS Song is a variable, high, somewhat buzzy, and ascending *churee-churee-churee-churee* or *chee-chee-chee-churry-churry*. Calls include a dry *chit* and a very high chip. ABUNDANCE An uncommon breeder from Southwest Central to Northwest NM and from North Central to Northeast AZ, most abundant in the Southern Rockies and in AZ's White Mountains. An uncommon migrant across both states. LOOKALIKE Mourning Warbler (rare in the Southwest). MacGillivray's distinguished by bold eye arcs and longer tail. ALTERNATIVE NAME Crescent-eyed Warbler. CONSERVATION STATUS Least Concern (Decreasing).

Adult male

Female

Common Yellowthroat

Geothlypis trichas

Densely vegetated marshes, wetlands, grasslands, forests; suburban yards and parks with suitable vegetation
LENGTH 4–5 in. WINGSPAN 6–7.5 in.

Little bandits of marsh, grassland, and forest edges. Small, stout-necked songbird with round body and rounded tail often held up like a wren. Male shows gold-washed, olive-gray back; bright yellow throat and belly that fade to straw color; and bold black mask (just a shadow of a mask on immature males) bordered by creamy white toward the crown. Female shows olive-gray overall, with golden yellow throat and undertail coverts; lacks black mask. BEHAVIOR These little floofs spend much of their time sheltered by thick vegetation in marsh and grassland habitat, where you can find them peeking out from time to time in the spring and summer months. More likely, you'll hear the male belting out his song before spotting him. Will forage with mixed flocks during migration, when they become more opportunistic and make use of a wider variety of habitat. VOCALIZATIONS Male song is a distinct, three-part (and incessant during breeding season) *witchity witchity witchity woot*, sounding off from cattail stands and dense grassland. Call is a sharp *chuk* or rapid, sputtering chatter. ABUNDANCE Breeds primarily along the Rio Grande River corridor in NM, and less commonly in wetland/marsh areas throughout the region. Resident in Southeast AZ; overwinters in Southwest AZ. LOOKALIKES Nashville, Orange-crowned warblers. Female Common Yellowthroat distinguished by olive-gray versus gray coloration and bright yellow throat with dark cheeks. Common Yellowthroat lacks exaggeratedly thin bill of Orange-crowned. CONSERVATION STATUS Least Concern (Decreasing).

Female

Adult male

Immature male

American Redstart

Setophaga ruticilla

Open deciduous woodlands; any forested area during migration
LENGTH 4–5 in. WINGSPAN 6–7.5 in.

Medium-sized warbler with broad, flat bill; long, fanned tail; and long wings. Adult male shows black plumage overall, with flame red–orange upper flanks, wing patches, and outer tail patches. Belly is white. Female mirrors this plumage pattern, but with olive-gray plumage overall accented by buttery yellow patches and a pale gray head. Immature male shows yellow patches with blotchy black bits on the face, back, and chest. BEHAVIOR Flitting and fluttering about the canopy, frantically foraging and flushing insects with a flash of their brilliant flame-orange tails. Clever predators, they fan their bright tails at insects, like a little matador, and then sally forth to snag them, flycatcher style. VOCALIZATIONS During breeding, male sings several variations on a theme: a high-pitched series of *su-ee su-ee su-ee sue* or *sue-sue-sue-sue* notes, sometimes with an ascending and descending quality, and ending abruptly. Call is a thin *tsip*. ABUNDANCE Uncommon but regularly observed during migration throughout AZ and NM, overwinters along lower Colorado River corridor. CONSERVATION STATUS Least Concern.

Adult male | Adult female

Yellow Warbler

Setophaga petechia

Habitat generalist of thickets, forest edges, streamsides; low-desert riparian to high-elevation mixed woodlands
LENGTH 4.5–5.5 in. WINGSPAN 6–7.5 in.

Small warbler with stout neck, round head, and short tail. Like a ball of singing sunshine, the male is a deep golden yellow with streaks of chestnut-brown radiating down its chest and sides. Back is olive-gray-green. Female shows paler, straw-yellow coloration and lacks streaking on chest. Immature shares paler coloration, sometimes showing pale gray with a wash of pale yellow. Beady eye is jet-black. BEHAVIOR One of the most cooperative species when it comes to humoring birdwatchers, often hanging out in low to mid-canopy, aloofly foraging for insects it gleans from branches and leaves. Often falling victim to nest parasitism, they are fated to raise giant, unceasingly hungry cowbird chicks. VOCALIZATIONS Breeding male song is the ubiquitous, distinct, and easy to recall "sweet sweet sweet, I'm so sweet," often the first warbler song learned by beginning birders. Calls range from a sharp, clear chip to a thin, metallic *tseet*. ABUNDANCE Common breeding resident throughout breeding range, with populations concentrated along riparian corridors. Wintering population along lower Colorado River corridor. Common migrant. LOOKALIKES Female Wilson's, Orange-crowned warblers. Yellow distinguished from Wilson's by solid yellow face, lacking hint of eyebrow and cap. Distinguished from Orange-crowned by heavier bill and brighter versus drab (sorry, Orange-crowneds) overall coloration. Song similar to Lucy's Warbler but less varied and longer, with cleaner, crisper notes. CONSERVATION STATUS
Least Concern (Decreasing).

Yellow-rumped Warbler

Setophaga coronata

Breeds in montane coniferous and mixed forests; in winter and migration, lower elevations with fruiting vegetation in nearly any shrubby, wooded habitat including deserts, riparian areas, open woodlands, suburban settings
LENGTH 4.75–5.5 in. WINGSPAN 7.5–9 in.

Referred to affectionately as "butterbutts." Somewhat large for a warbler, with a pointed bill, long tail, and often drooping wings. Adult male Audubon's form has a gray face and back, white underparts, a black breast, coarsely streaked or fully black flanks, messy white wingbars, yellow crown patch, a restricted and usually yellow but sometimes white throat patch, yellow patch on the flanks, and yellow rump. Female and immature birds lack yellow crown patch and are more brownish overall. In both sexes, Myrtle's form distinguished by dark mask, white throat patch that extends behind the eye, and pale eyebrow.

BEHAVIOR Forages for insects through gleaning and frequent flycatching. Wintering birds rely heavily on fruit and visit productive patches in large groups. VOCALIZATIONS Song is a variable, messy, accelerating, and sometimes two-parted *chewy-chewy-chewy-chewy* or *chewy-chewy-chewy-chee-chee* that may rise or fall in pitch toward the end. Common call is a distinctively harsh chip. ABUNDANCE Common. Breeds at high elevations from South Central to Northwest NM and from Southeast to Northwest AZ. Winters abundantly in Western and North Central to Southern AZ, in Southern NM, and northward along the Rio Grande. Found regionwide during migration. Audubon's form is far more abundant, but Myrtle's form does occur, especially in winter and migration.
CONSERVATION STATUS Least Concern.

Adult male

Female/immature

Grace's Warbler

Setophaga graciae

Mature, high-elevation, often montane pine and pine-oak forests with relatively open understories
LENGTH 5 in. WINGSPAN 8 in.

Small warbler with a needlelike bill and a medium-length tail. Adult male is blue-gray; white below; and has two white wingbars; a broad yellow eyebrow; a thin, black eyeline and broken mask; a small, yellow arc below the eye; a yellow throat, breast, and upper belly; and a white rump. Female and immature birds are similar but a bit drab, with less contrast. BEHAVIOR Usually remains frustratingly high in the pines, bouncing from branch to branch in search of insects and spiders to pluck from under bark or from within clusters of needles. VOCALIZATIONS Song, *chew-chew-chew-chee-chee-cheecheechee*, is a rising series of clear notes that accelerates into a fast trill. Compared to Virginia's Warbler song, is faster and less distinctly two-parted. Common call is a flat *tsip*. ABUNDANCE A common breeder at high elevations in Eastern and from Central to Northern AZ, in Western and Northern NM, and southward along the Sandia Crest and into Lincoln National Forest. LOOKALIKE Yellow-rumped Warbler. Grace's distinguished by yellow breast and eyebrow. ALTERNATIVE NAME Pine-oak Warbler. CONSERVATION STATUS Least Concern (Decreasing).

Adult male

Adult female

Black-throated Gray Warbler

Setophaga nigrescens

Chaparral, pine-oak, pinyon-juniper forests and canyons; urban and suburban forested areas throughout migration
LENGTH 4.5–5 in. WINGSPAN 7.5–8 in.

Small, round, and compact, with a short, broad bill and a big, round head. As this warbler's name suggests, both sexes show black throat, with gray back and boldly striped, black-and-white face with bright yellow spot in front of each eye. Crisp black streaking on sides contrasts with a clean, bright white belly. Female sometimes shows some patchiness on the throat and paler coloration overall. Immature shows a beautiful sooty gray and lacks black throat, but facial pattern is the same, complete with yellow spots. BEHAVIOR Moves slowly (for a warbler) throughout pine-oak canopy in search of insects and caterpillars. Look for them feeding on oakworm moth caterpillars, a favorite treat. Nests relatively low in large trees. VOCALIZATIONS Male sings buzzy *zeedle zeedle zeedle zeedle ZEET zee*, give or take a *zeedle* or two. Call (both sexes) is a quick, lackluster *tsup*. ABUNDANCE Common throughout extensive AZ and NM breeding range. Uncommon but regularly observed throughout winter in southernmost AZ. LOOKALIKES Black-and-white, Blackpoll warblers. Black-throated Gray distinguished by bright yellow spot in front of each eye, solid gray back, and solid black throat. CONSERVATION STATUS Least Concern (Decreasing).

Adult male

Female/immature

Townsend's Warbler

Setophaga townsendi

Breeds in mature, undisturbed coniferous and mixed forests with dense understories; migrates mostly through lush pine-oak woodlands, also variety of wooded habitats from low to high elevations including coniferous forests, riparian areas, desert oases, suburban settings **LENGTH** 5 in. **WINGSPAN** 7.5–8 in.

Small and full-bodied with a fairly long tail and a short, pointed bill. Adult male shows olive-yellow upper back with dark streaks, and gray wings with two white wingbars. Yellow face, breast, upper belly, and upper flanks contrast with black mask, cap, and throat; with coarse, black flank streaks; a white lower belly; white outer tail feathers; and white undertail coverts with black flecking. Female and immature birds are duller overall with grayish green-yellow replacing black, a dusky belly, and clean white undertail coverts. **BEHAVIOR** Stays high in the pines foraging for caterpillars and other arthropods during the breeding season but ventures lower in migration and winter to take insects and nectar from flowers. **VOCALIZATIONS** Song is a high and raspy *sweeda-sweeda-zee-zee-za* or *swee-swee-swee-swee-sweee-tsoo-tsoo* that ends on a high note. Chip can be thin and high or full and squeaky. **ABUNDANCE** A common migrant across both states and an uncommon winter resident at middle elevations from Central to Southeast AZ. **LOOKALIKE** Hermit Warbler. Townsend's distinguished by dark mask and yellow breast. **ALTERNATIVE NAMES** Black-throated Yellow Warbler, Black-eared Warbler. **CONSERVATION STATUS** Least Concern (Decreasing).

Adult male

Female/immature

Hermit Warbler

Setophaga occidentalis

Breeds in high-elevation conifer forests of spruce, fir, and pine; migrates through forested streamside/riparian woodlands
LENGTH 5.5 in. **WINGSPAN** 7.5–8 in.

Small songbird, short-necked and round-headed, with a longish tail and slender, sharp bill. Both male and female show gray-and-black coloration overall: two white wingbars on sooty gray back, pale grayish white belly with very faint streaking, and black throat. A pop of golden yellow extends from cheek to forehead, as if someone cut and pasted part of a Yellow Warbler onto this bird's head. Female and immature show more standard crayon-yellow versus deep gold of breeding male, with a dusky patch on the cheek; immature shows thin yellow eye ring. **BEHAVIOR** Get familiar with this warbler's underside, because that's likely what you'll see if you spot one. Spend most of their time moving busily high in conifer canopy, from trunk to tip and over again, gleaning insects. Will also join mixed flocks in migration. Known to hybridize with Townsend's Warbler where breeding ranges overlap. **VOCALIZATIONS** Breeding males sing two songs, most commonly a series of high notes followed by a descending raspier phrase. Slightly buzzy tone overall. Call is a dull chip. **ABUNDANCE** Range is limited to migration season in AZ and NM, with frequency of observations decreasing along northern boundaries of range. **LOOKALIKE** Townsend's Warbler. Hermit distinguished by all-yellow face, grayish white underparts with extremely faint streaks, and an absence of yellow on the throat and back. **CONSERVATION STATUS** Least Concern.

Adult male

Female/immature

Wilson's Warbler

Cardellina pusilla

Breeds in moist, brushy, high-elevation habitats including montane meadows and riparian areas, preferably rich in alder, willow, sometimes aspen; migrates through brushy, wooded habitats including riparian corridors, mesquite woodlands, suburban landscapes **LENGTH** 4–4.75 in. **WINGSPAN** 6–7 in.

Small and round-bodied, even for a warbler, with a small, sharp bill and a long, slender, and often raised tail. Adult male is rich yellow below, greenish yellow above and on the cheek, and has a small black cap, a barely visible yellow eye ring, a yellow eyebrow, a grayish undertail, and grayish flight feathers. Female and immature birds usually show greenish yellow crowns but can also show limited black caps. **BEHAVIOR:** Forages for small insects in low vegetation or on the ground. Fidgety and nervous, rarely spends much time at a single perch and flicks tail habitually. **VOCALIZATIONS** Song is a forceful, clear, whistled, and sweet *chee-chee-chee-chee-choo-choo-choo-choo*, sometimes accelerating and descending toward the end. Chip call is flat and nasal. **ABUNDANCE** An uncommon breeder near and north of Santa Fe and a common migrant across both states. **LOOKALIKE** Yellow Warbler. Wilson's is smaller and less uniformly yellow, has a daintier bill and grayish undertail, and sports a contrasting eyebrow and dark cap. **ALTERNATIVE NAMES** Toupeed Warbler, Black-capped Warbler, Alder Warbler. **CONSERVATION STATUS** Least Concern (Decreasing).

Adult male

Adult female

Red-faced Warbler

Cardellina rubrifrons

High elevation pine-oak, aspen, spruce-fir woodlands and canyons
LENGTH 5–5.5 in. WINGSPAN 8 in.

Small, slender, elongated warbler with stubby bill and long tail. Gray plumage on back, delicately accented by a thin white wingbar and nape, white underparts and rump, and a lipstick-red face and throat with black over the crown extending down over the ears (if a warbler had mutton chops), matching its large black eyes. Male shows brighter red face and throat than female, and immature shows rosy pink. BEHAVIOR In typical warbler fashion, flits through canopy gleaning insects, sometimes sallying forth to catch them in flight. Look for tail-flicking behavior. Unlike many of its relatives, males are relatively unaggressive when it comes to claiming breeding territory and will often join in the mating chorus with other males. They hide their cup-shaped nests in shallow depressions in the ground among woody plants or under overhangs. VOCALIZATIONS Clear and sweet, reminiscent of a Yellow or Lucy's warbler, with several *sweet-sweet-sweet* notes followed by two or three descending notes. Call is an abrupt chip. ABUNDANCE AZ/NM mountain specialty, regularly observed breeding in preferred high-elevation habitat. KEY SITES AZ: Madrean Sky Islands, Mogollon Rim; NM: Gila National Forest. CONSERVATION STATUS Least Concern (Decreasing).

Adult

Juvenile

Adult tail display

Painted Redstart

Myioborus pictus

Montane forests of oak, pine, juniper, frequently near canyons or riparian corridors; arid, open slopes
LENGTH 5.25–5.75 in. WINGSPAN 8.25–8.75 in.

Among our most unmistakable warblers. Small and stout with a pointed bill, relatively long wings, and a long tail. Adult is solid black overall with a bright red belly and small white arc below each eye. Deep black contrasts with mostly white undertail, bold white wing patches, and white tail edges. Juvenile is similar but grayer overall, lacking red belly. BEHAVIOR Bounces through the middle and lower canopies, frequently flicking its wings and fanning its tail in an attempt to scare up potential insect prey, which it captures primarily through gleaning and some flycatching. VOCALIZATIONS Song is sweet, warbling, whistled, clear, and variable, with two-part notes sandwiched between simpler whistles: *woo-woo-woo-weeda-weeda-weeda-we, wee-weeda-weeda-woo-woo-wee-wee*. Call is a harsh *CHEEUR*. ABUNDANCE Common breeder across the Colorado Plateau transition zone from West Central NM to Northwest AZ. Common breeder and uncommon winter resident in montane forests from Central AZ to the southeastern corner of the state and into Southwest NM. CONSERVATION STATUS Least Concern (Decreasing).

TANAGERS, CARDINALS, AND OTHERS

In the Southwest, the family *Cardinalidae* includes the tanagers, cardinals, grosbeaks, North American buntings, and Dickcissel. This diverse group is broadly midsized with hefty heads, long tails, and prominent bills that convey a good deal of information about the birds that carry them. Their size and shape are indicative of their diet, and birders can increase their odds of spotting birds in this family by taking their food into account. Watch for the parrot-billed Pyrrhuloxia in deserts rich with berries, gawk for grosbeaks where large nuts and seeds abound, and tease out tanagers in woodlands with plentiful large, flying insects such as bees and wasps.

ABOVE Female Western Tanager

Hepatic Tanager

Piranga flava

Montane pine-oak woodlands, oak highlands/canyons/slopes, wooded riparian corridors
LENGTH 7.5–8. in. WINGSPAN 12–13 in.

Sturdy bodied songbird with distinct, elongated form and heavy bill. Male coloration is a dusky, earthy, red-gray tone on the back (*hepatic* refers to the liver), with a clean red belly below, a silvery bill, and a distinct, dusky gray cheek patch. Female/immature shows maize-yellow coloration overall, with pale gray cheek patch and back. Individuals in AZ known to show sunnier, golden orange hue on throats and crowns. Plumage varies slightly among subspecies (fifteen total, two within AZ/NM range), with *P. flava hepatica* throughout AZ and Western NM and the slightly brighter *P. flava dextra* in Eastern NM. BEHAVIOR A treat for hardworking birders, this species is a slow mover, skulking through the canopy seeking insects, caterpillars, beetles, and berries. Pauses with an elongated posture, seeking out its next meal. Male defends the nest with song, often perching in open view on snags and outer branches. VOCALIZATIONS Song is a singsong, cheerful, and clear warble akin to that of a robin, grosbeak, or vireo. Call is a chattery *chuk* or, in flight, an inquisitive, wheezy *whait?* ABUNDANCE Another AZ/NM specialty (breeding range just peeks into Southern CO). Common throughout all but the northernmost extent of breeding range, where breeding populations dwindle. LOOKALIKE Summer Tanager. Hepatic distinguished by dusky gray wash on back, silvery bill, gray cheeks, and less saturation of color on both male and female. CONSERVATION STATUS Least Concern.

Summer Tanager

Piranga rubra

Breeds along low- to mid-elevation streams and rivers within relatively open patches of riparian forests of cottonwood, willow, mesquite, sometimes tamarisk; in migration, similarly wooded and well-watered habitats such as rivers, streams, lush suburban landscapes LENGTH 6.5–7.5 in. WINGSPAN 11–12 in.

Medium-sized and hefty with a slightly peaked head, long wings, a medium-length tail, and a fairly long, heavy, dull, and pinkish gray to yellow-gray bill. Adult male is bright red throughout, darkest in the wings. Female and immature birds are entirely yellow, brightest below and with a greenish cast above and in the wings. Subadult male may show awkwardly patchy red-and-yellow plumage. BEHAVIOR Hunts for bees, wasps, and other large insects from the high canopy through flycatching or careful gleaning. VOCALIZATIONS Song, a short series of three-syllable whistles, similar to that of American Robin but shorter, slower, and more distinctly phrased. Common call is a brief, dry *chit-it-tup*, usually with three or more notes. ABUNDANCE An increasingly uncommon but locally abundant breeder and migrant in Western and from North Central to Southern AZ, in Eastern and from Central to Southern NM, and farther north along major river corridors including the Colorado and Rio Grande. LOOKALIKES Female/immature Hepatic and Western tanagers. Summer distinguished from Hepatic by more strongly peaked head, more purely red (not grayish) plumage, lack of a grayish ear patch, and a pale bill; from female Western by less grayish wings lacking wingbars. CONSERVATION STATUS Least Concern.

Adult male

Female/immature

Western Tanager

Piranga ludoviciana

Breeds in open evergreen/conifer woodlands of various elevations, from pinyon-juniper to pine-spruce-fir
LENGTH 6–7 in. WINGSPAN 11–12 in.

A little flying flame, more petite and shorter bodied than other family members, but still larger than a warbler or sparrow. Breeding male plumage shows bright, flaming red-orange face, fading to banana-yellow on head, neck, underparts, and rump. Back, tail, and wings show black, with two wide wingbars—yellow on top, white below. Immature male shows fainter red coloration on face and scaled appearance on back, while nonbreeding male lacks flame coloration entirely. Female plumage is highly variable by individual, from pale gray–yellow to deep golden yellow, but with consistently dark wings and distinct wingbars. Sometimes shows faint wash of orange on forehead. BEHAVIOR Nests and forages high in canopy, slowly and steadily searching for insects, berries, and flower nectar. On occasion, will take flight to snatch an insect meal midair. Males are most vocal in late spring/early summer. VOCALIZATIONS Coarse but sweet melody, evoking a robin or grosbeak, typically repeating its up-down, up-down verse three or four times, sometimes immediately followed by its call. Call typically a two- or three-note, quick, chattery *pit-it* or *pit-it-eat*. ABUNDANCE Common throughout both migratory and breeding ranges. LOOKALIKES Female/immature Hepatic and Summer tanagers. Western distinguished from both by more slender bill, dark wings, and distinct wingbars. CONSERVATION STATUS Least Concern.

Adult male

Female/immature

Flame-colored Tanager

Piranga bidentata

Humid, montane oak and pine-oak forests, most frequently in shaded riparian canyons **LENGTH** 7–7.5 in. **WINGSPAN** 12 in.

Medium-sized and heavy bodied with a rounded head, long wings, a longish tail, and a dark slate-gray bill that is heavy, slightly decurved, and fairly long. Adult male is a bright red-orange that fades to a yellowish orange on the lower belly, with a blackish tail, blackish wings with prominent white wingbars and spots, smudgy brown ear patches, and a dingy orange back with black streaks and hints of olive-brown. Female and immature are more yellow overall, but display similar wing and back plumage, and gray ear patch. **BEHAVIOR** Forages for a variety of insects by gleaning from treetops, flycatching, and occasionally visiting the ground, where it will also take advantage of available fruit. **VOCALIZATIONS** Song is a quickly delivered series of two- or three-part whistled phrases, similar to that of American Robin or vireo. Call is a fast, slurred, and typically three-part *chur-di-dip*. **ABUNDANCE** First documented in the United States in 1985, remains an uncommon visitor, with summer detections restricted to the lush canyons of Southeast AZ's Chiricahua, Huachuca, and Santa Rita mountains. **KEY SITES** AZ: Cave Creek Canyon (Chiricahua), Miller Canyon (Huachuca), and Madera Canyon (Santa Rita). **LOOKALIKES** Other tanagers. Flame-colored differentiated by combination of bold, white wingbars and spots on dark wings; dark ear patch; and streaked upper back. **CONSERVATION STATUS** Least Concern.

Adult male

Female

Northern Cardinal

Cardinalis cardinalis

Relatively lush, brushy, wooded habitat including low- to mid-elevation riparian corridors, mesquite- and hackberry-lined desert washes, irrigated landscapes such as parks, golf courses, yards
LENGTH 8.5–9 in. WINGSPAN 10–12 in.

Long-tailed and medium-sized with a large, finchlike bill and a shaggy, exaggerated crest. Adult male is bright orange-red overall, darkest and with a faint wash of purple above, with a black mask and an orange bill. Adult female is also masked but is buffy to brownish gray overall, with only hints of red in the crest, tail, and wings. Juvenile has a gray bill and resembles a dull, disheveled, and maskless female. Subspecies observed in Arizona and New Mexico (*Cardinalis cardinalis superbus*) shows flame-orange coloration on male, taller crest, and more elongated build on both sexes than eastern counterparts. BEHAVIOR Forages inconspicuously and often in pairs for seed, fruit, and insects within low, dense vegetation or on the ground. Pairs give away their location with frequent calls and a habit of singing from exposed perches. A taste for sunflower seeds makes them frequent visitors to feeders. VOCALIZATIONS Song is a variable series of rich, clear, sometimes two-part, and often descending whistles: *peeeur-peeeur-peeeur-whit-whit-whit-whit-whit*, *weeda-weeda-weeda-pew-pew-pew-pew-pew*. Call is a sharp, metallic, and persistent chip. ABUNDANCE Common year round from Northwest Central to South Central and Southeast AZ. Less frequently encountered in the southwest corner and along the southern edge of NM. LOOKALIKE Pyrrhuloxia. Northern Cardinal distinguished by larger orange, finchlike bill with a less deeply curved upper mandible. CONSERVATION STATUS Least Concern.

Pyrrhuloxia

Cardinalis sinuatus

Low and dry desert scrub, sparse mesquite forests, columnar cactus gardens/forests; occasionally migrates locally to desert riparian areas to overwinter
LENGTH 8–9 in. **WINGSPAN** 11–13 in.

Oft referred to as the "desert cardinal"; shows markedly different characteristics than its northern relatives. If you're having trouble identifying a cardinal-like bird in the field, look for the Pyrrhuloxia's waxy yellow, parrotlike bill. Medium-sized overall, with long crest; rounded, chunky bill; and long tail. Adult male shows dusky gray overall with yellow bill; rosy red face, chest patch, and wings; and rosy wash down the tail. Female shows similar plumage with less red, particularly on the face and undersides. Immature shows gray overall with distinct rounded bill. **BEHAVIOR** Forages low in canopy or along the ground for insects and seeds, making short, buoyant flights between stands of cactus and mesquite. Will gather in small flocks in winter. Male will perch conspicuously on poles, tips of saguaros, or high atop mesquite to sing. **VOCALIZATIONS** Like a Northern Cardinal that's lost some oomph, song is a thinner rendition of its relative's laserlike, staccato melody. Call is a sweet but sharp chip, also reminiscent of the Northern Cardinal. **ABUNDANCE** Common resident in harsh, dry habitats of Southern AZ and NM, with overwintering population in limited Central NM range. **LOOKALIKE** Northern Cardinal. Several subtle differences can help you in identification, but the clearest distinguishing feature is the Pyrrhuloxia's rounded, parrotlike, yellow bill, versus the orange, conical bill of the cardinal. Note that Pyrrhuloxia and Cardinals are known to hybridize, just to confuse birders.
CONSERVATION STATUS Least Concern (Decreasing).

Rose-breasted Grosbeak

Pheucticus ludovicianus

Deciduous woodlands, forest edges, forested parks, suburbs
LENGTH 7–8.5 in. WINGSPAN 11.5–13 in.

A sturdy, barrel-chested, thick-necked songbird with heavy, triangular bill and squared tail. Adult male shows jet-black plumage on head and back, contrasted by a clean white belly, white wing and tail patches, and bright rose-red chest patch. In flight, underwings show deep red bordered by bright white. Bill is pale. Female shows stark contrast in plumage, brown overall with warm, peachy chest heavily streaked with brown; bold white eyebrows; and white wing and tail patches similar to male. Immature male shows a lovely combination of the two, with a pink chest patch and white eyebrows. BEHAVIOR Surprisingly tricky to spot, as they seem to disappear into the canopy, but if your echolocation is good, the male's crooning will give him away. Omnivores, they forage for insects, berries, and seeds and will frequent backyard feeders, lured by black oil sunflower seeds. VOCALIZATIONS Song is full, sweet, and long, like a highly talented, verbose American Robin. Call is a resonant, wheezy *chink*. ABUNDANCE Rare vagrant, but regularly observed during migration, often in wooded areas along riparian corridors or well-forested parks. CONSERVATION STATUS Least Concern (Decreasing).

Black-headed Grosbeak

Pheucticus melanocephalus

Variety of habitats at varied elevations, preferring wooded and relatively moist areas with brushy understories and large trees—most abundant in riparian forests; patchy coniferous, pine-oak, oak woodlands; lush desert washes; mature chaparral; suburban settings
LENGTH 7.25–8.25 in. **WINGSPAN** 12.5 in.

Medium-sized and chunky with an extremely heavy bill, short tail, large head, and long, rounded wings. Adult male, a jack-o'-lantern of a bird, has a grayish bill; a black head, face, back, tail, and wings; a pumpkin-orange nape, rump, breast, and belly; bold white wing spots; and yellow underwing coverts. Adult female is more subdued with brownish, white-spotted wings; a streaky brown back; buffy underparts with thin streaking in the lower belly and flanks; and a boldly patterned face with a white eyebrow, brown cheek, and white malar. Young birds resemble females but lack streaking below. **BEHAVIOR** Hops through the middle and high canopies with frequent vocalizations, gleaning seeds and insects, which it crunches with its heavy bill. **VOCALIZATIONS** Song, a warbling series of whistled phrases, is reminiscent of American Robin but hoarser, longer, and with less repetition. Call is a reedy and woodpecker-like *PIK*. Female and juvenile give a plaintive *wee-ohh*. **ABUNDANCE** A common breeding resident in Northern and from Central to Eastern AZ and throughout NM, except for in the dry grasslands on its eastern edge. A common migrant throughout both states. **LOOKALIKES** Female, juvenile Rose-breasted Grosbeak. Distinguished from female Rose-breasted by darker bill, buffier breast with finer streaks, and sometimes buffier eyebrow; from juvenile Rose-breasted by the absence of pinkish underwing coverts. **CONSERVATION STATUS** Least Concern.

Breeding male

Female/immature

Blue Grosbeak

Passerina caerulea

Grasslands and scrub, especially fallow agricultural fields with high grasses, forbs, shrubs; also shrubby growth along riparian corridors in arid habitat
LENGTH 6–6.5 in. WINGSPAN 10–11 in.

Heavy-chested songbird, with notably heavy, broad, triangular bill; elongated body; and medium-length tail. Breeding male shows cobalt-blue plumage accented by chestnut-brown wingbars and a small black mask just in front of the eyes. Female/immature displays apricot-cinnamon-brown overall, with head/scapulars showing deeper saturation of color, and faint wash of blue on rump and tail. Wingtips show ashy gray on both sexes; wingbars are cinnamon-brown on top, buffy below. Nonbreeding male shows scruffy, patchy cinnamon-brown and blue overall. BEHAVIOR Preys upon a wide variety of insects, as well as snails and arachnids. Feeds on seeds and spent grain in cooler seasons. Males will perch in the open to sing; otherwise not that showy, despite their brilliant plumage, sticking to low shrub canopy or ground foraging in the grass. Look for characteristic catlike tail twitching behavior when perched. Forages in flocks outside of breeding season. VOCALIZATIONS Male song is a resonant, coarse, but sweet warble, lasting roughly two to four seconds—like a shorter, less buzzy version of a singing House Finch. Call is a very metallic, tinny *tink*, and flight call is a quick, low buzz. ABUNDANCE Common throughout much of breeding range, with breeding populations dwindling in Northwest NM and Northern AZ. LOOKALIKES Female Indigo, Lazuli buntings. Blue Grosbeak distinguished from both by darker, clean belly; brown/cinnamon wingbars; and that big ol' cone of a bill that occupies most of its face. CONSERVATION STATUS Least Concern.

Breeding male

Nonbreeding adult male (top) with female/immature (bottom) at feeder

Lazuli Bunting

Passerina amoena

Brushy, wooded habitats, often near rivers or streams, including open riparian, pine-juniper, pine-oak, and oak woodlands; desert washes; scrubby agricultural fields; chaparral; suburban settings
LENGTH 5.5–6 in. **WINGSPAN** 8.75 in.

Small and finchlike with a conical bill, a slightly peaked head, a sloping forehead, and a medium-length, notched tail. Breeding male is luminescent azure-blue above with apricot-colored breast and flanks; white belly; blackish tail, wings, and lores; and white shoulder patches and wingbars. Nonbreeding male shows tawny blotches mixed with azure-blue. Female and immature are warm, peachy-brown above with off-white wingbars, hints of blue in the wings and tail, grayish throat, buffy breast, and pale belly. **BEHAVIOR** Forages on the ground, in shrubs, and in low trees for seeds and insects. During breeding, male will take an obvious perch to announce himself to potential mates and watch for intruding rivals. **VOCALIZATIONS** Song, a short, cheerful series of high chirps, whistles, and trills, is delivered repetitively with short pauses between bursts. Calls include high chips and short buzzes. **ABUNDANCE** An uncommon breeding resident in the northern portions of both states, an uncommon winter resident within a limited patch of Eastern South Central Arizona around Nogales, and a common migrant throughout the region. **LOOKALIKES** Female and immature buntings, bluebirds. Female Lazuli differentiated from all other brownish female and immature buntings by gray throat, unstreaked underparts, and whitish wingbars; smaller than bluebirds with a larger bill. **CONSERVATION STATUS** Least Concern.

Breeding male

Female/immature

Indigo Bunting

Passerina cyanea

Woodland edges and roadsides; streamside scrub; brushy, shrubby patches; overgrown grasses
LENGTH 4.5–5.5 in. WINGSPAN 7.5–9 in.

Petite songbird with conical bill and relatively short tail. Breeding males show spectrum of brilliant blues, some azure, some cerulean, some deep turquoise—oddly enough, not a true indigo. Strokes of sooty gray on the primaries and tail feathers. Bill is silvery gray. Nonbreeding male shows patches of dusky brown with flecks of blue all over. Female shows solid dusky brown with faintly streaked breast and pale wash of blue down rump and tail. BEHAVIOR Vocal and voracious, they sing from shrubby perches, poles, and fence posts, and forage for a variety of insects, berries, and seeds, sometimes frequenting backyard thistle/nyjer seed feeders. VOCALIZATIONS Song is a series of paired phrases/notes, sung with a sense of urgency: "Fire! Fire! There! There! See it? See it?" Song varies by "neighborhood," in that young males learn to sing by listening to and emulating the song of their elders, resulting in various dialects/variations on the song's theme. Call is a thin, high chip. ABUNDANCE Uncommon but regularly observed throughout breeding and migratory ranges, with primary migratory path running just east of NM border. LOOKALIKE Female Lazuli Bunting. Indigo distinguished by lack of distinct wingbars, faint streaking on breast, white throat, and faint blue wash on tail and/or wings (not always present). CONSERVATION STATUS Least Concern (Decreasing).

Adult male

Female/immature

Varied Bunting

Passerina versicolor

Thornscrub and desert canyons, streams, washes rich with oak, willow, mesquite, hackberry
LENGTH 4.5–5.5 in. WINGSPAN 7.75–8.25 in.

Finchlike and petite with a rounded head, medium-length and subtly forked tail, and conical bill with a slightly curved upper mandible. Adult male frequently appears simply dark, but in sunlight, shows rich blue and royal purple with a limited black mask and deep red in the nape, rear crown, throat, mantle, and breast. Female and immature are unmarked grayish brown, darkest above. BEHAVIOR Forages, often in pairs, for insects, seeds, and fruit while bouncing on fidgety wings through dense, thorny brush. Like other buntings, males will take to an obvious perch for a bit of showing off. VOCALIZATIONS Song, an up-and-down warble of whistles and chirps, is reminiscent of House Finch but less robust and more likely to be heard from a single bird. Calls include high, flat chips and in-flight buzzing. ABUNDANCE Most abundant in the foothills of the Madrean Sky Islands. An uncommon breeder in Southeast AZ and the extreme southwest and southeast corners of NM. LOOKALIKES Female, immature buntings. Varied differentiated by streaks, curved bill, grayish plumage, and lack of wingbars. CONSERVATION STATUS Least Concern.

Adult male

Female/immature

Painted Bunting

Passerina ciris

Dense thickets and edge habitat within low-elevation riparian areas
LENGTH 4.5–5 in. WINGSPAN 6–7.5 in.

Compact songbird with finchlike bill, heavy and conical for easy seed eating. Adult male dons a striking palette of cobalt-blue on the head and scapulars, bright sulfur/chartreuse on the back, and strawberry-red on the belly and rump. Wings show a beautiful combination of all three colors, accented with sooty gray, and a broken red eye ring for another touch of flair. Females and immature males show a distinct, muted yellow-green overall with pale yellow eye ring. Some immature males show a wash of rosy pink on belly and rump. BEHAVIOR Forages for seeds and insects in low, dense brush, occasionally venturing out to grasses. Will frequent bird feeders. Like many songbirds, males defend territory with song, but they won't back down from a physical scuffle if necessary. VOCALIZATIONS Song is musical and variable, reminiscent and easily mistaken for that of a House Finch. Call is a sharp, snappy, metallic *tsit*. ABUNDANCE Small population of breeding birds in limited range of Southeast NM, though frequently observed throughout migratory range. KEY SITES NM: Rattlesnake Canyon, Carlsbad Caverns National Park. CONSERVATION STATUS Least Concern.

Adult female

Juvenile

Adult male

Dickcissel

Spiza americana

Brushy grasslands and similar habitats including agricultural fields, weedy fields, roadsides
LENGTH 5.5–6.25 in. WINGSPAN 9.75–10.25 in.

Small, slender, and sparrow-like with long wings, a medium-length tail, and a long, heavy bill. Breeding male has a gray head, cheek, nape, and uppertail; a white eyebrow and malar, both with splashes of yellow; a gray mantle with black streaks; rufous shoulders; blackish flight feathers showing white edging at rest; a yellow breast; whitish underparts; and a black V shape on the chest that connects to thin malar stripes bordering a white chin. Female lacks black chest but shows thin malar stripes. Juvenile resembles female but with streaking below and only faded yellow tones. BEHAVIOR Hops along the ground or in low weeds and shrubs in search of seeds and insects. Mostly solitary when breeding but often observed in large flocks in migration.

VOCALIZATIONS Song, the source of this bird's name, is a varied and hoarse *dic-dic-SEE-SEE-SEE*. Common calls include clear chirps, thin chips, and a rattling in-flight buzz. ABUNDANCE An uncommon migrant in Eastern NM with limited breeding mostly in the northeast corner of the state. Increasingly rare farther west. LOOKALIKE Female House Sparrow. Dickcissel distinguished by longer bill, yellow tones, and dark malar stripes. CONSERVATION STATUS Least Concern.

RARITIES

Were we to include every species ever encountered in Arizona and New Mexico in this book, the resulting tome would not be something you'd want to carry into the field! As such, you may one day find yourself watching a bird for which there is no species account in this guide. In those cases, start with this list of regular and irregular rarities. Keep in mind, though, that birds don't care much for books, wildlife on the wing can show up in surprising places, and climate change is leading to shifting ranges. An unlikely sighting is never an impossible one!

FAMILY ANATIDAE: DUCKS, GEESE, AND SWANS
Fulvous Whistling-Duck (*Dendrocygna bicolor*)
Brant (*Branta bernicla*)
Trumpeter Swan (*Cygnus buccinator*)
Garganey (*Spatula querquedula*)
Eurasian Wigeon (*Mareca penelope*)
Tufted Duck (*Aythya fuligula*)
Harlequin Duck (*Histrionicus histrionicus*)
Long-tailed Duck (*Clangula hyemalis*)

FAMILY PHASIANIDAE: PHEASANTS, GROUSE, AND TURKEY
Gunnison Sage-Grouse (*Centrocercus minimus*) [Extirpated in AZ, NM]
Sharp-tailed Grouse (*Tympanuchus phasianellus*) [Extirpated in AZ, NM]

FAMILY PODICIPEDIDAE: GREBES
Red-necked Grebe (*Podiceps grisegena*)

FAMILY HELIORNITHIDAE:
Sungrebe (*Heliornis fulica*)

FAMILY COLUMBIDAE: PIGEONS AND DOVES
White-tipped Dove (*Leptotila verreauxi*)

FAMILY CUCULIDAE: CUCKOOS
Black-billed Cuckoo (*Coccyzus erythropthalmus*)

OPPOSITE Tricolored Heron

FAMILY CAPRIMULGIDAE: NIGHTJARS
Chuck-will's-widow (*Antrostomus carolinensis*)
Eastern Whip-poor-will (*Antrostomus vociferus*)

FAMILY APODIDAE: SWIFTS
Black Swift (*Cypseloides niger*)
Chimney Swift (*Chaetura pelagica*)

FAMILY TROCHILIDAE: HUMMINGBIRDS
Mexican Violetear (*Colibri thalassinus*)
Ruby-throated Hummingbird (*Archilochus colubris*)
Bumblebee Hummingbird (*Selasphorus heloisa*)
Cinnamon Hummingbird (*Amazilia rutila*)

FAMILY RALLIDAE: RAILS, GALLINULES, AND COOTS
Rufous-necked Wood-Rail (*Aramides axillaris*)
Clapper Rail (*Rallus crepitans*)
King Rail (*Rallus elegans*)
Purple Gallinule (*Porphyrio martinicus*)
Yellow Rail (*Coturnicops noveboracensis*)

FAMILY GRUIDAE: CRANES
Common Crane (*Grus grus*)

FAMILY CHARADRIIDAE: PLOVERS
European Golden-Plover (*Pluvialis apricaria*)
American Golden-Plover (*Pluvialis dominica*)
Pacific Golden-Plover (*Pluvialis fulva*)
Lesser Sand-Plover (*Charadrius mongolus*)
Piping Plover (*Charadrius melodus*)

FAMILY *SCOLOPACIDAE*: SANDPIPERS
Hudsonian Godwit (*Limosa haemastica*)
Ruddy Turnstone (*Arenaria interpres*)
Black Turnstone (*Arenaria melanocephala*)
Red Knot (*Calidris canutus*)
Ruff (*Calidris pugnax*)
Sharp-tailed Sandpiper (*Calidris acuminata*)
Curlew Sandpiper (*Calidris ferruginea*)
Little Stint (*Calidris minuta*)
American Woodcock (*Scolopax minor*)
Wandering Tattler (*Tringa incana*)
Red Phalarope (*Phalaropus fulicarius*)

FAMILY *JACANIDAEL*: JACANAS
Northern Jacana (*Jacana spinosa*)

**FAMILY *STERCORARIIDAE*:
SKUAS AND JAEGERS**
Pomarine Jaeger (*Stercorarius pomarinus*)
Parasitic Jaeger (*Stercorarius parasiticus*)
Long-tailed Jaeger (*Stercorarius longicaudus*)

FAMILY *ALCIDAE*: AUKS, MURRES, AND PUFFINS
Long-billed Murrelet (*Brachyramphus perdix*)
Ancient Murrelet (*Synthliboramphus antiquus*)

FAMILY *LARIDAE*: GULLS, TERNS, AND SKIMMERS
Black-legged Kittiwake (*Rissa tridactyla*)
Ivory Gull (*Pagophila eburnea*)
Sabine's Gull (*Xema sabini*)
Little Gull (*Hydrocoloeus minutus*)
Laughing Gull (*Leucophaeus atricilla*)
Black-tailed Gull (*Larus crassirostris*)
Heermann's Gull (*Larus heermanni*)
Short-billed Gull (*Larus brachyrhynchus*)
Western Gull (*Larus occidentalis*)
Yellow-footed Gull (*Larus livens*)
Iceland Gull (*Larus glaucoides*)
Lesser Black-backed Gull (*Larus fuscus*)
Glaucous-winged Gull (*Larus glaucescens*)
Glaucous Gull (*Larus hyperboreus*)
Sooty Tern (*Onychoprion fuscatus*)
Gull-billed Tern (*Gelochelidon nilotica*)
Arctic Tern (*Sterna paradisaea*)
Royal Tern (*Thalasseus maximus*)
Elegant Tern (*Thalasseus elegans*)
Black Skimmer (*Rynchops niger*)

FAMILY *PHAETHONTIDAE*: TROPICBIRDS
White-tailed Tropicbird (*Phaethon lepturus*)
Red-billed Tropicbird (*Phaethon aethereus*)

FAMILY *GAVIIDAE*: LOONS
Red-throated Loon (*Gavia stellata*)
Pacific Loon (*Gavia pacifica*)
Yellow-billed Loon (*Gavia adamsii*)

FAMILY *DIOMEDEIDAE*: ALBATROSSES
Laysan Albatross (*Phoebastria immutabilis*)

**FAMILY *HYDROBATIDAE*:
NORTHERN STORM-PETRELS**
Wedge-rumped Storm-Petrel
    (*Hydrobates tethys*)
Black Storm-Petrel (*Hydrobates melania*)
Least Storm-Petrel (*Hydrobates microsoma*)

**FAMILY *PROCELLARIIDAE*:
SHEARWATERS AND PETRELS**
Juan Fernandez Petrel (*Pterodroma externa*)
Hawaiian Petrel (*Pterodroma sandwichensis*)
Wedge-tailed Shearwater (*Ardenna pacifica*)
Sooty Shearwater (*Ardenna grisea*)
Black-vented Shearwater (*Puffinus opisthomelas*)

FAMILY *CICONIIDAE*: STORKS
Wood Stork (*Mycteria americana*)

FAMILY *FREGATIDAE*: FRIGATEBIRDS
Magnificent Frigatebird (*Fregata magnificens*)

**FAMILY *SULIDAE*:
GANNETS AND BOOBIES**
Blue-footed Booby (*Sula nebouxii*)
Brown Booby (*Sula leucogaster*)

FAMILY *ANHINGIDAE*: ANHINGAS
Anhinga (*Anhinga anhinga*)

FAMILY *PELECANIDAE*: PELICANS
Brown Pelican (*Pelecanus occidentalis*)

FAMILY *ARDEIDAE*: BITTERNS
AND HERONS
Little Blue Heron (*Egretta caerulea*)
Tricolored Heron (*Egretta tricolor*)
Reddish Egret (*Egretta rufescens*)
Yellow-crowned Night-Heron
    (*Nyctanassa violacea*)

FAMILY *THRESKIORNITHIDAE*:
IBISES AND SPOONBILLS
White Ibis (*Eudocimus albus*)
Glossy Ibis (*Plegadis falcinellus*)
Roseate Spoonbill (*Platalea ajaja*)

FAMILY *ACCIPITRIDAE*: KITES,
EAGLES, AND HAWKS
Swallow-tailed Kite (*Elanoides forficatus*)
White-tailed Hawk
    (*Geranoaetus albicaudatus*)
Short-tailed Hawk (*Buteo brachyurus*)

FAMILY *STRIGIDAE*: TYPICAL OWLS
Eastern Screech-Owl (*Megascops asio*)
Barred Owl (*Strix varia*)

FAMILY *ALCEDINIDAE*: KINGFISHERS
Ringed Kingfisher (*Megaceryle torquata*)

FAMILY *PICIDAE*: WOODPECKERS
Golden-fronted Woodpecker
    (*Melanerpes aurifrons*)
Red-bellied Woodpecker (*Melanerpes carolinus*)
Red-breasted Sapsucker (*Sphyrapicus ruber*)

FAMILY *FALCONIDAE*: FALCONS
AND CARACARAS
Aplomado Falcon (*Falco femoralis*)

FAMILY *PSITTACIDAE*: HOLOTROPICAL PARROTS
Thick-billed Parrot (*Rhynchopsitta pachyrhyncha*) [Extirpated in AZ, NM]

FAMILY *TITYRIDAE*: TITYRAS AND ALLIES
Gray-collared Becard (*Pachyramphus major*)

FAMILY *TYRANNIDAE*: TYRANT FLYCATCHERS
Nutting's Flycatcher (*Myiarchus nuttingi*)
Great Crested Flycatcher (*Myiarchus crinitus*)
Great Kiskadee (*Pitangus sulphuratus*)
Piratic Flycatcher (*Legatus leucophaius*)
Tufted Flycatcher (*Mitrephanes phaeocercus*)
Eastern Wood-Pewee (*Contopus virens*)
Yellow-bellied Flycatcher
    (*Empidonax flaviventris*)
Acadian Flycatcher (*Empidonax virescens*)
Pine Flycatcher (*Empidonax affinis*)

FAMILY *VIREONIDAE*: VIREOS
Black-capped Vireo (*Vireo atricapilla*)
White-eyed Vireo (*Vireo griseus*)
Yellow-throated Vireo (*Vireo flavifrons*)
Blue-headed Vireo (*Vireo solitarius*)
Philadelphia Vireo (*Vireo philadelphicus*)
Yellow-green Vireo (*Vireo flavoviridis*)

FAMILY *CORVIDAE*: JAYS, MAGPIES,
AND CROWS
California Scrub-Jay (*Aphelocoma californica*)

FAMILY *HIRUNDINIDAE*: SWALLOWS
Brown-chested Martin (*Progne tapera*)

FAMILY *TROGLODYTIDAE*: WRENS
Winter Wren (*Troglodytes hiemalis*)
Sedge Wren (*Cistothorus stellaris*)
Carolina Wren (*Thryothorus ludovicianus*)
Sinaloa Wren (*Thryophilus sinaloa*)

FAMILY *MUSCICAPIDAE*:
OLD WORLD FLYCATCHERS
Northern Wheatear (*Oenanthe oenanthe*)

FAMILY *TURDIDAE*: THRUSHES
Orange-billed Nightingale-Thrush
  (*Catharus aurantiirostris*)
Brown-backed Solitaire (*Myadestes occidentalis*)
Veery (*Catharus fuscescens*)
Gray-cheeked Thrush (*Catharus minimus*)
Wood Thrush (*Hylocichla mustelina*)
Clay-colored Thrush (*Turdus grayi*)
White-throated Thrush (*Turdus assimilis*)
Varied Thrush (*Ixoreus naevius*)
Aztec Thrush (*Ridgwayia pinicola*)

**FAMILY *MIMIDAE*: MOCKINGBIRDS
AND THRASHERS**
Blue Mockingbird (*Melanotis caerulescens*)
Long-billed Thrasher (*Toxostoma longirostre*)

FAMILY *BOMBYCILLIDAE*: WAXWINGS
Bohemian Waxwing (*Bombycilla garrulus*)

FAMILY *MOTACILLIDAE*: WAGTAILS AND PIPITS
White Wagtail (*Motacilla alba*)
Red-throated Pipit (*Anthus cervinus*)

FAMILY *FRINGILLIDAE*: FINCHES
Purple Finch (*Haemorhous purpureus*)
Common Redpoll (*Acanthis flammea*)
White-winged Crossbill (*Loxia leucoptera*)
Smith's Longspur (*Calcariidae pictus*)
Snow Bunting (*Calcariidae nivalis*)
Little Bunting (*Emberizidae pusilla*)

**FAMILY *PASSERELLIDAE*:
NEW WORLD SPARROWS**
Worthen's Sparrow (*Spizella wortheni*)
  [Extirpated in AZ, NM]
Golden-crowned Sparrow
  (*Zonotrichia atricapilla*)
Harris's Sparrow (*Zonotrichia querula*)
LeConte's Sparrow (*Ammospiza leconteii*)
Nelson's Sparrow (*Ammospiza nelsoni*)
Henslow's Sparrow (*Centronyx henslowii*)
Eastern Towhee (*Pipilo erythrophthalmus*)

FAMILY *ICTERIDAE*: NEW WORLD BLACKBIRDS
Bobolink (*Dolichonyx oryzivorus*)
Eastern Meadowlark (*Sturnella magna*)
Black-vented Oriole (*Icterus wagleri*)
Baltimore Oriole (*Icterus galbula*)
Rusty Blackbird (*Euphagus carolinus*)

FAMILY *PARULIDAE*: NEW WORLD WARBLERS
Worm-eating Warbler
  (*Helmitheros vermivorum*)
Louisiana Waterthrush (*Parkesia motacilla*)
Golden-winged Warbler
  (*Vermivora chrysoptera*)
Blue-winged Warbler (*Vermivora cyanoptera*)
Swainson's Warbler (*Limnothlypis swainsonii*)
Prothonotary Warbler (*Protonotaria citrea*)
Crescent-chested Warbler
  (*Oreothlypis superciliosa*)
Tennessee Warbler (*Leiothlypis peregrina*)
Connecticut Warbler (*Oporornis agilis*)
Mourning Warbler (*Geothlypis philadelphia*)
Kentucky Warbler (*Geothlypis formosa*)
Hooded Warbler (*Setophaga citrina*)
Cape May Warbler (*Setophaga tigrina*)
Cerulean Warbler (*Setophaga cerulea*)
Northern Parula (*Setophaga americana*)
Tropical Parula (*Setophaga pitiayumi*)
Magnolia Warbler (*Setophaga magnolia*)
Bay-breasted Warbler (*Setophaga castanea*)
Blackburnian Warbler (*Setophaga fusca*)
Chestnut-sided Warbler
  (*Setophaga pensylvanica*)
Blackpoll Warbler (*Setophaga striata*)
Black-throated Blue Warbler
  (*Setophaga caerulescens*)
Palm Warbler (*Setophaga palmarum*)
Pine Warbler (*Setophaga pinus*)
Yellow-throated Warbler (*Setophaga dominica*)
Prairie Warbler (*Setophaga discolor*)
Golden-cheeked Warbler
  (*Setophaga chrysoparia*)
Black-throated Green Warbler
  (*Setophaga virens*)

RARITIES | 513

Bobolink

Fan-tailed Warbler (*Basileuterus lachrymosus*)
Rufous-capped Warbler (*Basileuterus rufifrons*)
Canada Warbler (*Cardellina canadensis*)
Golden-crowned Warbler
  (*Basileuterus culicivorus*)
Slate-throated Redstart (*Myioborus miniatus*)

FAMILY *CARDINALIDAE*: CARDINALS AND ALLIES
Scarlet Tanager (*Piranga olivacea*)
Yellow Grosbeak (*Pheucticus chrysopeplus*)

FAMILY *THRAUPIDAE*: TANAGERS
Blue-black Grassquit (*Volatinia jacarina*)

RECOMMENDED RESOURCES

BIRDING AND CONSERVATION ORGANIZATIONS

American Birding Association: ABA.org
American Ornithological Society: AmericanOrnithology.org
Arizona Field Ornithologists: AZFO.org
Bird Names for Birds: BirdNamesforBirds.Wordpress.com
Birdability: Birdability.org
Black AF in STEM and Black Birders Week: BlackAFinSTEM.com
Cornell Lab of Ornithology: Birds.Cornell.edu
Female Bird Day/The Galbatross Project: FemaleBirdDay.Wordpress.com
Feminist Bird Club: FeministBirdClub.org
International Dark-Sky Association: DarkSky.org
Local Audubon Society Chapters in AZ and NM
Bird Alliance of Central New Mexico: BACNM.org
Desert Rivers: DesertRiversAudubon.org
Maricopa: MaricopaAudubon.org
Mesilla Valley: MVASAudubon.org
Northern Arizona: NorthernArizonaAudubon.org
Prescott: PrescottAudubon.org
Sangre de Cristo: AudubonSantaFe.org
Sonoran: SonoranAudubon.org
Southwestern New Mexico: SWNMAudubon.org
Sun Devil Audubon Student Conservation Chapter: Facebook.com/groups/459870621350177
Tucson: TucsonAudubon.org
White Mountain: WhiteMountainAudubon.org
Yuma: AudubonYuma.org
National Audubon Society/Audubon Southwest: Audubon.org, Southwest.Audubon.org
New Mexico Ornithological Society: NMBirds.org
Western Rivers Conservancy: WesternRivers.org

DIGITAL RESOURCES

Audubon Bird Guide App, National Audubon Society: Audubon.org/app
eBird Mobile, Cornell Lab of Ornithology: Ebird.org/about/ebird-mobile
iNaturalist: iNaturalist.org
Merlin Bird ID, Cornell Lab of Ornithology: Merlin.AllAboutBirds.org

SOCIAL MEDIA ACCOUNTS WE FOLLOW

Rosemary Mosco, Bird and Moon Comics: Instagram [@rosemarymosco]
Alexis Nikole Nelson, Black Forager: TikTok [@alexisnikole] and Instagram [@blackforager]
Kaeli Swift, PhD, corvid researcher: TikTok and Instagram [@corvidresearch]
Joshua Barkman, False Knees webcomic: Instagram [@falseknees]

OPPOSITE Killdeer

Maynard Okereke, Hip Hop Science Show:
YouTube and Instagram
[@hiphopscienceshow]
Corina Newsome, Hood Naturalist:
Instagram [@hood__naturalist]
Joey Santore, Crime Pays But Botany Doesn't:
YouTube and Instagram [@crime_pays_but_botany_doesnt]

PODCASTS
American Birding Podcast: ABA.org/podcast
BirdNote: BirdNote.org
The Field Guides:
TheFieldGuidesPodcast.com
Ologies with Alie Ward:
AlieWard.com/ologies
Talkin' Birds with Ray Brown:
TalkinBirds.com

ARIZONA FESTIVALS
Wings Over Willcox Birding & Nature Festival, January: WingsOverWillcox.com
Friends of the Verde River, Verde Valley Birding and Nature Festival, April: VerdeRiver.org
Southeast Arizona Birding Festival, August: TucsonAudubon.org/festival

NEW MEXICO FESTIVAL
Festival of the Cranes, December:
FriendsofBosquedelApache.org/festival

FURTHER READING
Ackerman, Jennifer. 2016. *The Genius of Birds*. New York: Penguin Press.

OPPOSITE Great-tailed Grackle

Brown, D. E., and C. Lowe. 1980. Biotic Communities—Southwestern United States and Northwestern Mexico [Map]. General Technical Report RM-78, Rocky Mountain Forest and Range Experiment Station, Forest Service, US Department of Agriculture.

Cantú, Francisco. 2018. *The Line Becomes a River: Dispatches from the Border*. New York: Riverhead Books.

Cooper, Christian. 2023. *Better Living Through Birding: Notes from a Black Man in the Natural World*. New York: Random House.

Craig, Mya-Rose. 2023. *Birdgirl: Looking to the Skies in Search of a Better Future*. New York: Celadon Books.

Kimmerer, Robin Wall. 2015. *Braiding Sweetgrass: Indigenous Wisdom, Scientific Knowledge, and the Teachings of Plants*. Minneapolis: Milkweed Editions.

Lanham, J. Drew. 2017. *The Home Place: Memoirs of a Colored Man's Love Affair with Nature*. Minneapolis: Milkweed Editions.

Macdonald, Helen. 2021. *Vesper Flights: New and Collected Essays*. Waterville, Maine: Thorndike Press.

Mosco, Rosemary. 2021. *A Pocket Guide to Pigeon Watching: Getting to Know the World's Most Misunderstood Bird*. New York: Workman Publishing Company.

Pyle, Peter. 1997. *Identification Guide to North American Birds, Part I: Columbidae to Ploceidae*. Point Reyes Station, California: Slate Creek Press.

Pyle, Peter. 2008. *Identification Guide to North American Birds, Part II: Anatidae to Alcidae*. Point Reyes Station, California: Slate Creek Press.

Sibley, David Allen. 2020. *What It's Like to be a Bird: From Flying to Nesting, Eating to Singing—What Birds Are Doing, and Why*. New York: Alfred A. Knopf.

Strycker, Noah. 2015. *The Thing with Feathers: The Surprising Lives of Birds and What They Reveal About Being Human.* New York: Riverhead Books.

Tallamy, Douglas W. 2020. *Nature's Best Hope: A New Approach to Conservation that Starts in Your Yard.* Portland, Oregon: Timber Press.

Weidensaul, Scott, 2021. *A World on the Wing: The Global Odyssey of Migratory Birds.* London: Picador Publishing.

Weidensaul, Scott. 1999. *Living on the Wind: Across the Hemisphere with Migratory Birds.* New York: North Point Press.

Zarankin, Julia. 2020. *Field Notes from an Unintentional Birder: A Memoir.* Madeira Park, BC: Douglas and McIntyre.

OPPOSITE Northern Shoveler

ACKNOWLEDGMENTS

WHERE TO BEGIN? Where to end? From the support and enthusiasm (even if feigned) of friends and family, to the colleagues and mentors whose encouragement and shared knowledge led us to this place—thank you, from the bottom of our hearts.

Without the contributions of dozens of talented photographers and researchers, this would be a far less attractive, far less accurate work. We are endlessly grateful for your talent and generosity.

A special debt of gratitude to Nick, Linda, Joan, Jen, Nicole, Jess, Char, Josh, Sarah, Debbie, Helen, Anthony, Sam, Felice, Jeff, Cathy Wise, Tice Supplee, Damon Praefke, Gail Riley, Tom Kerr, Lauren Makeyenko, Loren Smith, Linda Kennedy, Benjamin Beal, Suzanne Wilcox, Greg Clark, Jim and Robin Winters, Deeohn Ferris, David Gordon, Taddy Dann, Bill Michalek, Stef Ecker, Bridget Baker, Heidi Romer, Riana Johnson, Jennie Duberstein, and many others for supporting, encouraging, and tolerating our endless curiosity and love for birds, plants, and all other critters (maybe not mosquitoes).

Finally, our sincere gratitude to the lovely folks at Timber Press, with special thanks to Ryan Harrington, Matthew Burnett, and Sarah Milhollin.

OPPOSITE Chihuahuan Meadowlark

PHOTO AND ILLUSTRATION CREDITS

PHOTOGRAPHS
All photographs by Melissa Fratello and Steven Prager with the exception of the following:

Brady Karg, 219, 220 (left), 221 (left), 224 (left), 262, 265, 270 (right), 365, 379, 382 (left), 386, 391 (left), 393, 427, 448, 469 (left, top), 471 (right), 474, 483 (right), 500 (left), 504
Felice Prager, 151 (right)
Greg Ralbovsky, 37, 151 (left), 217 (right, bottom), 237 (right, top), 275 (right)
Jack Little, 361
Jeff Ritz, 118 (main), 138 (left), 183, 185 (inset), 200 (right, bottom), 202 (right, top),
Josh Kanuchok, 99 (left)
Manon Ringuette, 419 (left, top)
Paul Tashjian, 39 (top), 74 (right, top), 168
Richard Fray, 350 (main)
Sky Jacobs, 258
TJ Winkler, 131 (right), 139 (right), 442 (right)
Tom Mangelsdorf, 148 (left), 152 (left), 290, 294, 376 (left)

Alamy
All Canada Photos, 328
Brian Small, 440 (left, second)

Dreamstime
Agami Photo Agency, 186 (right)
© Elba Cabrera Abdecoral, 87 (right)
Grian12, 342 (left)
Harold Stiver, 342 (right)
Jonathan Chancasana, 215 (left)
Kaido Rummel, 497 (left)
Martha Marks, 430 (right, bottom), 499 (left, top)
Mexitographer, 215 (right, bottom)

OPPOSITE Downy Woodpecker

Peterll, 160
Rachel Hopper, 142, 158 (right), 437
Rinus Baak, 116 (right), 286 (left)
Shawn Mason, 167
Shayne Kaye, 285 (right)
Steve Byland, 147
Taviphoto, 343 (right)
Tracy Immordino, 419 (right)
Vitalii Otroshko, 351 (right, top)
Wirestock, 440 (right)

Flickr
Public Domain
Abbeyprivate, 64
Laura Wolf, 21 (right)
Neal Herbert/National Park Service, 113 (left)
Sandra Uecker\U.S. Fish and Wildlife Service, 119 (left), 263 (left)
Shawn Taylor, 26
Susan Young, 81 (main), 220 (right, top)

Public Domain 1.0
abbeyprivate, 466 (right, bottom)
Adam Jackson, 163
Alan Schmierer, 104 (right, top), 109 (right), 113 (right, top), 121 (right), 126, 148 (right), 152 (right), 154 (right, top), 155 (right, bottom), 175 (right), 189 (right, bottom), 203 (right, bottom), 210 (right, bottom), 243 (right), 246 (left), 247 (left), 271 (right), 297, 300, 309, 311, 340, 353 (right), 368, 377 (right and left, top), 395, 410 (right), 411 (right), 412 (right), 421 (right), 425, 426, 428 (right), 429 (right), 431 (right, bottom), 434 (left), 440 (left, third), 445 (right), 460 (right), 461 (right), 481 (right), 486 (right), 487 (right), 492 (right), 494 (right), 499 (left, bottom), 500 (right, bottom)
Andy Rankin/National Park Service, 43

Courtney Celley/U.S. Fish and Wildlife Service, 333 (right), 409 (right), 413 (left)
Dan Pawlak/National Park Service, 367
Eve Turek/U.S. Fish and Wildlife Service, 83 (left, top), 485 (right, bottom)
Glacier National Park, 339 (right)
Grayson Smith/U.S. Fish and Wildlife Service, 430 (right, top)
Jacob W. Frank/National Park Service, 416 (left)
Jeff Ritz, 121 (left)
Jerry Ting, 454
Jesus Moreno/U.S. Fish and Wildlife Service, 307 (right)
Kegen Benson/Bureau of Land Management, 234 (left)
Ken Sturm/U.S. Fish and Wildlife Service, 41
ksblack99, 236 (right, bottom)
Lisa Hupp/U.S. Fish and Wildlife Service, 189 (left), 200 (right, top), 410 (left)
Liz Julian/U.S. Fish and Wildlife Service, 203 (left), 209 (bottom)
Mark Lellouch/National Park Service, 29 (right, top)
Mathias Appel, 405 (right)
Mike Budd/U.S. Fish and Wildlife Service, 222 (right, bottom)
Mike Carlo/U.S. Fish & Wildlife Service, 207 (right)
N. Lewis/National Park Service, 331, 363 (right), 375 (right), 380, 382 (right, bottom), 413 (right)
Peter Pearsall/U.S. Fish & Wildlife Service, 184 (left), 187 (left), 370, 420 (left)
Rick Bohn/U.S. Fish and Wildlife Service, 421 (left), 447
Rita Wiskowski, 187 (right, top), 202 (left)
Sandra Uecker\U.S. Fish and Wildlife Service, 236 (right, top),
Shiloh Schulte/U.S. Fish and Wildlife Service, 187 (right, bottom)
Smithsonian Institute, 60
Susan Young, 239 (left, top), 382 (right, bottom), 467 (left, top), 506 (right)
Tom Koerner/U.S. Fish and Wildlife Service, 82 (left and right, top), 154 (left), 156 (right), 199 (left, top), 235 (right, bottom), 371, 444, 490 (right), 496 (left)
U.S. Fish and Wildlife Service, 459 (right)

CC-BY 2.0

Alan Shearman, 211 (right, bottom)
Alyenaa Buckles, 91 (right)
Amit Patel, 229 (left)
Andrew Cannizzaro, 189 (right, top)
Andrey Gulivanov, 76 (right, top), 238 (left and right, top)
Andy Morffew, 225 (right), 239 (left, bottom), 288 (left and right, top), 289 (left), 319, 392 (left), 498 (right), 499 (right)
Andy Reago and Chrissy McClarren, 77, 120 (right, bottom), 128, 134, 171 (inset), 177, 186 (left), 195 (main), 197 (right), 198 (right), 214 (inset), 217 (right, top), 226 (right), 240 (right and left, top), 241 (left), 247 (right), 249 (left), 278 (left), 284 (left), 305, 325, 345 (right), 348 (right, bottom), 358, 375 (left), 377 (left, bottom), 403 (right), 408 (right), 432 (right, top), 435, 462 (right, bottom), 466 (left), 468 (left), 475, 479 (left), 483 (left, bottom), 485 (right, top), 486 (left), 494 (left), 495 (left, bottom), 507 (left, top)
Becky Matsubara, 114 (left), 123, 162, 194 (right), 199 (right), 223 (right, bottom), 235 (left), 255, 270 (right), 281 (right), 302, 316, 347, 348 (left), 352 (left), 372, 387 (left), 402 (left), 439 (left), 440 (left, bottom), 443 (left), 482 (left), 485 (left, bottom), 488 (left), 490 (left), 503 (left)
Bettina Arrigoni, 206 (bottom), 223 (left), 241 (right, bottom), 254, 266, 267 (right), 268 (left), 277 (left), 280 (right), 315, 317, 394, 422 (left)
Brad Lanam, 318
Brent Myers, 122
Brett Morrison, 206 (top)
capt_tain Tom, 288 (right, bottom)

PHOTO AND ILLUSTRATION CREDITS | 525

Channel City Camera Club/Steve Colwell, 95 (inset), 166 (right), 179 (main), 233 (right, top), 237 (right, bottom), 244 (left), 273, 470 (left), 489 (right), 500 (right, top), 507 (left, bottom)
Charles Gates, 76 (left and right, bottom), 246 (right, top), 250 (right, bottom), 274 (left), 277 (right), 279 (left), 292 (left), 313, 314, 326, 346 (left), 348 (right, top), 383 (left, top), 385 (left), 414, 415 (left), 436 (left), 439 (right), 446, 455 (right), 458, 459 (left, bottom), 469 (left, bottom), 476 (left), 477, 479 (right), 481 (left), 488 (right), 489 (left), 495 (right)
Chrissy McClarren, 95 (main), 349 (right)
Colby Stopa, 496 (right)
Colin Durfee, 281 (left), 283 (left)
Danielle Brigg, 443 (right)
David A Mitchell, 257, 356, 362, 409 (left), 433 (right), 469 (right)
dfaulder, 182 (left), 191 (main), 204 (left), 210 (left)
Don Debold, 196 (right)
Don Henise, 433 (left)
Don Owens, 91 (left)
eGuide Travel, 57
Ellen Urbanski/U.S. Fish and Wildlife Service, 222 (left)
Eric Ellingson, 100 (left)
Eric Gropp, 308
Etienne Gosse, 252 (top)
Fyn Kynd, 103 (right), 174, 198 (left), 419 (left, bottom)
Greg Kramos/U.S. Fish and Wildlife Service, 116 (right)
Indiana Ivy Nature Photographer, 467 (left, bottom)
Jason Crotty, 84 (left), 172 (right, top), 176 (right)
jeffhutchison, 471 (left, bottom)
Jeremy Meyer, 483 (left, top)
Joanna Gilkeson/U.S. Fish and Wildlife Service, 244 (right, top)
Jerry Kurkhart, 179 (inset)
John Brighenti, 78 (right, top)
Jordan Walmsley, 214 (main)
Juan Zamora, 156 (left)
Judy Gallagher, 99 (right, top), 207 (left, bottom)
Kaaren Perry, 93 (left), 154 (right, bottom), 456 (right, bottom)
Katja Schulz, 113 (right, bottom)
Kim Taylor Hull, 336
Larry Lamsa, 335, 408 (left), 440m (left, top), 461 (left), 465, 493
Laura Wolf, 91 (left), 106 (left), 157 (right), 180 (right), 230 (right), 240 (left, bottom), 307 (left), 330, 389 (left, bottom), 463 (right)
Laurie Shaull, 484 (left)
Lily Douglas, 243 (left)
Logan Ward, 432 (right, bottom)
Lorie Shaull, 46 (top)
Luiz Lapa, 246 (right)
Marshal Hedin, 182 (right)
Martien Brand, 208 (left)
Mary Madigan, 51 (bottom)
Matt MacGillivray, 279 (right)
Matt Tillett, 341 (right)
Melissa McMasters, 147, 296 (right), 495 (left, top)
Mndoci, 327
Moore Laboratory, 268 (right)
Nick Varvel, 431 (left)
Nicole Desnoyers/Institute for Wildlife Studies, 445 (left)
Nigel Winnu, 172 (left), 192 (left)
Niki Robertson, 261
Oregon Department of Transportation, 248 (left)
Peterichman, 345 (left)
Renee Grayson, 227 (right), 485 (left, top)
Richard Hurd, 86 (right)
Rinus Baak/U.S. Fish and Wildlife Service, 212 (inset)
Rodney Campbell, 96 (main)
Ron Knight, 112 (right, top), 175 (left), 205, 403 (left)

Russ W., 73 (right, bottom)
Russimages, 197 (left), 216 (right, bottom)
Ryan Mandelbaum, 184 (right), 199 (left, bottom), 280 (left), 337, 357, 364, 415 (right), 451 (right), 452 (right), 502 (right)
sasastro, 110 (right)
Shawn Taylor, 341 (left), 349 (left), 353 (left), 428 (left)
Shiva Shenoy, 101
Silver Leapers, 178
Steve Jurvetson, 111 (main)
Steve Maslowski/U.S. Fish and Wildlife Service, 106 (right)
Sue Cook/Channel City Camera Club, 84 (right)
Sunny Lo, 194 (left), 236 (left), 378
Tom Koerner/U.S. Fish and Wildlife Service, 75, 91 (right),189 (left), 234 (right, bottom), 246 (right, bottom), 249 (right), 384 (left and right, top), 397
Tom Shockey 99 (right, bottom)
Tony Hisgett, 242 (right, bottom)
Trish Hartmann, 74 (right, bottom)
U.S. Fish and Wildlife Service, 451 (left)
Under the Same Moon, 129, 216 (left)
Watts, 104 (left), 200 (left)
Wouter Koch, 130

CC-BY-ND 2.0

Allan Hack, 89, 111 (inset), 171 (main)
Ashley Wahlberg, 114 (right)
Brian Holsclaw, 263 (right)
Chris Parker, 252 (bottom)
Cristian Gonzalez G., 405 (left, top)
D. Fletcher, 405 (left, bottom)
Don Johnson 395, 232 (inset)
Emelie Chen, 73 (left), 250 (left), 376 (right, bottom)
Florida Fish and Wildlife, 176 (left)
Heather Paul, 127
Kelly Colgan Azar, 312, 401 (right), 484 (right)
Kjetil Rimolsrønning, 203 (left)
Mick Sway, 224 (right)

Mr.TinMD, 207 (left, top)
Pacific Power, 250 (right, top)
Rick Bergstrom, 359, 396
Shawn McCready, 94 (inset), 164, 196 (left), 292 (right), 324
Tara Lemezis, 153 (left)
Tim Spouge, 78 (right, bottom)
Tom Lee, 291 (right)
Veit Irtenkauf, 166 (left), 233 (left and right, bottom), 248 (right, top)

CC-BY-SA 2.0

Alastair Rae, 497 (right)
Allan Hack, 223 (right, top),234 (right, top)
Amado Demesa, 464 (left), 487 (left), 491 (left)
Andrew Weitzel, 181 (right), 462 (right, top), 482 (right)
Becky Matsubara, 97 (left)
Bengt Nyman, 211 (left)
Bertknot, 209 (middle)
cuatrok77, 221 (right)
Darren Kirby, 339 (left), 406 (left)
Deborah Freeman, 103 (left), 120 (right, top), 190 (left), 232 (main)
Dennis Jarvis, 463 (left, top)
Dominic Sherony, 79 (main), 149 (left), 241 (right, top), 267 (left), 285 (left), 287, 350 (inset), 478 (left)
Don Faulkner, 96 (inset), 441 (right)
Don Graham, 229 (right, bottom)
Enoch Leung, 132
Félix Uribe, 195 (inset)
Francesco Veronesi, 141, 158 (left), 343 (left)
Frank Schulenberg, 88 (left)
gardener41, 40
Gary Leavens, 93 (right, top), 149 (left), 189 (right, top), 191 (inset), 202 (right, bottom), 203 (right, top), 204 (right, bottom), 244 (right, bottom), 245 (left)
Gregory "Slobirdr" Smith, 411 (left)
Imran Shah, 115
Jerry Friedman, 333 (left), 384 (right, bottom)
Joel Huntress, 438 (left)

Jordan Jones, 217 (left)
Joseph Gage, 248 (right, bottom), 274 (right)
Laurie Boyle, 209 (top)
Len Worthington, 150 (right)
marneejill, 456 (right, top)
Mary Shattock, 291 (left)
Maureen Leong-Kee, 220 (right, bottom)
Mike's Birds, 271 (left), 296 (left), 376 (right, top), 383 (left, bottom), 418, 491 (right)
Panegyrics of Granovetter, 388
Paul Hartado, 431 (right, top)
Scott Heron, 185 (main)
Shanthanu Bhardwaj, 110 (left)
Simon Wray/Oregon Department of Fish and Wildlife, 283 (right)
Tim Sackton, 93 (right, bottom)
Tracie Hall, 155 (right, top)

iStock
Dennis Swena, 32
Saeedatun, 79 (inset)

Shutterstock
A. Viduetsky, 329
Agami Photo Agency, 144, 149 (right), 181 (left), 210 (right, top), 211 (right, top), 334, 352 (right), 436 (right), 506 (left)
Albert Barr, 286 (right)
Andrew M. Allport, 269
auldscot, 464 (right)
C. Hamilton, 245 (right)
Chase D'animulls, 73 (right, top), 139 (left)
Daniel Dunca, 264
Danita Delimont, 118 (inset)
Feng Yu, 253, 310
Gregory Johnston, 153 (right)
Harry Collins Photography, 291 (left)
James Chen, 102
Joe McDonald, 462 (left)
Jukka Jantunen, 438 (right)
Kamil Srubar, 189 (right, bottom)
Keneva Photography, 505 (left)
Kern Freesland, 412 (left)
Kerry Hargrove, 112 (left and right, bottom), 422 (right)
Lev Frid, 407
Linda Burek, 208 (right)
Mark Caunt, 188
Marwin Pongprayoon, 449 (right)
Michael J. Thompson, 434 (right)
Mike Jackson, 235 (right, top)
Moosehenderson, 216 (right, top)
Nick Bossenbroek, 385 (right)
Ray Hennessy, 172 (right, bottom)
Ray vizgirdas, 455 (left)
rck_953, 150 (left)
Richard Fitzer, 212 (main)
Russ Jenkins, 238 (right, bottom)
Sandra Standbridge, 420 (right)
Sen Yang, 100 (right)
Serge Miles, 387 (right)
SergeyCo, 351 (right, bottom)
Susan E. Viera, 259
Susan Hodgson, 82 (right, bottom)
vagabond54, 193, 452 (left)
Viduetsky, 97 (right)

Vireo
B. Steele/VIREO, 138 (right)
Brian E. Small/VIREO, 360 (right)
J. Fuhrman/VIREO, 301, 480 (left)
Laure Neish/VIREO, 143
R.&N. Bowers/VIREO, 140 (right), 480 (right)
T. Vezo, 505 (right)

ILLUSTRATIONS
Range maps, habitat graphics, and bird illustrations by Melissa Fratello
Habitat map adapted from Brown and Lowe, 1980
Tucson Audubon Society, 63 (bottom)

INDEX

Abert's towhee, 33, 66, 453
abundance of species, 66–67
Acadian flycatcher, 511
Acanthis flammea, 512
accessibility, 7, 13, 16, 33, 34, 36, 38, 40, 49, 50, 53, 58–59, 63
Accipiter atricapillus, 238
Accipiter cooperii, 237
Accipiter striatus, 236
Accipitridae, 228
acorn woodpecker, 275
Actitis macularius, 194
Adams, Molly, 61
Aechmophorus clarkii, 123
Aechmophorus occidentalis, 122
Aegithalidae, 354
Aegolius acadicus, 265
Aegolius funereus, 264
Aeronautes saxatalis, 144
Agapornis roseicollis, 294
Agelaius phoeniceus, 467
Aimophila ruficeps, 454
Aix sponsa, 80
Alaudidae, 344
Alcedinidae, 269
Alectoris chukar, 115
Allen's hummingbird, 155
Amazilia rutila, 509
American avocet, 169, 171
American bittern, 219
American coot, 161, 166
American crow, 341
American dipper, 374, 378
American golden-plover, 509
American goldfinch, 419
American goshawk, 38, 238
American kestrel, 287, 289

American Ornithological Society (AOS), 69
American Ornithological Union (AOU), 65
American pipit, 404, 406
American redstart, 483
American robin, 389
American three-toed woodpecker, 280
American tree sparrow, 438
American white pelican, 213, 217
American wigeon, 72, 85
American woodcock, 510
Ammodramus savannarum, 427
Ammospiza leconteii, 512
Ammospiza nelsoni, 512
Amphispiza bilineata, 429
Amphispizopsis quinquestriata, 428
Anas acuta, 88
Anas crecca, 89
Anas diazi, 87
Anas platyrhynchos, 86
ancient murrelet, 510
Anderson Mesa, 36
anhinga, 511
Anhinga anhinga, 511
Anna's hummingbird, 22, 151
Anser albifrons, 76
Anser caerulescens, 74
Anser rossii, 75
Anthus cervinus, 512
Anthus rubescens, 406
Anthus spragueii, 407
Antigone canadensis, 168
Antrostomus arizonae, 142
Antrostomus carolinensis, 509
Antrostomus ridgwayi, 141
Antrostomus vociferus, 509
Aphelocoma californica, 511
Aphelocoma wollweberi, 338

Aphelocoma woodhouseii, 337
aplomado falcon, 287, 511
Appleton-Whittell Research Ranch, 53
apps, 13
Aquila chrysaetos, 234
Aramides axillaris, 509
Archilochus alexandri, 150
Archilochus colubris, 509
Arctic tern, 510
Ardea alba, 222
Ardea herodias, 221
Ardeidae, 218
Ardenna grisea, 510
Ardenna pacifica, 510
Arenaria interpres, 510
Arenaria melanocephala, 510
Arizona
    central, 36
    habitat map, 8
    Indigenous communities, 11
    Indigenous lands, 52–53
    northern, 36
    public lands, 52
    southeast, 33–34
    southwest, 36
    state bird, 15, 366
Arizona woodpecker, 34, 284
Artemisiospiza belli, 445
Artemisiospiza nevadensis, 444
ash-throated flycatcher, 299
Asio flammeus, 263
Asio otus, 262
Athene cunicularia, 260
attracting birds, 45
Audubon, John James, 60
auricular, 24
Auriparus flaviceps, 360
avocets, 169
Aythya affinis, 94
Aythya americana, 91
Aythya collaris, 92

OPPOSITE Gila Woodpecker

Aythya fuligula, 509
Aythya marila, 93
Aythya valisineria, 90
Aztec thrush, 512

backroads, 57
Baeolophus ridgwayi, 359
Baeolophus wollweberi, 358
Bailey, Florence Augusta Merriam, 60
Baird's sandpiper, 184
Baird's sparrow, 447
bald eagle, 21, 36, 40, 239
Baltimore oriole, 512
band-tailed pigeon, 126
bank swallow, 346
barn owl, 41, 251, 252
barn swallow, 351
barred owl, 511
Barrow's goldeneye, 100
Bartramia longicauda, 177
Basileuterus culicivorus, 513
Basileuterus lachrymosus, 513
Basileuterus rufifrons, 513
Basilinna leucotis, 158
bay-breasted warbler, 512
behavioral cues, 16, 20, 66
Bell's sparrow, 445
Bell's vireo, 43, 322, 325
belted kingfisher, 269, 270
Bendire's thrasher, 394
berylline hummingbird, 160
Bewick's wren, 372
Bill Williams River wildlife refuge, 36
binoculars, 12–13, 49
biodiversity, 7
biomes, 33
BIPOC birding community, 61
bird houses, 45
Birds Through an Opera Glass (Bailey), 60

birdwatching. *see also* identification
 ethics and etiquette, 47–54
 gear, 12–13, 28
 inclusivity, 48–50, 58–61
 on Indigenous lands, 52–53
 private land, 54
 on public lands, 52
 right place, right time, 27–29
 safety considerations, 54–57
Bitter Lake National Wildlife Refute, 42
bitterns, 218
Black AF in STEM, 61
black-and-white warbler, 476
black-bellied plover, 172
black-bellied whistling-duck, 33, 73
black-billed cuckoo, 509
black-billed magpie, 340
Black Birders Week, 61
Blackburnian warbler, 512
black-capped chickadee, 41, 47, 354, 355
black-capped gnatcatcher, 70, 377
black-capped vireo, 511
black-chinned hummingbird, 150
black-chinned sparrow, 434
black-crowned night-heron, 226
black-headed grosbeak, 501
black-legged kittiwake, 510
black-necked stilt, 169, 170
black phoebe, 318
blackpoll warbler, 512
black rail, 167
black rosy-finch, 411
black scoter, 97

black skimmer, 510
black storm-petrel, 510
black swift, 509
black-tailed gnatcatcher, 376
black-tailed gull, 510
black tern, 41, 210
black-throated blue warbler, 512
black-throated gray warbler, 68, 487
black-throated green warbler, 512
black-throated sparrow, 429
black turnstone, 510
black-vented oriole, 512
black-vented shearwater, 510
black vulture, 228, 230
blue-footed booby, 510
blue-gray gnatcatcher, 375
blue grosbeak, 68, 502
blue-headed vireo, 511
blue jay, 332, 336
blue mockingbird, 512
blue-throated mountain-gem, 34, 148
blue-winged teal, 81
blue-winged warbler, 512
bobolink, 512
Bohemian waxwing, 512
Bombycilla cedrorum, 401
Bombycilla garrulus, 512
Bombycillidae, 399
Bonaparte's gull, 202
book features
 abundance, 66–67
 alternative names, 68–69
 behavior, 66
 conservation status, 69
 habitats, 66
 key sites, 68
 lookalikes, 68
 naming conventions, 66

photos, 65–66
range map, 69–70
vocalizations, 66
boreal owl, 251, 264
Bosque del Apache National Wildlife Refuge, 29, 38, 67
Botaurus lentiginosus, 219
Botteri's sparrow, 66, 425
Boyce Thompson Arboretum, 36
Brachyramphus perdix, 510
brant, 509
Branta bernicla, 509
Branta canadensis, 78
Branta hutchinsii, 77
breeding grounds, 29
Brewer's blackbird, 33, 470
Brewer's sparrow, 436
bridled titmouse, 358
broad-billed hummingbird, 157
broad-tailed hummingbird, 156
broad-winged hawk, 245
bronzed cowbird, 468
brown-backed solitaire, 512
brown booby, 510
brown-capped rosy-finch, 412
brown-chested martin, 511
brown creeper, 365
brown-crested flycatcher, 17, 300
brown-headed cowbird, 469
brown pelican, 511
brown thrasher, 393
brush piles, 47
Bubo virginianus, 256
Bubulcus ibis, 224
Bucephala albeola, 98
Bucephala clangula, 99
Bucephala islandica, 100
buff-breasted flycatcher, 317

buff-breasted sandpiper, 187
buff-collared nightjar, 141
bufflehead, 98
Bullock's oriole, 465
bumblebee hummingbird, 509
buntings, 16, 21
burrowing owl, 40, 41, 260
bushtit, 20, 361
Buteo albonotatus, 247
Buteo brachyurus, 511
Buteogallus anthracinus, 241
Buteo jamaicensis, 248
Buteo lagopus, 249
Buteo lineatus, 244
Buteo plagiatus, 243
Buteo platypterus, 245
Buteo regalis, 250
Buteo swainsoni, 246
Butorides virescens, 225

cackling goose, 77
cactus wren, 15, 366, 373
cairns, 50
Calamospiza melanocorys, 431
Calcariidae, 404
Calcarius lapponicus, 420
Calcarius ornatus, 421
Calcarius pictus, 512
Calidris acuminata, 510
Calidris alba, 182
Calidris alpina, 183
Calidris bairdii, 184
Calidris canutus, 510
Calidris ferruginea, 510
Calidris fuscicollis, 186
Calidris himantopus, 181
Calidris mauri, 190
Calidris melanotos, 188
Calidris minuta, 510
Calidris minutilla, 185
Calidris pugnax, 510

Calidris pusilla, 189
Calidris subruficollis, 187
California condor, 229
California gull, 205
California quail, 110
California scrub-jay, 511
callbacks, 47
calliope hummingbird, 153
Callipepla californica, 110
Callipepla gambelii, 108
Callipepla squamata, 107
calls, 16
Calothorax lucifer, 149
Calypte anna, 151
Calypte costae, 152
Camptostoma imberbe, 297
Campylorhynchus brunneicapillus, 373
Canada goose, 78
Canada jay, 333
Canada warbler, 513
canvasback, 90
canyon towhee, 452
canyon wren, 368
Cape May warbler, 512
Caprimulgidae, 137
Caracara plancus, 288
Cardellina canadensis, 513
Cardellina pusilla, 490
Cardellina rubrifrons, 491
Cardinalidae, 493
Cardinalis cardinalis, 498
Cardinalis sinuatus, 499
Carlsbad Caverns National Park, 43
Carolina wren, 511
Caspian tern, 209
Cassin's finch, 414
Cassin's kingbird, 303
Cassin's sparrow, 426
Cassin's vireo, 328
Catalina Highway, 33

catbirds, 390
Cathartes aura, 231
Cathartidae, 228
Catharus aurantiirostris, 512
Catharus fuscescens, 512
Catharus guttatus, 387
Catharus minimus, 512
Catharus ustulatus, 386
Catherpes mexicanus, 368
Cave Creek Canyon, 34
cave swallow, 353
cedar waxwing, 401
Centrocercus minimus, 509
Centronyx bairdii, 447
Centronyx henslowii, 512
ceres, 24
Certhia americana, 365
Certhiidae, 354
cerulean warbler, 512
Chaetura pelagica, 509
Chaetura vauxi, 143
Charadriidae, 169
Charadrius melodus, 509
Charadrius mongolus, 509
Charadrius montanus, 175
Charadrius nivosus, 176
Charadrius semipalmatus, 174
Charadrius vociferus, 173
checklists, 13
Cherokee people, 69
chestnut-collared longspur, 41, 421
chestnut-sided warbler, 512
chickadees, 354
Chihuahuan Desert, 38, 40
Chihuahuan meadowlark, 460
Chihuahuan raven, 342
chimney swift, 509
chin, 24
chipping sparrow, 423, 432
Chiricahua Mountains, 34

Chiricahua Wilderness/Portal, 34
Chlidonias niger, 210
Chloroceryle americana, 271
Chondestes grammacus, 430
Chordeiles acutipennis, 138
Chordeiles minor, 139
Chroicocephalus philadelphia, 202
Chuck-will's-widow, 509
chukar, 115
Cibola wildlife refuge, 36
Cinclidae, 374
Cinclus mexicanus, 378
cinnamon hummingbird, 509
cinnamon teal, 33, 82
Circus hudsonius, 235
Cistothorus palustris, 371
Cistothorus stellaris, 511
Clangula hyemalis, 509
clapper rail, 509
Clark's grebe, 41, 117, 123
Clark's nutcracker, 339
clay-colored sparrow, 433
clay-colored thrush, 512
cliff swallow, 344, 352
climate change, 7, 62–63
clothing recommendations, 12, 28
Coccothraustes vespertinus, 408
Coccyzus americanus, 136
Coccyzus erythropthalmus, 509
Colaptes auratus, 285
Colaptes chrysoides, 286
Colibri thalassinus, 509
Colinus virginianus, 106
Columba livia, 125
Columbiadae, 124
Columbina inca, 128
Columbina passerina, 129
Columbina talpacoti, 130
common black hawk, 241

common crane, 509
common gallinule, 161, 165
common goldeneye, 99
common grackle, 471
common ground dove, 129
common loon, 214
common merganser, 72, 102
common nighthawk, 139
common poorwill, 137, 140
common raven, 332, 343
common redpoll, 512
common tern, 201, 211
common yellowthroat, 482
community science, 62–63
Connecticut warbler, 512
conservation, 62–63
   community involvement, 62–63
   species status, 69
   women in, 60–61
conspecifics, 20–23
Contopus cooperi, 308
Contopus pertinax, 309
Contopus sordidulus, 310
Contopus virens, 511
Cooper's hawk, 66, 237
coots, 161
Coragyps atratus, 230
cordilleran flycatcher, 33
cormorants, 213
Corthylio calendula, 379
Corvidae, 332
corvids, 332
Corvus brachyrhynchos, 341
Corvus corax, 343
Corvus cryptoleucus, 342
Costa's hummingbird, 47, 152
Coturnicops noveboracensis, 509
cranes, 161
creepers, 354
crescent-chested warbler, 512

crest, 24
crested caracara, 288
crissal thrasher, 396
Crotophaga sulcirostris, 134
crown, 23, 24
crows, 332
Cuculidae, 133
curlew sandpiper, 510
curve-billed thrasher, 390, 392
Cyanocitta cristata, 336
Cyanocitta stelleri, 335
Cygnus buccinator, 509
Cygnus columbianus, 79
Cynanthus latirostris, 157
Cypseloides niger, 509
Cyrtonyx montezumae, 109

dark-eyed junco, 21, 439
Dendragapus obscurus, 113
Dendrocygna autumnalis, 73
Dendrocygna bicolor, 509
dickcissel, 507
dippers, 374
dogs on trails, 50
Dolichonyx oryzivorus, 512
double-crested cormorant, 216
doves, 21, 124
downy woodpecker, 281
Driver, Robert, 69
drones, 48
drought, 7
drug trafficking, 55
Dryobates arizonae, 284
Dryobates pubescens, 281
Dryobates scalaris, 282
Dryobates villosus, 283
ducks, 42
Dumetella carolinensis, 391
dunlin, 183
dusky-capped flycatcher, 21, 295, 298

dusky flycatcher, 315
dusky grouse, 105, 113

eagles, 228
eared grebe, 121
eared quetzal, 268
eastern bluebird, 382
eastern kingbird, 306
eastern meadowlark, 512
eastern phoebe, 319
eastern screech-owl, 511
eastern towhee, 512
eastern whip-poor-will, 509
eastern wood-pewee, 511
ecotones, 27–28
Egretta caerulea, 511
Egretta rufescens, 511
Egretta thula, 223
Egretta tricolor, 511
Elanoides forficatus, 511
Elanus leucurus, 233
elegant tern, 510
elegant trogon, 34, 52, 65, 66, 256, 267
elf owl, 63, 259
Emberiza pusilla, 512
Empidonax affinis, 511
Empidonax difficilis, 316
Empidonax flaviventris, 511
Empidonax fulvifrons, 317
Empidonax hammondii, 313
Empidonax minimus, 312
Empidonax oberholseri, 315
Empidonax traillii, 311
Empidonax virescens, 511
Empidonax wrightii, 314
Eremophila alpestris, 345
ethics, 47–54, 68–69
etiquette
 callbacks, 47
 inclusivity, 48–50
 photography, 47–48

proximity, 47
trail, 50, 52
Eudocimus albus, 511
Eugenes fulgens, 146
Euphagus carolinus, 512
Euphagus cyanocephalus, 470
Euptilotis neoxenus, 268
Eurasian collared-dove, 127
Eurasian wigeon, 509
European cuckoo, 133
European golden-plover, 509
European starling, 400
evening grosbeak, 408
eyeline, 23, 24
eye ring, 24

Falco columbarius, 290
Falco femoralis, 511
Falco mexicanus, 292
Falconidae, 287
falcons, 228, 287
Falco peregrinus, 291
Falco sparverius, 289
fan-tailed warbler, 513
feeders, 45
Feminist Bird Club, 61
ferruginous hawk, 250
ferruginous pygmy-owl, 258
field guides, 13, 20, 60
field marks, 15, 20, 23
field sparrow, 435
finches, 47, 404
five-striped sparrow, 428
flame-colored tanager, 497
flammulated owl, 253
flanks, 24
flooding, 55, 62
flycatchers, 15, 41, 66, 295
forehead, 24
Forster's tern, 212
fox sparrow, 437
Franklin's gull, 203

Fregata magnificens, 510
Fringillidae, 404
Fulica americana, 166
fulvous whistling-duck, 509

gadwall, 84
Gallinago delicata, 193
Gallinula galeata, 165
gallinules, 161
Gambel's quail, 105, 108
garganey, 509
Gavia adamsii, 510
Gavia immer, 214
Gavia pacifica, 510
Gavia stellata, 510
Gaviidae, 213
Gelochelidon nilotica, 510
Gene C. Reid Park, 45
Geococcyx californianus, 135
geographic variations, 21
Geothlypis formosa, 512
Geothlypis philadelphia, 512
Geothlypis tolmiei, 481
Geothlypis trichas, 482
Geranoaetus albicaudatus, 511
Gila Cliff Dwellings, 38
Gila National Forest, 38
Gila woodpecker, 272, 276
gilded flicker, 272, 286
Glaucidium brasilianum, 258
Glaucidium gnoma, 257
glaucous gull, 510
glaucous-winged gull, 510
glossy ibis, 511
gnatcatchers, 374
golden-cheeked warbler, 512
golden-crowned kinglet, 380
golden-crowned sparrow, 512
golden-crowned warbler, 513
golden eagle, 7, 33, 38, 66–67, 234

golden-fronted woodpecker, 511
golden-tailed oriole, 69
golden-winged warbler, 512
gorget, 24
Gould's wild turkey, 34
Grace's warbler, 486
grackles, 66
Grand Canyon National Park, 36
grasshopper sparrow, 427
gray catbird, 391
gray-cheeked thrush, 512
gray-collared becard, 511
gray-crowned rosy-finch, 410
gray flycatcher, 66, 314
gray hawk, 243
gray vireo, 43, 326
great blue heron, 221
great crested flycatcher, 511
great egret, 218, 222
greater pewee, 38, 309
greater roadrunner, 15, 33, 133, 135
greater scaup, 72, 93
greater white-fronted goose, 76
greater yellowlegs, 198
great horned owl, 41, 251, 256
great kiskadee, 511
Great Plains, 40
great-tailed grackle, 472
grebes, 117
green heron, 36, 225
green kingfisher, 7, 34, 269, 271
green-tailed towhee, 455
green-winged teal, 89
groove-billed ani, 133, 134
Gruidae, 161
Grus grus, 509
gull-billed tern, 510
gulls, 15, 201

Gunnison sage-grouse, 509
Gymnogyps californianus, 229
Gymnorhinus cyanocephalus, 334

habitats
  book descriptions, 66
  convergence of, 27–28
  creating, 45, 47
  diversity of, 7, 27–28
  identification, 19–20
  map, 8
  productivity of, 27
  urban, 45
Haemorhous cassinii, 414
Haemorhous mexicanus, 413
Haemorhous purpureus, 512
hairy woodpecker, 283
Haliaeetus leucocephalus, 239
Hall, Minna, 60
Hammond's flycatcher, 313
harlequin duck, 509
Harris's hawk, 8, 42, 242
Harris's sparrow, 512
Havasu wildlife refuge, 36
Hawaiian petrel, 510
hawks, 228
heat exhaustion and stroke, 54, 57
Heermann's gull, 510
Heliomaster constantii, 147
Heliornis fulica, 509
Helmitheros vermivorum, 512
Hemenway, Harriet, 60
Henslow's sparrow, 512
hepatic tanager, 494
hermit thrush, 381, 387
hermit warbler, 489
herons, 218
herring gull, 207
Himantopus mexicanus, 170
Hirundo rustica, 351

Histrionicus histrionicus, 509
hooded merganser, 101
hooded oriole, 68, 463
hooded warbler, 512
horned grebe, 120
horned lark, 344, 345
house finch, 66–67, 413
house sparrow, 405
house wren, 369
how to use this book, 65–70
Hudsonian godwit, 510
human trafficking, 55
hummingbirds, 21, 137, 145
Hutton's vireo, 327
Hydrobates microsoma, 510
Hydrocoloeus minutus, 510
Hydroprogne caspia, 209
Hylocichla mustelina, 512

ibises, 218
Iceland gull, 510
Icteria virens, 458
icterids, 457
Icteriidae, 457
Icterus bullockii, 465
Icterus cucullatus, 463
Icterus galbula, 512
Icterus parisorum, 466
Icterus pustulatus, 464
Icterus spurius, 462
Icterus wagleri, 512
Ictinia mississippiensis, 240
identification
    auditory, 16
    behavioral cues, 16
    habitat, 19–20
    how to use this book, 65–70
    range, 19–20
    seasonality, 19, 21
    tips, 20
    tools, 13
    visual, 15

illegal activity, 55, 57
Imperial wildlife refuge, 36
Important Bird Areas (IBAs), 36
Inca dove, 128
inclusivity, 48–50, 58–61
Indigenous communities, 11, 69
Indigenous lands, 52–53
indigo bunting, 504
International Union for the Conservation of Nature (IUCN), 67, 69
invasive species, 7
ivory gull, 510
Ixobrychus exilis, 220
Ixoreus naevius, 512

Jacana spinosa, 510
jays, 332
Juan Fernandez petrel, 510
Junco hyemalis, 439
Junco phaeonotus, 441
juniper titmouse, 359

Kaibab Plateau, 7
Kentucky warbler, 512
key sites, 68
killdeer, 173
kingfishers, 269
kinglets, 374
king rail, 509
kites, 228

Lacey Act (1900), 60
ladder-backed woodpecker, 282
Lagopus leucura, 112
Lake Mary, 36
Lampornis clemenciae, 148
land acknowledgement, 11
Laniidae, 322

Lanius borealis, 324
Lanius ludovicianus, 323
Lapland longspur, 420
Laridae, 201
lark bunting, 431
larks, 344
lark sparrow, 430
Larus argentatus, 207
Larus brachyrhynchus, 510
Larus californicus, 205
Larus crassirostris, 510
Larus delawarensis, 204
Larus fuscus, 510
Larus glaucescens, 510
Larus glaucoides, 510
Larus heermanni, 510
Larus hyperboreus, 510
Larus livens, 510
Larus occidentalis, 510
Las Vegas National Wildlife Refuge, 40
Laterallus jamaicensis, 167
laughing gull, 510
Lawrence's goldfinch, 418
Laysan albatross, 510
lazuli bunting, 503
least bittern, 220
least flycatcher, 312
least grebe, 117, 118
least sandpiper, 185
least storm-petrel, 510
least tern, 208
LeConte's sparrow, 512
LeConte's thrasher, 395
Legatus leucophaius, 511
Leiothlypis celata, 477
Leiothlypis luciae, 478
Leiothlypis peregrina, 512
Leiothlypis ruficapilla, 479
Leiothlypis virginiae, 480
Leptotila verreauxi, 509
lesser black-backed gull, 510

lesser goldfinch, 417
lesser nighthawk, 138
lesser prairie-chicken, 41, 116
lesser sand-plover, 509
lesser scaup, 72, 94
lesser yellowlegs, 196
Leucolia violiceps, 159
Leucophaeus atricilla, 510
Leucophaeus pipixcan, 203
Leucosticte atrata, 411
Leucosticte australis, 412
Leucosticte tephrocotis, 410
Lewis's woodpecker, 273
LGBTQIA birding community, 61
Limnodromus griseus, 191
Limnodromus scolopaceus, 192
Limnothlypis swainsonii, 512
Limosa fedoa, 180
Limosa haemastica, 510
Lincoln's sparrow, 450
little blue heron, 511
little bunting, 512
little gull, 510
little stint, 510
loggerhead shrike, 323
logs, moving, 50
long-billed curlew, 40, 179
long-billed dowitcher, 192
long-billed murrelet, 510
long-billed thrasher, 512
long-eared owl, 40, 262
longspurs, 404
long-tailed duck, 509
long-tailed jaeger, 510
long-tailed tits, 354
lookalikes, 16, 68
loons, 213
Lophodytes cucullatus, 101
lores, 24
Los Alamos Nature Center, 58–59

Louisiana waterthrush, 512
Lower Colorado River, 29, 36
Loxia curvirostra, 415
Loxia leucoptera, 512
lucifer hummingbird, 149
Lucy's warbler, 478

MacGillivray's warbler, 481
Madera Canyon, 7, 34
Madrean Sky Islands, 7
magnificent frigatebird, 510
magnificent hummingbird, 34
magnolia warbler, 512
magpies, 332
malar stripe, 24
mallard, 86
mallards, 66, 72
mandible, 24
mantle, 24
marbled godwit, 180
Mareca americana, 85
Mareca penelope, 509
Mareca strepera, 84
marsh wren, 371
Massachusetts Audubon, 60
Maxwell National Wildlife Refuge, 41
McLaughlin, Jessica, 69
Megaceryle alcyon, 270
Megaceryle torquata, 511
Megascops asio, 511
Megascops kennicottii, 255
Megascops trichopsis, 254
Melanerpes aurifrons, 511
Melanerpes carolinus, 511
Melanerpes erythrocephalus, 274
Melanerpes formicivorus, 275
Melanerpes lewis, 273
Melanerpes uropygialis, 276
Melanitta americana, 97
Melanitta deglandi, 96
Melanitta perspicillata, 95

Melanotis caerulescens, 512
Meleagris gallopavo, 111
Melospiza georgiana, 451
Melospiza lincolnii, 450
Melospiza melodia, 449
Melozone aberti, 453
Melozone fusca, 452
Melrose Woods migrant trap, 41
Mergus merganser, 102
Mergus serrator, 103
merlin, 290
Mesilla Valley Bosque State Park, 40
Mexican chickadee, 357
Mexican duck, 87
Mexican jay, 34, 332, 338
Mexican violetear, 509
Mexican whip-poor-will, 142
Micrathene whitneyi, 259
migrant traps, 41, 43
migration, 19, 23, 29
Migratory Bird Treaty Act (1918), 60
Milnesand Prairie Reserve, 41
Mimidae, 390
mimids, 390
Mimus polyglottos, 398
misidentification, 20
Mississippi kite, 240
Mitrephanes phaeocercus, 511
Mniotilta varia, 476
mockingbirds, 390
Molothrus aeneus, 468
Molothrus ater, 469
molting, 21, 23
monsoon season, 55
Montezuma quail, 38, 105, 109
Mormon Lake, 36
Motacilla alba, 512
Motacillidae, 404
mountain bluebird, 384

mountain chickadee, 356
mountain plover, 175
Mount Lemmon, 33
mourning dove, 132
mourning warbler, 512
mustachial stripe, 24
Myadestes occidentalis, 512
Myadestes townsendi, 385
Mycteria americana, 510
Myiarchus cinerascens, 15, 299
Myiarchus crinitus, 511
Myiarchus nuttingi, 511
Myiarchus tuberculifer, 298
Myiarchus tyrannulus, 300
Myioborus miniatus, 513
Myioborus pictus, 492
Myiodynastes luteiventris, 301

naming conventions, 65, 68–69
Nannopterum auritum, 216
Nannopterum brasilianum, 215
nape, 23, 24
Nashville warbler, 479
National Audubon Society, 53, 60
native plants, 45, 47
natural hazards, 55
The Nature Conservancy, 41
Nelson's sparrow, 512
neotropic cormorant, 215
New Mexico
  central, 40
  eastern/southeast, 41–43
  habitat map, 8
  Indigenous communities, 11
  Indigenous lands, 52–53
  northern, 41
  public lands, 52
  state bird, 15, 133
  western/southwest, 38–40
nightjars, 28, 137

Nikole, Alexis, 61
northern beardless-tyrannulet, 34, 297
northern bobwhite, 106
northern cardinal, 498
northern flicker, 285
northern harrier, 235
northern jacana, 510
northern mockingbird, 398
northern parula, 512
northern pintail, 88
northern pygmy-owl, 257
northern rough-winged swallow, 349
northern saw-whet owl, 265
northern shoveler, 83
northern shrike, 324
northern waterthrush, 475
northern wheatear, 511
Nucifraga columbiana, 339
Numenius americanus, 179
Numenius phaeopus, 178
nutcrackers, 332
nuthatches, 354
Nutting's flycatcher, 511
Nyctanassa violacea, 511
Nycticorax nycticorax, 226

observations
  auditory, 16, 20
  behavioral cues, 16, 20
  how to use this book, 65–70
  plumage variation, 20–23
  right place, right time, 27–29
  sharing, 48
  visual, 15, 20
Oceanodroma melania, 510
Oceanodroma tethys, 510
Odontophoridae, 105
Oenanthe oenanthe, 511

olive-sided flycatcher, 308
olive warbler, 33, 399, 403
Onychoprion fuscatus, 510
Oporornis agilis, 512
optics, 12–13, 49
orange-billed nightingale-thrush, 512
orange-crowned warbler, 477
orbital ring, 24
orchard oriole, 462
Oreoscoptes montanus, 397
Oreothlypis superciliosa, 512
orioles, 41
osprey, 228, 232
ovenbird, 474
owls, 28, 48, 251
Oxyura jamaicensis, 104

Pachyramphus aglaiae, 296
Pachyramphus major, 511
Pacific golden-plover, 509
Pacific loon, 510
Pacific wren, 366, 370
Pagophila eburnea, 510
painted bunting, 43, 506
painted redstart, 38, 473, 492
Pajarito Environmental Education Center (PEEC), 59
palm warbler, 512
Pandion haliaetus, 232
Pandionidae, 228
Parabuteo unicinctus, 242
parasitic jaeger, 510
Paridae, 354
Parkesia motacilla, 512
Parkesia noveboracensis, 475
parrots, 293
parts of a bird, 23–25
Parulidae, 473
Passerculus sandwichensis, 448
Passer domesticus, 405

Passerella iliaca, 437
Passerellidae, 423
Passeridae, 404
Passerina amoena, 503
Passerina caerulea, 502
Passerina ciris, 506
Passerina cyanea, 504
Passerina versicolor, 505
Patagioenas fasciata, 126
Patagonia, 7, 34
Patagonia Lake State Park, 34
Patagonia Mountains, 34
Paton Center for Hummingbirds, 7, 34
Pattie Gonia, 61
Pecos River, 42
pectoral sandpiper, 188
Pelecanidae, 213
Pelecanus erythrorhynchos, 217
Pelecanus occidentalis, 511
pelicans, 213
penduline tits, 354
Percha Dam State Park, 38
peregrine falcon, 287, 291
Perisoreus canadensis, 333
Petrochelidon fulva, 353
Petrochelidon pyrrhonota, 352
Peucaea botterii, 425
Peucaea carpalis, 424
Peucaea cassinii, 426
Peucedramidae, 399
Peucedramus taeniatus, 403
Phaethon aethereus, 510
Phaethon lepturus, 510
Phainopepla, 19, 399, 402
Phainopepla nitens, 402
Phalacrocoracidae, 213
Phalaenoptilus nuttallii, 140
Phalaropus fulicarius, 510
Phalaropus lobatus, 200
Phalaropus tricolor, 199
Phasianidae, 105

Phasianus colchicus, 114
Pheucticus chrysopeplus, 513
Pheucticus ludovicianus, 500
Pheucticus melanocephalus, 501
Philadelphia vireo, 511
Phoebastria immutabilis, 510
photography, 47–48, 50
photos of birds (in book), 65–66
Pica hudsonia, 340
Picidae, 272
Picoides dorsalis, 280
pied-billed grebe, 117, 119
pigeons, 124
pine flycatcher, 511
pine grosbeak, 409
pine siskin, 416
pine warbler, 512
Pinicola enucleator, 409
pinyon jay, 36, 332, 334
Pipilo chlorurus, 455
Pipilo erythrophthalmus, 512
Pipilo maculatus, 456
piping plover, 509
pipits, 404
Piranga bidentata, 497
Piranga flava, 494
Piranga ludoviciana, 496
Piranga olivacea, 513
Piranga rubra, 495
piratic flycatche, 511
Pitangus sulphuratus, 511
plain-capped starthroat, 147
Platalea ajaja, 511
Plectrophenax nivalis, 512
Plegadis chihi, 227
Plegadis falcinellus, 511
plovers, 169
plumage
  terms, 24
  variations, 20–23
Plumbeous Vireo, 329

Pluvialis apricaria, 509
Pluvialis dominica, 509
Pluvialis fulva, 509
Pluvialis squatarola, 172
Podiceps auritus, 120
Podiceps grisegena, 509
Podiceps nigricollis, 121
Podicipedidae, 117
Podilymbus podiceps, 119
Poecile atricapillus, 355
Poecile gambeli, 356
Poecile sclateri, 357
Polioptila caerulea, 375
Polioptila melanura, 376
Polioptila nigriceps, 377
Polioptilidae, 374
pomarine jaeger, 510
Pooecetes gramineus, 446
Porphyrio martinica, 509
Porzana carolina, 164
prairie falcon, 292
prairie warbler, 512
primaries, 24
primary coverts, 24
private land, 54, 55
Progne subis, 350
Progne tapera, 511
prothonotary warbler, 512
Protonotaria citrea, 512
proximity, 47
Psaltriparus minimus, 361
Psiloscops flammeolus, 253
Psittacidae, 293
Psittaculidae, 293
Pterodroma externa, 510
Pterodroma sandwichensis, 510
Ptiliogonatidae, 399
public lands, 50, 52
Puffinus opisthomelas, 510
purple finch, 512
purple gallinule, 509
purple martin, 36, 350

pygmy nuthatch, 59, 364
Pyrocephalus rubinus, 321
pyrrhuloxia, 40, 42, 493, 499

Quiscalus mexicanus, 472
Quiscalus quiscula, 471

racism, 59, 61
Rallidae, 161
Rallus crepitans, 509
Rallus elegans, 509
Rallus limicola, 163
Rallus obsoletus, 162
range, 19–20
range maps, 69–70
raptors, 28, 228
rarities, 509–513
Rattlesnake Springs, 43
ravens, 332
Recurvirostra americana, 171
Recurvirostridae, 169
red-bellied woodpecker, 511
red-billed tropicbird, 510
red-breasted merganser, 103
red-breasted nuthatch, 362
red-breasted sapsucker, 511
red crossbill, 59, 415
reddish egret, 511
red-eyed vireo, 331
red-faced warbler, 33, 36, 38, 491
redhead, 91
red-headed woodpecker, 274
red knot, 510
red-naped sapsucker, 279
red-necked grebe, 509
red-necked phalarope, 200
red phalarope, 510
red-shouldered hawk, 244
red-tailed hawk, 7, 248
red-throated loon, 510
red-throated pipit, 512

red-winged blackbird, 33, 467
Regulus satrapa, 380
Remizidae, 354
representation, 61
resident breeders, 7
Rhynchophanes mccownii, 422
Rhynchopsitta pachyrhyncha, 511
Ridgwayia pinicola, 512
Ridgway's rail, 36, 162
Ring-billed Gull, 204
ringed kingfisher, 511
ring-necked duck, 92
ring-necked pheasant, 114
Rio Fernando Wetlands, 41
Rio Grande Nature Center State Park, 40, 58
Riparian Preserve at Water Ranch, 36, 45
Riparia riparia, 346
Rissa tridactyla, 510
Rivoli's hummingbird, 146
rock pigeon, 124, 125
rocks, moving, 50
rock wren, 367
Rocky Mountains, 7, 38, 40
roseate spoonbill, 36, 511
rose-breasted grosbeak, 65, 67, 70, 500
rose-throated becard, 295, 296
Ross's goose, 42, 75
rosy-faced lovebird, 70, 293, 294
rough-legged hawk, 249
royal tern, 510
ruby-crowned kinglet, 374, 379
ruby-throated hummingbird, 509
Ruddy Duck, 104
ruddy ground dove, 124, 130
ruddy turnstone, 510

ruff, 510
rufous-backed robin, 388
rufous-capped warbler, 513
rufous-crowned sparrow, 454
rufous hummingbird, 154
rufous-necked wood-rail, 509
rufous-winged sparrow, 17, 424
rusty blackbird, 512
Rynchops niger, 510

Sabine's gull, 510
safety considerations, 54–57
sagebrush sparrow, 444
sage thrasher, 397
Salpinctes obsoletus, 367
sanderling, 182
sandhill crane, 19, 29, 33, 38, 40, 42, 67, 161, 168
sandpipers, 169
Santa Rita Lodge, 7
Santa Rita Mountains, 34
Saucerottia beryllina, 160
Savannah sparrow, 448
Sayornis nigricans, 318
Sayornis phoebe, 319
Sayornis saya, 320
Say's phoebe, 320
scaled quail, 67, 107
scapulars, 24
scarlet tanager, 513
scissor-tailed flycatcher, 307
Scolopacidae, 169
Scolopax minor, 510
Scott, Winfield, 69
Scott's oriole, 65, 68–69, 466
screech owl, 251
seasonality, 19, 21, 23, 29
secondaries, 24
secondary coverts, 24
sedge wren, 511
Seiurus aurocapilla, 474

Selasphorus calliope, 153
Selasphorus heloisa, 509
Selasphorus platycercus, 156
Selasphorus rufus, 154
Selasphorus sasin, 155
semipalmated plover, 174
semipalmated sandpiper, 189
Setophaga americana, 512
Setophaga caerulescens, 512
Setophaga castanea, 512
Setophaga cerulea, 512
Setophaga chrysoparia, 512
Setophaga citrina, 512
Setophaga coronata, 485
Setophaga discolor, 512
Setophaga dominica, 512
Setophaga fusca, 512
Setophaga graciae, 486
Setophaga magnolia, 512
Setophaga nigrescens, 487
Setophaga occidentalis, 489
Setophaga palmarum, 512
Setophaga pensylvanica, 512
Setophaga petechia, 484
Setophaga pinus, 512
Setophaga pitiayumi, 512
Setophaga ruticilla, 483
Setophaga striata, 512
Setophaga tigrina, 512
Setophaga townsendi, 488
Setophaga virens, 512
sexism, 59
sexual dimorphism, 21
sharp-shinned hawk, 66, 236
sharp-tailed grouse, 509
sharp-tailed sandpiper, 510
shorebirds, 15, 29, 38, 66
short-billed dowitcher, 191
short-billed gull, 510
short-eared owl, 251, 263
short-tailed hawk, 511
shrikes, 322

Sialia currucoides, 384
Sialia mexicana, 383
Sialia sialis, 382
Sierra Estrella Mountains, 36
silky-flycatchers, 399
Sinaloa wren, 511
Sitta canadensis, 362
Sitta carolinensis, 363
Sitta pygmaea, 364
Sittidae, 354
Sky Island, 34
slate-throated redstart, 513
Smith's longspur, 512
Smithsonian Institution, 61
snow bunting, 512
snow goose, 29, 33, 38, 42, 74
snowy egret, 223
snowy plover, 176
solitary sandpiper, 195
songbirds, 28
songs, 16, 20, 66
song sparrow, 449
sooty shearwater, 510
sooty tern, 510
sora, 164
Southwestern Research Station, 34
sparrows, 15, 21, 41, 47, 404, 423
Spatula clypeata, 83
Spatula cyanoptera, 82
Spatula discors, 81
Spatula querquedula, 509
Sphyrapicus nuchalis, 279
Sphyrapicus ruber, 511
Sphyrapicus thyroideus, 277
Sphyrapicus varius, 278
Spinus lawrencei, 418
Spinus pinus, 416
Spinus psaltria, 417
Spinus tristis, 419
Spiza americana, 507

Spizella atrogularis, 434
Spizella breweri, 436
Spizella pallida, 433
Spizella passerina, 432
Spizella pusilla, 435
Spizella wortheni, 512
Spizelloides arborea, 438
spotted owl, 261
spotted sandpiper, 16, 66, 194
spotted towhee, 456
Sprague's pipit, 407
Stelgidopteryx serripennis, 349
Steller's jay, 33, 335
Stercorarius longicaudus, 510
Stercorarius parasiticus, 510
Stercorarius pomarinus, 510
Sterna forsteri, 212
Sterna hirundo, 211
Sterna paradisaea, 510
Sternula antillarum, 208
stilts, 169
stilt sandpiper, 181
storms, 54–55
streak-backed oriole, 464
Streptopelia decaocto, 127
Strigidae, 251
Strix occidentalis, 261
Strix varia, 511
Sturnella lilianae, 460
Sturnella magna, 512
Sturnella neglecta, 461
Sturnidae, 399
Sturnus vulgaris, 400
subterminal band, 24
Sula leucogaster, 510
Sula nebouxii, 510
sulphur-bellied flycatcher, 301
summer tanager, 16, 21, 495
sungrebe, 509
sun protection, 54
supercilium, 24
supreloral, 24

surf scoter, 95
Swainson's hawk, 246
Swainson's thrush, 386
Swainson's warbler, 512
swallows, 344
swallow-tailed kite, 511
swamp sparrow, 451
Sweetwater Wetlands, 33, 45, 58
swifts, 137
Synthliboramphus antiquus, 510

Tachybaptus dominicus, 118
Tachycineta bicolor, 347
Tachycineta thalassina, 348
tail coverts, 24
taxonomy, 15
temperature variations, 54–55
Tennessee warbler, 512
terns, 201
tertials, 24
Thalasseus elegans, 510
Thalasseus maximus, 510
thick-billed kingbird, 304
thick-billed longspur, 422
thick-billed parrot, 293, 511
thrashers, 21, 390
Threskiornithidae, 218
throat, 24
thrushes, 19, 381
Thryomanes bewickii, 372
Thryophilus sinaloa, 511
Thryothorus ludovicianus, 511
time of day, 28–29
titmice, 354
Tityridae, 295
Townsend's solitaire, 385
Townsend's warbler, 488
Toxostoma bendirei, 394
Toxostoma crissale, 396
Toxostoma curvirostre, 392

Toxostoma lecontei, 395
Toxostoma longirostre, 512
Toxostoma rufum, 393
trail etiquette, 50, 52
Trail of Tears, 69
tree swallow, 347
Tres Rios Wetlands, 36
tricolored heron, 511
Tringa flavipes, 196
Tringa incana, 510
Tringa melanoleuca, 198
Tringa semipalmata, 197
Tringa solitaria, 195
Trochilidae, 137, 145
Troglodytes aedon, 369
Troglodytes hiemalis, 511
Troglodytes pacificus, 370
Troglodytidae, 366
Trogon elegans, 267
Trogonidae, 266
trogons, 256
tropical kingbird, 302
tropical parula, 512
trumpeter swan, 509
tufted duck, 509
tufted flycatcher, 511
tundra swan, 79
Turdidae, 381
Turdus assimilis, 512
Turdus grayi, 512
Turdus migratorius, 389
Turdus rufopalliatus, 388
turkey vulture, 228, 231
Tympanuchus pallidicinctus, 116
Tympanuchus phasianellus, 509
Tyranniadae, 295
Tyrannus crassirostris, 304
Tyrannus forficatus, 307
Tyrannus melancholicus, 302
Tyrannus tyrannus, 306
Tyrannus verticalis, 305

Tyrannus vociferans, 303
tyrant flycatchers, 295
Tyto alba, 252
Tytonidae, 251

undertail coverts, 24
upland sandpiper, 177
urban birding, 45

varied bunting, 43, 505
varied thrush, 512
Vaux's swift, 143
veery, 512
venomous reptiles, 55, 57
verdin, 20, 354, 360
vermilion flycatcher, 13, 15, 295, 321
Vermivora chrysoptera, 512
Vermivora cyanoptera, 512
vesper sparrow, 446
violet-crowned hummingbird, 34, 159
violet-green swallow, 348
Vireo atricapilla, 511
Vireo bellii, 325
Vireo cassinii, 328
Vireo flavifrons, 511
Vireo flavoviridis, 511
Vireo gilvus, 330
Vireo griseus, 511
Vireo huttoni, 327
Vireonidae, 322
Vireo olivaceus, 331
Vireo philadelphicus, 511
Vireo plumbeus, 329
vireos, 16, 41, 322
Vireo solitarius, 511
Vireo vicinior, 326
Virginia rail, 163
Virginia's warbler, 36, 480
vocalizations, 16, 23, 47, 66
vultures, 228

wandering tattler, 510
warblers, 16, 19, 38, 41, 43, 473
warbling vireo, 330
waste disposal, 52
water features, 47
water resources, 7, 62
water treatment plants, 45
Watson and Willow Lakes Ecosystem, 36
waxwings, 399
weather conditions, 29, 54–55
wedge-rumped storm-petrel, 510
wedge-tailed shearwater, 510
Weeks-Mclean Act (1913), 60
western bluebird, 383
western cattle egret, 224
western flycatcher, 316
western grebe, 41, 117, 122
western gull, 510
western kingbird, 305
western meadowlark, 461
western sandpiper, 190
western screech-owl, 255
western tanager, 16, 41, 493, 496
western wood-pewee, 310
wetlands, 161
whimbrel, 178
whiskered screech-owl, 251, 254
white-breasted nuthatch, 363
white-crowned sparrow, 21, 442
white-eared hummingbird, 158
white-eyed vireo, 511
white-faced ibis, 227
white ibis, 511
white pelican, 36
white-rumped sandpiper, 186
white-tailed hawk, 511

white-tailed kite, 233
White-tailed Ptarmigan, 112
white-tailed tropicbird, 510
white-throated sparrow, 443
white-throated swift, 144
white-throated thrush, 512
white-tipped dove, 509
white wagtail, 512
Whitewater Draw Wildlife Area, 33, 67
white-winged crossbill, 512
white-winged dove, 131
white-winged scoter, 96
wildfire, 62, 63
wildlife refuges, 36
wild turkey, 105, 111
Willcox Playa, 29
willet, 197
Williamson's sapsucker, 277
willow flycatcher, 36, 311
Wilson, Alexander, 69
Wilson's phalarope, 41, 42, 199
Wilson's snipe, 193
Wilson's warbler, 68, 69, 490
wing anatomy, 25
wingbars, 24
wingtips, 24
winter wren, 511
women in birding and conservation, 59–61
wood duck, 36, 80
Woodhouse's scrub-jay, 337
woodpeckers, 16, 21, 272
wood stork, 510
wood thrush, 512
worm-eating warbler, 512
Worthen's sparrow, 512
wrens, 366

Xanthocephalus xanthocephalus, 459
Xema sabin, 510

yellow-bellied flycatcher, 511
yellow-bellied sapsucker, 278
yellow-billed cuckoo, 27, 36, 67, 133, 136
yellow-billed loon, 510
yellow-breasted chat, 19, 457, 458
yellow-crowned night-heron, 511
yellow-eyed junco, 441
yellow-footed gull, 510
yellow-green vireo, 511
yellow grosbeak, 513
yellow-headed blackbird, 15, 33, 457, 459
yellow rail, 509
yellow-rumped warbler, 45, 485
yellow-throated vireo, 511
yellow-throated warbler, 512
yellow warbler, 36, 484
yucca oriole, 65, 68, 69

Zenaida asiatica, 131
Zenaida macroura, 132
zone-tailed hawk, 247
Zonotrichia albicollis, 443
Zonotrichia atricapilla, 512
Zonotrichia leucophrys, 442
Zonotrichia querula, 512

OPPOSITE Rufous-crowned Sparrow

ABOUT THE AUTHORS

Melissa Fratello is an ardent naturalist, photographer, tinkerer, and writer. She has worked to advance a more equitable and approachable birding community, supporting Feminist Bird Club chapters in Tucson and beyond, and currently directs Tucson Audubon Society, which she is guiding through a renaming process. She feels strongly that birders owe it to the birds they observe to protect their precious and quickly disappearing habitats, and that birding can serve as a most wonderful gateway to conservation action and as a cure to the consumerism that ails us. When she's not working, she's gardening, cooking, or exploring the far flung reaches of the Sky Islands peeping at plants, listening for birds, and always hoping to happen upon a snake.

Steven Prager is a field biologist, science communicator, conservation advocate, and lifelong Arizonan. He may have co-written a book about birds, but the avian world is just a small part of what keeps this naturalist busy. He's always happy to go birding, but he's likely to miss that passing rarity as he struggles to not let his thoughts and eyes wander to the snakes, lizards, amphibians, fish, and invertebrates (especially ants) that also call the Southwest home. He thinks that birds are best appreciated as just one part of the habitats they occupy, and he believes that the ethical enjoyment of birds and their habitats requires a commitment to their conservation. Whether Steven is down in the dirt admiring the efforts of his backyard leafcutter ants or eyeball-deep in thornscrub searching (so far, unsuccessfully) for a Brown Vinesnake, he's just happy to be outside.